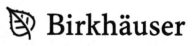

Operator Theory: Advances and Applications
Volume 243

Founded in 1979 by Israel Gohberg

Alberto Cialdea • Vladimir Maz'ya

Semi-bounded Differential Operators, Contractive Semigroups and Beyond

 Birkhäuser

Alberto Cialdea
Dept. of Mathematics, Computer Science
   and Economics
University of Basilicata
Potenza, Italy

Vladimir Maz'ya
Department of Mathematics
Linköping University
Linköping, Sweden

ISSN 0255-0156   ISSN 2296-4878 (electronic)
ISBN 978-3-319-35670-9   ISBN 978-3-319-04558-0 (eBook)
DOI 10.1007/978-3-319-04558-0
Springer Cham Heidelberg New York Dordrecht London

Printed on acid-free paper

Springer Basel is part of Springer Science+Business Media (www.birkhauser-science.com)

To our wives Flavia and Tatyana

Contents

Introduction

Motivation. In the present book we study conditions for the semi-boundedness of partial differential operators which is interpreted in different ways. The semi-boundedness, i.e., the boundedness from above or from below of quadratic forms generated by operators acting in Hilbert spaces has clear physical meaning. This explains why such operators play an exceptional role in applications of mathematical analysis.

In particular the definiteness of the sign of quadratic forms leads to theorems of invertibility for the corresponding operators. It also gives quantitative information on inverse operators. Operators with definite sign of the quadratic form are naturally divided in the classes of positive and negative operators.

From the mathematical point of view the semi-boundedness is closely connected with such properties of differential operators as strong ellipticity, Gårding inequality, maximum norm principles, and L^2-contractivity of the semigroup generated by the operator.

Needless to say an operator A is bounded from below if and only if $-A$ is bounded from above and vice versa. Hence we choose one of these classes in different contexts without any serious reason, just according to our taste.

Nowadays one knows rather much about L^2-semi-bounded differential and pseudo-differential operators, although their complete characterization in analytic terms causes difficulties even for rather simple operators, like, for example, the Schrödinger operator $u \mapsto -\Delta u + c(x)u$ (see Maz'ya [72] and references therein, and Jaye, Maz'ya, Verbitsky [43, 44]).

Until recently almost nothing was known about analytic characterizations of semi-boundedness for differential operators in other Hilbert function spaces and in Banach function spaces.

The goal of the present book is to partially fill this gap. We consider various types of semi-boundedness and give some relevant conditions which are either necessary and sufficient or best possible in a certain sense.

The material included here has been either unpublished or published in journal articles, so that the book has no significant intersections with the huge existing monographic literature on partial differential operators in Hilbert and Banach spaces.

Most of the results reported in this book reflect our joint work on this subject [11, 12, 13, 14]. We have included also theorems due to Stefan Eilertsen [22, 24] and to the second author and Ovidiu Costin [15], Gershon Kresin [50, 51], Mikael Langer [54, 55], Guo Luo [62, 63] and Svitlana Mayboroda [64, 65].

Structure of the book. There are seven chapters in the present book.

In Chapter 1 we study a certain subclass of semi-bounded operators in a general Banach space, called dissipative. One is interested in such operators because of their close relation with the question of contractivity of the generated semigroup.

In Chapter 2 we consider a scalar second-order partial differential operator whose coefficients are complex-valued measures. For a class of such operators we find algebraic necessary and sufficient conditions for the L^p-dissipativity. We remark that such conditions characterize the L^p-dissipativity individually, for each p. Previous known results dealt with the L^p-dissipativity for any $p \in [1, +\infty)$, simultaneously.

Generally speaking we show that our conditions are necessary and sufficient for the so-called quasi-dissipativity. Algebraic necessary and sufficient conditions are obtained also for scalar operators with complex constant coefficients.

Chapter 3 is devoted to the elasticity system. In the two-dimensional case we give an inequality involving p and the Poisson ratio, which provides a necessary and sufficient condition for the L^p-dissipativity of the Lamé operator.

In Chapter 4 we study some classes of systems for which we give necessary and sufficient conditions.

The angle of L^p-dissipativity of an operator A, i.e., the largest angle $\vartheta_- \leqslant \arg z \leqslant \vartheta_+$ such that zA is L^p-dissipative, is the subject of Chapter 5. Thanks to the necessary and sufficient conditions we have obtained, we are able to characterize such an angle for different operators.

In Chapter 6 we have considered higher-order partial differential operators. It is shown that we cannot have L^p-dissipativity for higher-order operators if $p \neq 2$. We may have L^p-contractivity only for fourth-order operators on the cone of nonnegative functions and only for particular values of p.

In Chapter 7 we collect a series of results concerning the positivity of various differential operators in different L^2-weighted spaces.

More detailed information on the material included here can be found in the introductions to chapters. We collect bibliographical notes in comments to each chapter.

Readership. The volume is addressed to mathematicians working in partial differential equations and applications.

Prerequisites for reading this book are undergraduate courses in theory of partial differential equations and functional analysis.

Acknowledgment. The final version of the book was prepared during our stay in the Institut Mittag-Leffler in July, 2013. We are grateful to the staff of this Institute who created excellent conditions for our work.

We would ask Sylvia Lotrovski and Thomas Hempfling to accept our gratitude for deciding to publish this book.

Chapter 1

Preliminary Facts on Semi-boundedness of Forms and Operators

Let us recall that, in the classical theory of Hilbert spaces, the class of symmetric operators A, semi-bounded below is selected by the inequality

$$(Au, u) \geqslant c \, \|u\|^2$$

valid for any u belonging to $D(A)$, the domain of A ($c \in \mathbb{R}$). Similarly an operator A is said to be semi-bounded above if the previous inequality holds with $\geqslant$ replaced by $\leqslant$.

More generally, a not necessarily symmetric operator A on a Hilbert space is said to be semi-bounded below (above) if

$$\mathrm{Re}(Au, u) \geqslant c \, \|u\|^2 \qquad (\mathrm{Re}(Au, u) \leqslant c \, \|u\|^2) \tag{1.1}$$

for any u belonging to $D(A)$.

If (1.1) holds with $c = 0$, the operator A is said to be positive (negative). There are alternative terms, accretive and dissipative, which come from physical applications.

In the present book we are led by analogy with these Hilbert space notions to speak about semi-bounded, positive, negative, accretive and dissipative operators in more general spaces (see Section 1.2).

The importance of dissipative operators lies, in particular, in the fact that they generate contractive semigroups. Sections 1.3, 1.4 are devoted to basic properties of dissipative operators in Banach spaces and to the contractivity of the generated semigroups.

The last section 1.5 deals with time dependent semi-bounded operators acting in general Banach spaces.

1.1   Duality set

Let X be a (complex) Banach space and denote by X^* its (topological) dual space.

Given $x \in X$, denote by i(x) the set

$$\mathrm{i}(x) = \{x^* \in X^* \mid \langle x^*, x \rangle = \|x\|^2 = \|x^*\|^2\}.$$

The set i(x) is called *the duality set of x*.

It is not difficult to prove that, for any $x \in X$, i(x) is not empty. In fact, if $x = 0$, i$(0) = \{0\}$. If $x \neq 0$, in view of the Hahn–Banach Theorem, there exists $f \in X^*$ such that

$$\langle f, x \rangle = \|x\|, \qquad \|f\| = 1.$$

Thus $x^* = \|x\| f$ belongs to i(x), since

$$\langle x^*, x \rangle = \|x\| \langle f, x \rangle = \|x\|^2, \qquad \|x^*\| = \|x\| \|f\| = \|x\|.$$

Generally speaking, the set i(x) can contain more than one element. This does not happen if X^* is strictly convex, in particular if X is a Hilbert space.

Lemma 1.1. *If X^* is strictly convex, for any $x \in X$ the set* i(x) *contains only one element.*

Proof. Since i$(0) = \{0\}$, the result is true if $x = 0$.

Let now $x \neq 0$ and let f, g be in i(x); let us prove that

$$\frac{f + g}{2} \in \mathrm{i}(x). \tag{1.2}$$

In fact, we have

$$\langle f, x \rangle = \|x\|^2 = \|f\|^2, \qquad \langle g, x \rangle = \|x\|^2 = \|g\|^2$$

and therefore

$$\left\langle \frac{f + g}{2}, x \right\rangle = \frac{1}{2}\langle f, x \rangle + \frac{1}{2}\langle g, x \rangle = \|x\|^2.$$

This implies

$$\|x\|^2 \leqslant \frac{1}{2}\|f + g\| \|x\|$$

from which

$$\|x\| \leqslant \frac{1}{2}\|f + g\|.$$

On the other hand, since $\|f + g\| \leqslant \|f\| + \|g\| = 2 \|x\|$, we have

$$\frac{1}{2}\|f + g\| = \|x\|$$

and this proves (1.2).

We have thus
$$\|f + g\| = 2\,\|x\| = \|f\| + \|g\|\,.$$

Because of the strict convexity of X^*, this implies that f and g are linearly dependent, i.e., there exists $(a, b) \neq (0, 0)$ such that $af + bg = 0$ a.e..

Supposing $a \neq 0$, we find $f = \lambda\, g$. Since
$$\langle g, x \rangle = \|x\|^2 = \langle f, x \rangle = \lambda \langle g, x \rangle$$

we have $\lambda = 1$ (note that $x \neq 0$ and then $\langle g, x \rangle \neq 0$), i.e., $f = g$. □

Let us determine the duality set i(x) in the particular case of the L^p spaces. Since the spaces $L^p(\Omega)$ $(1 < p < \infty)$ are strictly convex, the duality set i(f) contains only one element f^*. Let us look for f^* in the form

$$f^*(x) = \begin{cases} c_f \overline{f(x)} |f(x)|^\alpha & \text{if } f(x) \neq 0 \\ 0 & \text{if } f(x) = 0 \end{cases}$$

where c_f and α are to be determined.

Since f^* has to belong to L^q, and since
$$|f^*(x)|^q = c_f^q |f(x)|^{q(\alpha+1)}$$

where $f \neq 0$, we must have $q(\alpha + 1) = p$, i.e., $\alpha = \frac{p}{q} - 1 = p - 2$.

Imposing the condition $\langle f^*, f \rangle = \|f\|_p^2$ leads to

$$\|f\|_p^2 = \langle f^*, f \rangle = c_f \int_{f \neq 0} |f|^{\alpha+2} dx = c_f \int_\Omega |f|^p dx = c_f \|f\|_p^p$$

and then
$$c_f = \|f\|_p^{2-p}.$$

Let us prove that we have also $\|f^*\|_q = \|f\|_p$.

In fact, since $|f^*|^q = c_f^q |f|^{q(\alpha+1)} = c_f^q |f|^p$ (where $f \neq 0$), we have

$$\int_\Omega |f^*|^q dx = c_f^q \int_{f \neq 0} |f|^p dx = \|f\|_p^{q(2-p)+p} = \|f\|_p^q.$$

We have then proved

Lemma 1.2. *Let $X = L^p(\Omega)$ $(1 < p < \infty)$. The duality set i(f) contains only the element f^*, where*

$$f^*(x) = \begin{cases} = \|f\|_p^{2-p} \overline{f(x)}\, |f(x)|^{p-2} & \text{if } f(x) \neq 0 \\ 0 & \text{if } f(x) = 0. \end{cases}$$

Remark 1.3. If $p = 1$, we can take

$$f^*(x) = \begin{cases} \|f\|_1 \dfrac{\overline{f(x)}}{|f(x)|} & \text{if } f(x) \neq 0 \\ \psi(x) & \text{if } f(x) = 0 \end{cases}$$

where ψ is any measurable function such that $|\psi(x)| \leqslant \|f\|_1$ a.e.

We leave the proof to the reader. This remark shows that there are infinitely many functions f^* in $i(f)$, provided that the set $\{x \in \Omega \mid f(x) = 0\}$ has positive measure and $\|f\|_1 > 0$.

1.2   Semi-bounded operators on Banach spaces

Let X be a (complex) Banach space and X^* its (topological) dual space. Let $\mathscr{L}$ be a sesquilinear form

$$D \times D' \ni (u, \varphi) \mapsto \mathscr{L}(u, \varphi) \in \mathbb{C},$$

where D' and D are two linear dense subspaces in B^* and B, respectively. This means that

$$\mathscr{L}(\alpha u + \beta v, \varphi) = \alpha \, \mathscr{L}(u, \varphi) + \beta \, \mathscr{L}(u, \psi),$$
$$\mathscr{L}(u, \alpha \varphi + \beta \psi,) = \overline{\alpha} \, \mathscr{L}(u, \varphi) + \overline{\beta} \, \mathscr{L}(v, \varphi).$$

We give now the first basic definition we shall use everywhere.

Definition 1.4. We say that $\mathscr{L}$ is semi-bounded above in D if there exists $c \in \mathbb{R}$ such that, for any $u \in D$ with $i(u) \cap D' \neq \emptyset$, there exists $u^* \in i(u) \cap D'$ for which

$$\operatorname{Re} \mathscr{L}(u, \overline{u^*}) \leqslant c \|u\|^2. \tag{1.3}$$

In particular, if for any $u \in D$ with $i(u) \cap D' \neq \emptyset$, there exists $u^* \in i(u) \cap D'$ such that

$$\operatorname{Re} \mathscr{L}(u, \overline{u^*}) \leqslant 0,$$

we say that $\mathscr{L}$ is dissipative (negative) in D.

If (1.3) holds with a constant $c \geqslant 0$, we say that $\mathscr{L}$ is quasi-dissipative.

Similarly the semi-boundedness below is defined as follows.

Definition 1.5. We say that $\mathscr{L}$ is semi-bounded below in D if there exists $c \in \mathbb{R}$ such that, for any $u \in D$ with $i(u) \cap D' \neq \emptyset$, there exists $u^* \in i(u) \cap D'$ for which

$$\operatorname{Re} \mathscr{L}(u, \overline{u^*}) \geqslant c \|u\|^2. \tag{1.4}$$

In particular, if for any $u \in D$ with $i(u) \cap D' \neq \emptyset$, there exists $u^* \in i(u) \cap D'$ such that

$$\operatorname{Re} \mathscr{L}(u, \overline{u^*}) \geqslant 0, \tag{1.5}$$

we say that $\mathscr{L}$ is accretive (positive) in D.

If (1.4) holds with a constant $c \leqslant 0$, we say that $\mathscr{L}$ is quasi-accretive.

Example 1.6. Consider $X = L^p(\Omega)$, $X^* = L^{p'}(\Omega)$, $D = D' = C_0^1(\Omega)$,

$$\mathscr{L}(u, \varphi) = \int_\Omega \nabla u \cdot \nabla \bar{\varphi} \, dx.$$

Now, if $p \geqslant 2$, u^* belongs to $C_0^1(\Omega)$ for any $u \in C_0^1(\Omega)$. Therefore condition (1.5) means

$$\operatorname{Re} \int_\Omega \nabla u \cdot \nabla(|u|^{p-2}\bar{u}) \, dx \geqslant 0$$

for any $u \in C_0^1(\Omega)$.

If $1 < p < 2$, we prove the following simple fact: $u \in C_0^1(\Omega)$ is such that u^* belongs to $C_0^1(\Omega)$ if and only if we can write $u = \|v\|_{p'}^{2-p'} |v|^{p'-2}\bar{v}$, with $v \in C_0^1(\Omega)$.

In fact, if v is any function in $C_0^1(\Omega)$, then setting $u = \|v\|_{p'}^{2-p'} |v|^{p'-2}\bar{v}$, we have $u \in C_0^1(\Omega)$ and $u^* = v$ belongs to $C_0^1(\Omega)$ too. Conversely, if u is such that u^* belongs to $C_0^1(\Omega)$, set $v = u^*$. We have $v \in C_0^1(\Omega)$ and $u = \|v\|_{p'}^{2-p'} |v|^{p'-2}\bar{v}$. The proof is complete.

Therefore, if $1 < p < 2$, condition (1.5) for any $u \in D$ such that u^* belongs to D', means

$$\operatorname{Re} \int_\Omega \nabla(|v|^{p'-2}v) \cdot \nabla \bar{v} \, dx \geqslant 0$$

for any $v \in C_0^1(\Omega)$.

Now, suppose we have a linear operator $A : D(A) \subset X \to X$. Setting

$$\mathscr{L}(u, \overline{u^*}) = \langle u^*, Au \rangle, \quad D = D(A), \quad D' = B^*$$

in Definitions 1.5 or 1.4, we obtain the analogous definitions for operators.

Definition 1.7. Let $A : D(A) \subset X \to X$ be a linear operator, X being a (complex) Banach space. A is said to be semi-bounded above if there exists $c \in \mathbb{R}$ such that, for any $u \in D(A)$ there exists $u^* \in i(u)$ for which

$$\operatorname{Re} \langle u^*, Au \rangle \leqslant c \|u\|^2.$$

In particular, if this condition holds with $c = 0$ ($c \geqslant 0$) we say that A is dissipative (quasi-dissipative).

Definition 1.8. Let $A : D(A) \subset X \to X$ be a linear operator, X being a (complex) Banach space. A is said to be semi-bounded below if there exists $c \in \mathbb{R}$ such that, for any $u \in D(A)$ there exists $u^* \in i(u)$ for which

$$\operatorname{Re} \langle u^*, Au \rangle \geqslant c \|u\|^2.$$

In particular, if this condition holds with $c = 0$ ($c \leqslant 0$) we say that A is accretive (quasi-accretive).

Remark 1.9. Let A be a linear operator defined on a subspace $D(A)$ contained in $L^p(\Omega)$, Ω being a domain in $\mathbb{R}^n$, $p \in (1, \infty)$. Thanks to Lemma 1.2, the operator A is dissipative with respect to the L^p-norm, briefly is L^p-dissipative, if, and only if,

$$\text{Re} \int_\Omega \langle Au, u \rangle |u|^{p-2} dx \leqslant 0, \quad \forall\, u \in D(A),$$

where the integral is extended on the set $\{x \in \Omega \mid u(x) \neq 0\}$.

1.3   Criteria for the contractivity of a semigroup

1.3.1   Strongly continuous semigroups

Let X be a Banach space. A *semigroup of linear operators on X* is a family of linear and continuous operators $T(t)$ $(0 \leqslant t < \infty)$ from X into itself such that

$$T(0) = I$$
$$T(t + s) = T(t)T(s) \qquad (s, t \geqslant 0).$$

The linear operator

$$Ax = \lim_{t \to 0^+} \frac{T(t)x - x}{t} \tag{1.6}$$

is the *infinitesimal generator of the semigroup $T(t)$*.

The domain $D(A)$ of the operator A is the set of $x \in X$ such that the following limit exists:

$$\lim_{t \to 0^+} \frac{T(t)x - x}{t}.$$

We remark that the linear operator A does not need to be continuous. We say that $T(t)$ is a strongly continuous semigroup (briefly, a C^0-semigroup) if

$$\lim_{t \to 0^+} T(t)x = x, \qquad \forall\, x \in X.$$

The operator A is said to be the generator of the C^0-semigroup if (1.6) holds for any $x \in D(A)$.

The following inequalities hold for any C^0-semigroup.

Theorem 1.10. *Let $T(t)$ be a C_0 semigroup. There exist two constants $\omega \geqslant 0$, $M \geqslant 1$ such hat*

$$\|T(t)\| \leqslant M\, e^{\omega t} \qquad 0 \leqslant t < \infty. \tag{1.7}$$

Proof. First let us show that there exist constants M and $\eta > 0$ such that

$$\|T(t)\| \leqslant M \qquad \forall\, t \in [0, \eta]. \tag{1.8}$$

If (1.8) is false, we can find a sequence of real numbers $t_n > 0$ such that $\|T(t_n)\| > n$, $t_n \to 0$. It follows that there exists $x \in X$ such that

$$\sup_{n \in \mathbb{N}} \|T(t_n)x\| = \infty.$$

If not, we would have

$$\sup_{n \in \mathbb{N}} \|T(t_n)x\| < \infty \qquad \forall\, x \in X;$$

in view of the Banach–Steinhaus Theorem, this implies

$$\sup_{n \in \mathbb{N}} \|T(t_n)\| < \infty$$

and this is absurd. Formula (1.8) is proved.

Since $\|T(0)\| = 1$, we have $M \geqslant 1$. Let now t be a nonnegative number; we can write $t = n\eta + \delta$, where n is a natural number and $0 \leqslant \delta < \eta$. Therefore

$$\|T(t)\| = \|T(\delta)T(\eta)^n\| \leqslant M^{n+1} = M^{1+\frac{t-\delta}{\eta}} \leqslant M^{1+\frac{t}{\eta}} = M\,e^{\omega t}$$

where $\omega = (\log M)/\eta$. $\qquad\square$

A first consequence is that $T(t)x$ is continuous.

Theorem 1.11. *Let $T(t)$ be a C_0-semigroup. For any $x \in X$, $T(t)x$ is a continuous function on X of the real variable $t \geqslant 0$.*

Proof. The continuity from the right at $t = 0$ is obvious. Let us fix $t > 0$ and take $h \geqslant 0$; we have

$$\|T(t+h)x - T(t)x\| \leqslant \|T(t)\|\,\|T(h)x - x\| \leqslant Me^{\omega t}\|T(h)x - x\|$$

and then

$$\lim_{h \to 0+} \|T(t+h)x - T(t)x\| = 0.$$

On the other hand, if $t - h \geqslant 0$, we have also

$$\|T(t-h)x - T(t)x\| \leqslant \|T(t-h)\|\,\|x - T(h)x\| \leqslant Me^{\omega\,(t-h)}\|x - T(h)x\|\,.$$

It follows that

$$\lim_{h \to 0-} \|T(t+h)x - T(t)x\| = 0$$

and the result is proved. $\qquad\square$

The next theorem shows some properties of C^0-semigroups.

Theorem 1.12. *Let $T(t)$ be a C_0-semigroup and A its generator. Then*

a) $\displaystyle \lim_{h \to 0} \frac{1}{h} \int_t^{t+h} T(s)x \, ds = T(t)x \qquad \forall\, x \in X;$

b) $\displaystyle x \in X \implies \int_0^t T(s)x \, ds \in D(A)$ *e*

$$A\left(\int_0^t T(s)x \, ds \right) = T(t)x - x; \tag{1.9}$$

c) $x \in D(A) \implies T(t)x \in D(A)$ *e*

$$\frac{d}{dt} T(t)x = AT(t)x = T(t)Ax; \tag{1.10}$$

d) *for any $x \in D(A)$ we have*

$$T(t)x - T(s)x = \int_s^t T(\tau)Ax \, d\tau = \int_s^t AT(\tau)x \, d\tau. \tag{1.11}$$

Proof. Fix $x \in X$ and $t > 0$; given $\varepsilon > 0$, in view of Theorem 1.11, there exists $\delta_\varepsilon > 0$ such that

$$\|T(s)x - T(t)x\| < \varepsilon \qquad |s - t| < \delta_e.$$

It follows that, if $|s - t| < \delta_\varepsilon$,

$$\left\| \frac{1}{h} \int_t^{t+h} (T(s)x - T(t)x) \, ds \right\| \leqslant \frac{1}{|h|} \left| \int_t^{t+h} \|T(s)x - T(t)x\| ds \right| < \varepsilon,$$

i.e., a) (it is obvious how to change the proof for $t = 0$).

As far as b) is concerned, fix $x \in X$ and $h > 0$. One has

$$\frac{T(h) - I}{h} \int_0^t T(s)x \, ds = \frac{1}{h} \int_0^t (T(s + h)x - T(s)x) ds$$

$$= \frac{1}{h} \int_t^{t+h} T(s)x \, ds - \frac{1}{h} \int_0^h T(s)x \, ds.$$

The last term tending to $T(t)x - x$ as $h \to 0^+$ because of a), the integral in b) belongs to $D(A)$ and b) holds.

Let now $x \in D(A)$ and $h > 0$; we have

$$\frac{T(h) - I}{h} T(t)x = T(t) \left(\frac{T(h) - I}{h} \right) x \ \to \ T(t)Ax.$$

This shows that $T(t)x$ belongs to $D(A)$ and moreover $AT(t)x = T(t)Ax$.

We have also proved that

$$\frac{d^+}{dt}T(t)x = AT(t)x = T(t)Ax.$$

Let us consider now the left derivative. We can write

$$\frac{T(t-h)x - T(t)x}{-h} - T(t)Ax = T(t-h)\left[\frac{x - T(h)x}{-h}\right] - T(t)Ax$$

$$= T(t-h)\left[\frac{T(h)x - x}{h} - Ax\right] + [T(t-h) - T(t)]Ax.$$

Since $x \in D(A)$, we have

$$\lim_{h \to 0^+} \frac{T(h)x - x}{h} = Ax.$$

The norm $\|T(s)\|$ being bounded on the compact sets, in view of Theorem 1.10 (note that $T(t)$ does not need to be continuous!), we find

$$\lim_{h \to 0^+} T(t-h)\left[\frac{T(h)x - x}{h} - Ax\right] = 0.$$

Moreover

$$\lim_{h \to 0^+} [T(t-h) - T(t)]Ax = 0 \quad \text{and thus} \quad \frac{d^-}{dt}T(t)x = T(t)Ax.$$

This proves the statement c).
Finally, (1.11) is obtained by integrating (1.10). □

We recall that the operator A is closed if its graph is closed, which means that the implication holds

$$\begin{cases} x_n \in D(A) \\ x_n \to x \\ Ax_n \to y \end{cases} \implies \begin{cases} x \in D(A) \\ Ax = y. \end{cases} \tag{1.12}$$

Theorem 1.13. *Let A be the generator of the C_0-semigroup $T(t)$. Then A is a densely defined closed operator.*

Proof. We start by proving that $D(A)$ is dense in X. Let $x \in X$ and define

$$x_t = \frac{1}{t}\int_0^t T(s)x\,ds.$$

From b) of Theorem 1.12, $x_t \in D(A)$ and from a) $x_t \to x$. This means that $\overline{D(A)} = X$.

In order to prove that A is a closed operator, we have to show that (1.12) holds. Since $x_n \in D(A)$, (1.11) implies

$$T(t)x_n - x_n = \int_0^t T(s)Ax_n ds.$$

Letting $n \to \infty$, one has

$$T(t)x - x = \int_0^t T(s)y\, ds$$

from which

$$\frac{T(t)x - x}{t} = \frac{1}{t}\int_0^t T(s)y\, ds.$$

As $t \to 0^+$, the right-hand side tends to y and thus $x \in D(A)$, $Ax = y$. □

The next result shows that a C_0-semigroup is uniquely determined by its generator.

Theorem 1.14. *Let A and B two generators of the C_0-semigroups $T(t)$ and $S(t)$ respectively. If $A = B$ then the two semigroups coincide, i.e., $T(t) = S(t)$ for any $t \geqslant 0$.*

Proof. Let $x \in D(A) = D(B)$. From (1.10) it follows that

$$\frac{d}{ds}T(t - s)S(s)x = -AT(t - s)S(s)x + T(t - s)BS(s)x$$

$$= -T(t - s)AS(s)x + T(t - s)BS(s)x = 0 \quad (0 < s < t)$$

and then the function $T(t - s)S(s)x$ of the real variable s is constant. In particular $T(t)x = S(t)x$, i.e., $T(t) = S(t)$ on $D(A)$. The domain $D(A)$ being dense in X (see Theorem 1.13), it follows that $T(t) = S(t)$. □

Properties (1.7) and (1.10) imply that for any given $u_0 \in D(A)$ the function $u(t) = T(t)u_0$ is the only solution of the abstract Cauchy problem

$$\begin{cases} \dfrac{du}{dt} = Au, \quad (t > 0) \\[2mm] u(0) = u_0. \end{cases} \tag{1.13}$$

Remark 1.15. It is still possible to solve the Cauchy problem (1.13) where u_0 is an arbitrary element of X. In order to do that, it is necessary to introduce a concept of generalized solution. For this we refer to Pazy [87, Ch. 4].

Example 1.16. An example of C_0-semigroup.

Let $X = C^0([0, \infty])$, where this symbol means the space of the complex-valued functions defined in $[0, \infty)$ such that there exists the limit

$$\lim_{x \to +\infty} f(x).$$

The space X, equipped with the norm

$$\|f\|_\infty = \sup_{x \in [0,+\infty)} |f(x)| \,,$$

is a Banach space. Define the family of operators $T(t)$ $(t \geqslant 0)$ by

$$[T(t)f](x) = f(x + t).$$

Obviously, for any $t \geqslant 0$, it makes sense to consider $f(x+t)$. Moreover $T(t)f$ is a continuous function and since

$$\lim_{x \to +\infty} [T(t)f](x) = \lim_{x \to +\infty} f(x),$$

$T(t)$ maps X into itself. Let us remark that

$$\|T(t)f\|_\infty \leqslant \|f\|_\infty \,.$$

It is clear that $T(t)$ is a semigroup. Let us prove that it is a C_0-semigroup, i.e., that

$$\lim_{t \to 0^+} \|T(t)f - f\|_\infty = 0 \,. \tag{1.14}$$

By hypothesis, there exists $\alpha \in \mathbb{C}$ to which $f(x)$ tends as $x \to +\infty$. Given $\varepsilon > 0$, there exists $K_\varepsilon > 0$ such that

$$|f(x) - \alpha| < \varepsilon \qquad \forall \, x \geqslant K_\varepsilon.$$

This implies

$$|f(x+t) - f(x)| \leqslant |f(x+t) - \alpha| + |\alpha - f(x)| < 2\varepsilon \quad \forall \, x \geqslant K_\varepsilon, t \geqslant 0. \tag{1.15}$$

On the other hand f is uniformly continuous on $[0, K_\varepsilon + 1]$ and then there exists $\delta_\varepsilon > 0$ (which can be supposed to be less than 1) such that

$$|f(x+t) - f(x)| < \varepsilon \qquad \forall \, x \in [0, K_\varepsilon], \ 0 \leqslant t < \delta_\varepsilon \,.$$

Keeping in mind (1.15), we find

$$|f(x+t) - f(x)| < 2\varepsilon \qquad \forall \, x \in [0, \infty), \ 0 \leqslant t < \delta_\varepsilon$$

and (1.14) is proved.

What is the generator A of $T(t)$ and its domain $D(A)$?

The function f belongs to $D(A)$ if and only if there exists in X the limit

$$Af = \lim_{t \to 0^+} \frac{T(t)f - f}{t} = \lim_{t \to 0^+} \frac{f(t + \cdot) - f(\cdot)}{t}.$$

In particular

$$Af(x) = \lim_{t \to 0^+} \frac{f(t + x) - f(x)}{t} \qquad \forall\, x \in [0, \infty)$$

and then f admits the right derivative for any $x \geqslant 0$ and the right derivative, Af, is continuous everywhere. But then, in view of a well-known result in the theory of functions of one real variable (see, e.g., Pazy [87, pp. 42–43]), f is differentiable for any $x > 0$.

Moreover, since $Af \in X$, there exists also

$$\lim_{x \to +\infty} f'(x).$$

Therefore $D(A)$ is contained in the space of the functions $f \in C^1([0, \infty))$ such that $f' \in X$ and $Af = f'$.

Viceversa, if $f \in C^1([0, \infty))$ and $f' \in X$, then $f \in D(A)$. In fact, we have

$$\frac{f(x + t) - f(x)}{t} - f'(x) = \frac{1}{t} \int_x^{x+t} [f'(u) - f'(x)]\, du.$$

But, since $f' \in X$, f' is uniformly continuous and then $|f'(u) - f'(x)| < \varepsilon$ for $|x - u|$ less than a certain δ_ε. Thus

$$\left| \frac{f(x + t) - f(x)}{t} - f'(x) \right| \leqslant \frac{1}{t} \int_x^{x+t} |f'(u) - f'(x)|\, du < \varepsilon \qquad 0 < t < \delta_\varepsilon$$

and this shows that

$$\lim_{t \to 0^+} \left\| \frac{f(\cdot + t) - f}{t} - f' \right\|_\infty = 0,$$

i.e., $f \in D(A)$, $Af = f'$. We have thus proved that

$$D(A) = \{ f \in X \mid \exists\, f',\ f' \in X \}, \qquad Af = f'.$$

1.3.2   Main result

The case when $\omega = 0$ and $M = 1$ in the inequality (1.7) is of particular interest. We have

$$\| T(t) \| \leqslant 1$$

and the semigroup is said to be a contraction semigroup or a semigroup of contractions. If the operator A is the generator of a C^0-semigroup of contractions, the solution of the Cauchy problem (1.13) satisfies the estimate

$$\|u(t)\| \leqslant \|u_0\| \quad (t \geqslant 0).$$

In order to find characterizations of such operators we prove some lemmas.

We recall that the resolvent of a linear operator A, $\varrho(A)$, is the set of complex numbers λ such that there exists the resolvent operator $(\lambda I - A)^{-1}$ and it is continuous. The spectrum $\sigma(A)$ of the operator A is defined as $\mathbb{C} \setminus \varrho(A)$. By $R(\lambda : A)$ ($\lambda \in \varrho(A)$), or shortly R_λ, we denote the operator $(\lambda I - A)^{-1}$.

Lemma 1.17. *Let A be a linear operator. If $\varrho(A) \neq \emptyset$ then A is closed.*

Proof. Suppose that the sequence $\{x_n\}$, contained in the domain of A, is such that $x_n \to x$ and $Ax_n \to y$.

Given $\lambda \in \varrho(A)$, we get

$$\lambda x_n - A x_n \to \lambda x - y \quad \text{and then} \quad x_n \to R_\lambda(\lambda x - y).$$

Because of the uniqueness of the limit, we find

$$x = R_\lambda(\lambda x - y).$$

This shows that $x \in D(\lambda I - A) = D(A)$ and $(\lambda I - A)x = \lambda x - y$, i.e., $Ax = y$ and the lemma is proved (see (1.12)). $\square$

Lemma 1.18. *If $\lambda, \mu \in \varrho(A)$ we have the resolvent identity*

$$R_\lambda - R_\mu = (\mu - \lambda)R_\lambda R_\mu. \tag{1.16}$$

Moreover the operators R_λ and R_μ commute: $R_\lambda R_\mu = R_\mu R_\lambda$.

Proof. We have

$$(\lambda I - A)[R_\lambda - R_\mu](\mu I - A) = [I - (\lambda I - A)R_\mu](\mu I - A)$$
$$= (\mu I - A) - (\lambda I - A) = (\mu - \lambda)I$$

and (1.16) follows. By exchanging λ and μ we prove the commutativity. $\square$

Let A be an operator such that[1] $\mathbb{R}^+ \subset \varrho(A)$; we can then consider R_λ for any $\lambda > 0$. The operator $A_\lambda = \lambda A R_\lambda$ is called the Yosida approximation of A. Even if A is unbounded, the operator A_λ is a linear and continuous operator defined all over X. The linearity is obvious, while A_λ is continuous, because

$$(\lambda I - A)R_\lambda = I \qquad \Longleftrightarrow \qquad A R_\lambda = \lambda R_\lambda - I$$

---

[1] By $\mathbb{R}^+$ we denote the set $\{\lambda \in \mathbb{R} \mid \lambda > 0\}$.

and then
$$A_\lambda = \lambda^2 R_\lambda - \lambda I. \tag{1.17}$$

The next lemma shows in which sense A_λ is an approximation of A.

Lemma 1.19. *Let A be a densely defined operator such that $\mathbb{R}^+ \subset \varrho(A)$ and*

$$\|R_\lambda\| \leqslant \frac{1}{\lambda} \qquad \forall\, \lambda > 0.$$

Then

$$\lim_{\lambda \to 0^+} \lambda R_\lambda x = x \qquad \forall\, x \in X, \tag{1.18}$$

$$\lim_{\lambda \to 0^+} A_\lambda x = Ax \qquad \forall\, x \in D(A). \tag{1.19}$$

Proof. Suppose first $x \in D(A)$; since

$$R_\lambda(\lambda I - A)x = x \quad \text{we get} \quad \lambda R_\lambda x = x + R_\lambda Ax.$$

Limit (1.18) for any $x \in D(A)$ follows from the inequality

$$\|R_\lambda Ax\| \leqslant \frac{1}{\lambda}\,\|Ax\|\,.$$

Let now $x \in X$; given $\varepsilon > 0$, by hypothesis there exists $y \in D(A)$ such that $\|x - y\| < \varepsilon$. Since

$$\|\lambda R_\lambda x - x\| \leqslant \|\lambda R_\lambda x - \lambda R_\lambda y\| + \|\lambda R_\lambda y - y\| + \|y - x\|$$
$$\leqslant 2\,\|x - y\| + \|\lambda R_\lambda y - y\|$$

we have

$$\limsup_{\lambda \to \infty} \|\lambda R_\lambda x - x\| \leqslant 2\varepsilon\,.$$

Because of the arbitrariness of ε, (1.18) is proved for any $x \in X$.

As far as (1.19) is concerned, formula (1.18) clearly implies

$$\lim_{\lambda \to 0^+} \lambda R_\lambda Ax = Ax \qquad \forall\, x \in D(A)$$

and (1.19) follows, because R_λ commute with A on $D(A)$.[2]   □

Lemma 1.20. *Let $U(t)$ and $V(t)$ be two contraction semigroups whose generators are C and D respectively. Suppose that $U(t)$ and $V(t)$ commute, i.e., $U(t)V(s) = V(s)U(t)$ for any $s,t \geqslant 0$. Then*

$$\|U(t)x - V(t)x\| \leqslant t\,\|Cx - Dx\| \qquad \forall\, x \in D(C) \cap D(D). \tag{1.20}$$

---

[2]In fact $(\lambda I - A)R_\lambda = R_\lambda(\lambda I - A) = I$ on $D(A)$ and then $AR_\lambda x = R_\lambda Ax$ for any $x \in D(A)$.

Proof. First observe that from the commutativity of $U(t)$ and $V(t)$ it follows that also the generator of $U(t)$, C, commutes with $V(t)$. Specifically, let $x \in D(C)$; we can write

$$\frac{U(h) - I}{h} V(t)x = V(t) \frac{U(h)x - x}{h} ;$$

therefore $x \in D(C) \Rightarrow V(t)x \in D(C)$ and

$$CV(t)x = V(t)Cx$$

(for any $t \geqslant 0$). Keeping in mind (1.10), we have for any $x \in D(C) \cap D(D)$,

$$U(t)x - V(t)x = \int_0^t \frac{d}{ds}[U(s)V(t - s)x]\,ds$$

$$= \int_0^t [U(s)CV(t - s)x - U(s)V(t - s)Dx]\,ds$$

and then

$$U(t)x - V(t)x = \int_0^t [U(s)V(t - s)Cx - U(s)V(t - s)Dx]\,ds\,.$$

This implies

$$\|U(t)x - V(t)x\| \leqslant \int_0^t \|U(s)\|\,\|V(t - s)\|\,\|Cx - Dx\|\,ds$$

$$\leqslant t\,\|Cx - Dx\|\,. \qquad \square$$

We are now in a position to prove the following characterization of generators of contractive semigroups, which provides the main result of this section.

Theorem 1.21 (Hille–Yosida). *A linear operator A generates a C^0 semigroup of contractions $T(t)$ if, and only if,*

(i) *A is closed and $D(A)$ is dense in X;*
(ii) *$\varrho(A) \supset \varrho^+$ and*

$$\|R_\lambda\| \leqslant \frac{1}{\lambda}, \qquad \forall\,\lambda > 0. \tag{1.21}$$

Proof. Suppose that A is the generator of a contraction semigroup. We know already that A is a densely defined and closed operator (see Theorem 1.13).

In order to prove (ii), observe that for any $\lambda > 0$, $e^{-\lambda t}T(t)$ is a contraction semigroup, because

$$\|e^{-\lambda t}T(t)\| = e^{-\lambda t}\|T(t)\| \leqslant 1.$$

The generator of $e^{-\lambda t}T(t)$ is $A - \lambda I$; in fact

$$\frac{e^{-\lambda t}T(t)x - x}{t} = e^{-\lambda t}\frac{T(t)x - x}{t} + \frac{e^{-\lambda t} - 1}{t}\,x$$

and then

$$\lim_{t \to 0^+} \frac{e^{-\lambda t}T(t)x - x}{t} = Ax - \lambda x, \qquad \forall\, x \in D(A) = D(A - \lambda I).$$

We can apply (1.9) and (1.11) to $A - \lambda I$, obtaining

$$e^{-\lambda t}T(t)x - x = (A - \lambda I)\left(\int_0^t e^{-\lambda s}T(s)x\,ds\right), \qquad \forall\, x \in X;$$

$$e^{-\lambda t}T(t)x - x = \int_0^t e^{-\lambda s}T(s)(A - \lambda I)x\,ds, \qquad \forall\, x \in D(A).$$

Letting $t \to +\infty$ we find

$$x = (\lambda I - A)\left(\int_0^\infty e^{-\lambda s}T(s)x\,ds\right) \qquad \forall\, x \in X,$$

$$x = \int_0^\infty e^{-\lambda s}T(s)(\lambda I - A)x\,ds \qquad \forall\, x \in D(A).$$

The first equality shows that the range of $\lambda I - A$ is all of X, while the second one implies that $\lambda I - A$ is injective. Then there exists $(\lambda I - A)^{-1}$ and, setting $y = (\lambda I - A)x$,

$$(\lambda I - A)^{-1}y = \int_0^\infty e^{-\lambda s}T(s)y\,ds.$$

This leads to

$$\|(\lambda I - A)^{-1}y\| \leqslant \int_0^\infty e^{-\lambda s}\|T(s)\|\,\|y\|\,ds \leqslant \|y\|\int_0^\infty e^{-\lambda s}ds = \frac{\|y\|}{\lambda}.$$

Then we have proved that $(\lambda I - A)^{-1}$ is continuous (i.e., $\lambda \in \varrho(A)$) and (1.21) holds.

Viceversa, let A satisfy (i) and (ii). For any $\lambda > 0$ we can consider the Yosida approximation A_λ and the limit (1.19) holds. The operator A_λ, being linear and continuous, is the generator of a semigroup uniformly continuous e^{tA_λ}. This is a contractive semigroup, because, keeping in mind (1.17), we have

$$\|e^{tA_\lambda}\| = \|e^{-\lambda t I}e^{\lambda^2 tR_\lambda}\| \leqslant e^{-\lambda t}e^{\lambda^2 t\|R_\lambda\|} \leqslant e^{-\lambda t}e^{\lambda t} = 1$$

(here we used that $\lambda\|R_\lambda\| \leqslant 1$).

Define

$$T(t)x = \lim_{\lambda \to \infty} e^{tA_\lambda}x \qquad \forall\, x \in X. \tag{1.22}$$

To see that this definition makes sense, we have to show that this limit does exist for any $x \in X$.

Let us start by first showing that this limit does exist for any $x \in D(A)$: given $\lambda, \mu > 0$, it is easy to show that the contraction semigroups e^{tA_λ} and e^{tA_μ} commute and then we can apply Lemma 1.20. From (1.20) it follows that

$$\|e^{tA_\lambda}x - e^{tA_\mu}x\| \leqslant t\|A_\lambda x - A_\mu x\| \qquad \forall\, x \in D(A).$$

But, if $x \in D(A)$, (1.19) shows that $A_\lambda x \to Ax$ and then, given $\varepsilon > 0$, there exists $\lambda_\varepsilon > 0$ such that, for $\lambda, \mu > \lambda_\varepsilon$, one has $\|A_\lambda x - A_\mu x\| < \varepsilon$. Thus

$$\|e^{tA_\lambda}x - e^{tA_\mu}x\| \leqslant t\varepsilon.$$

This shows that the limit (1.22) does exist for any $x \in D(A)$. Let now $x \in X$; given $\varepsilon > 0$ let $y \in D(A)$ such that $\|x - y\| < \varepsilon$. We have

$$\|e^{tA_\lambda}x - e^{tA_\mu}x\| \leqslant \|e^{tA_\lambda}x - e^{tA_\lambda}y\| + \|e^{tA_\lambda}y - e^{tA_\mu}y\| + \|e^{tA_\mu}y - e^{tA_\mu}x\|$$
$$\leqslant 2\|x - y\| + \|e^{tA_\lambda}y - e^{tA_\mu}y\|$$

and then there exists λ_ε such that, for $\mu > \lambda_\varepsilon$, one has

$$\|e^{tA_\lambda}x - e^{tA_\mu}x\| \leqslant 3\varepsilon'.$$

Thus there exists the limit (1.22) for any $x \in X$.

(1.22) implies also

$$\lim_{\lambda \to \infty} \|T(t)x\| = \lim_{\lambda \to \infty} \|e^{tA_\lambda}x\| \leqslant \|x\| \qquad \forall\, x \in X$$

and then $\|T(t)\| \leqslant 1$.

From (1.22) it easily follows that $T(0) = I$, $T(t + s) = T(t)T(s)$, i.e., $T(t)$ is a semigroup. Let us show that $T(t)$ is continuous:

$$\lim_{t \to 0+} T(t)x = x \qquad \forall\, x \in X \tag{1.23}$$

We prove first (1.23) when $x \in D(A)$. In fact, if $x \in D(A)$, we have

$$e^{tA_\lambda}x - x = \int_0^t \frac{d}{ds}[e^{sA_\lambda}x]\,ds = \int_0^t e^{sA_\lambda}A_\lambda x\,ds,$$

from which, letting $\lambda \to \infty$ and keeping in mind (1.19), it follows that

$$T(t)x - x = \int_0^t T(s)Ax\,ds \tag{1.24}$$

(invoking the dominated convergence theorem, we can pass the limit under the integral sign, because $\|T(s)Ax\| \leqslant \|Ax\|$). We have then

$$\|T(t)x - x\| \leqslant \int_0^t \|T(s)Ax\|\,ds \leqslant t\|Ax\|$$

and (1.23) follows for any $x \in D(A)$. The density of $D(A)$ in X implies the result for any $x \in X$.

We have shown that $T(t)$ is a contractive C^0-semigroup. To complete the proof, we have to show that A is the generator of $T(t)$.

Let B be the generator of $T(t)$; we have to show that $A = B$.

Dividing (1.24) by t we get

$$\frac{T(t)x - x}{t} = \frac{1}{t} \int_0^t T(s) Ax \, ds \qquad \forall \, x \in D(A)$$

and then the limit (1.6) exists and

$$\lim_{t \to 0^+} \frac{T(t)x - x}{t} = Ax \qquad \forall \, x \in D(A).$$

This shows that $D(A) \subset D(B)$ and $Bx = Ax$ on $D(A)$. On the other hand $1 \in \varrho(A)$ (by hypothesis) and $1 \in \varrho(B)$ (because of the necessity part of the present theorem). Therefore $(I - A)^{-1}$ and $(I - B)^{-1}$ do exist and are continuous. In particular, $I - A$ and $I - B$ are surjective operators. Thus

$$(I - B)D(A) = (I - A)D(A) = X = (I - B)D(B)$$

from which follows: $D(A) = D(B)$ and then $A = B$. $\qquad\qquad\square$

Strictly speaking, the definition of e^{tA} doesn't make sense, because A can be unbounded. Nevertheless, we can still consider this exponential, provided it is understood in a generalized sense. This is shown by the next result.

Lemma 1.22. *If A is the generator of a C^0-semigroup, then $T(t) = e^{tA}$, where this exponential is understood as*

$$e^{tA}x = \lim_{\lambda \to \infty} e^{tA_\lambda}x, \qquad x \in X.$$

Proof. In the proof of the Hille–Yosida Theorem, we have seen that there exists the limit

$$\lim_{\lambda \to \infty} e^{tA_\lambda}x, \qquad \forall \, x \in X$$

and that it defines a semigroup $S(t)$, whose generator is A.

Both the semigroups $T(t)$ and $S(t)$ being generated by A, Theorem 1.14 implies $T(t) \equiv S(t)$. $\qquad\qquad\square$

Another interesting formula is the following one, which shows that the resolvent can be considered as the Laplace transform of the semigroup

$$R_\lambda u = \int_0^\infty e^{-\lambda t}[T(t)u] \, dt \qquad (\mathrm{Re}\,\lambda > \omega).$$

From the Hille–Yosida Theorem, one can obtain also the following characterization of the generators of C^0-semigroups, where M and ω are the constants appearing in (1.7)

Theorem 1.23. *A linear operator A generates a C^0 semigroup $T(t)$ if, and only if,*

(i) *A is closed and $D(A)$ is dense in X;*
(ii) *$\varrho(A) \supset \{\lambda \in \varrho \mid \lambda > \omega\}$ and*

$$\|R_\lambda^n\| \leqslant \frac{M}{(\lambda - \omega)^n}, \quad \forall\, \lambda > \omega, \ n = 1, 2, \dots.$$

1.4 Dissipativity in abstract setting

1.4.1 Dissipative operators on Banach spaces

In this section we prove some lemmas concerning dissipative operators (see Definition 1.7). Such results will permit us to prove a new characterization of generators of contractive semigroups.

Lemma 1.24. *Let $x, y \in X$. The inequality*

$$\|x\| \leqslant \|x - \alpha y\| \tag{1.25}$$

holds for any $\alpha > 0$ if, and only if, there exists $\varphi \in \mathrm{i}(x)$ such that

$$\mathrm{Re}\langle \varphi, y \rangle \leqslant 0. \tag{1.26}$$

Proof. If $x = 0$ the result is trivial, since $\mathrm{i}(0) = \{0\}$. Let $x \neq 0$.

If (1.26) is true, for any $\alpha > 0$ we may write

$$\|x\|^2 = \langle \varphi, x \rangle \leqslant \langle \varphi, x \rangle - \alpha\, \mathrm{Re}\langle \varphi, y \rangle = \mathrm{Re}\langle \varphi, x - \alpha y \rangle \leqslant \|x\|\,\|x - \alpha y\|$$

and (1.25) follows.

Conversely, let us suppose that (1.25) holds for any $\alpha > 0$. Let φ_α be an element of $\mathrm{i}(x - \alpha y)$ and define $\psi_\alpha = \varphi_\alpha / \|\varphi_\alpha\|$. Note that $\varphi_\alpha \neq 0$, because $\|x\| \leqslant \|x - \alpha y\| = \|\varphi_\alpha\|$ and we are assuming $x \neq 0$.

Moreover

$$\langle \psi_\alpha, x - \alpha y \rangle = \langle \varphi_\alpha, x - \alpha y \rangle / \|\varphi_\alpha\| = \|x - \alpha y\| \geqslant \|x\|. \tag{1.27}$$

Because of the Banach–Alaoglu Theorem, we can find a sequence $\{\alpha_n\}$ of positive numbers such that $\alpha_n \to 0$ and $\psi_{\alpha_n} \overset{*}{\rightharpoonup} \psi_0$, with

$$\|\psi_0\| \leqslant 1. \tag{1.28}$$

Putting $\alpha = \alpha_n$ in (1.27) and letting $n \to \infty$, we find[3]

$$\|x\| = \langle \psi_o, x \rangle \leqslant \|\psi_0\|\,\|x\|$$

---

[3] Note that, if $x_n \to x_0$ and $\psi_n \overset{*}{\rightharpoonup} \psi_0$, then $\langle \psi_n, x_n \rangle \to \langle \psi, x \rangle$.

from which, keeping in mind (1.28), we find

$$\|\psi_o\| = 1.$$

Define $\varphi = \|x\|\psi_0$. Since

$$\langle \varphi, x \rangle = \|x\| \langle \psi_o, x \rangle = \|x\|^2 = \|\varphi\|^2$$

we have $\varphi \in i(x)$. Moreover, the inequality

$$\|x\| \leqslant \|x - \alpha y\| = \operatorname{Re}\langle \psi_a, x \rangle - \alpha \operatorname{Re}\langle \psi_a, y \rangle \leqslant \|x\| - \alpha \operatorname{Re}\langle \psi_a, y \rangle$$

shows that

$$\operatorname{Re}\langle \psi_a, y \rangle \leqslant 0.$$

Putting $\alpha = \alpha_n$ and letting $n \to \infty$, we find (1.26). $\qquad\square$

This lemma has some interesting consequences.

Corollary 1.25. *The linear operator A is dissipative if and only if, for any $x \in D(A)$, we have*

$$\|x\| \leqslant \|x - \alpha A x\| \tag{1.29}$$

for any $\alpha > 0$.

Proof. The operator A is dissipative if, and only if, for any $x \in D(A)$, there exists $\varphi \in i(x)$ such that $\operatorname{Re}\langle \varphi, Ax \rangle \leqslant 0$. For a fixed $x \in D(A)$, Lemma 1.24 (where $y = Ax$) shows that this happens if and only if (1.29) holds for any $\alpha > 0$. $\qquad\square$

The operator A is said to be m-dissipative if A is dissipative and $\varrho(A) \cap \mathbb{R}^+ \neq \emptyset$. We denote by $\mathscr{R}(A)$ the range of A.

Corollary 1.26. *The operator A is m-dissipative if, and only if, A is dissipative and there exists $\lambda > 0$ such that $\mathscr{R}(\lambda I - A) = X$.*

Proof. If A is dissipative and $\mathscr{R}(\lambda I - A) = X$, then $(\lambda I - A)^{-1}$ does exist and is continuous, in view of (1.29). This shows that A is m-dissipative. The converse is obvious. $\qquad\square$

Corollary 1.27. *If A is closed and dissipative, then for any $\lambda > 0$ the range $\mathscr{R}(\lambda I - A)$ is closed.*

Proof. Let y_n be a sequence in $\mathscr{R}(\lambda I - A)$ such that $y_n \to y_0$. We can write $y_n = \lambda x_n - A x_n$, for some $x_n \in D(\lambda I - A) = D(A)$.

Because of Corollary 1.25, we have

$$\|x_{n+p} - x_n\| \leqslant \|(\lambda x_{n+p} - A x_{n+p}) - (\lambda x_n - A x_n)\| = \|y_{n+p} - y_n\|$$

and then $\{x_n\}$ is a Cauchy sequence in X. Let x_0 be its limit. We have $A x_n = \lambda x_n - y_n \to \lambda x_0 - y_0$. Since A is a closed operator, x_0 belongs to $D(A)$ and $A x_0 = \lambda x_0 - y_0$, i.e., $y_0 = \lambda x_0 - A x_0$. This shows that y_0 belongs to $\mathscr{R}(\lambda I - A)$, i.e., that $\mathscr{R}(\lambda I - A)$ is closed. $\qquad\square$

Lemma 1.28. *Let A be a linear operator and let $\mu \in \varrho(A)$. Then $\lambda \in \varrho(A)$ if and only if U^{-1} belongs to $\mathscr{B}(X)$, where*

$$U = I + (\lambda - \mu)(\mu I - A)^{-1}.$$

In this case

$$(\lambda I - A)^{-1} = (\mu I - A)^{-1} U^{-1}.$$

Proof. If U^{-1} does exist and is continuous, we have

$$
\begin{aligned}
(\lambda I - A)(\mu I - A)^{-1} U^{-1} &= [(\lambda - \mu)I + (\mu I - A)](\mu I - A)^{-1} U^{-1} \\
&= [(\lambda - \mu)(\mu I - A)^{-1} + I] U^{-1} = U U^{-1} = I.
\end{aligned}
$$

In the same way

$$
\begin{aligned}
(\mu I - A)^{-1} U^{-1}(\lambda I - A) &= (\mu I - A)^{-1} U^{-1}[(\lambda - \mu)I + \mu I - A] \\
&= (\mu I - A)^{-1} U^{-1}[(\lambda - \mu)(\mu I - A)^{-1} + I](\mu I - A) \\
&= (\mu I - A)^{-1}(\mu I - A) = I_{D(A)}.
\end{aligned}
$$

This means that $(\lambda I - A)^{-1}$ exists, is given by $(\mu I - A)^{-1} U^{-1}$ and thus it belongs to $\mathscr{B}(X)$.

The proof of the converse is similar. $\qquad\square$

1.4.2 Another characterization of generators of contractive semigroups

The Hille–Yosida Theorem 1.21 characterizes the generators of contractive semigroups. The next results provide different necessary and sufficient conditions under which A generates a contractive semigroup. Such conditions are related to the concept of dissipativity.

Theorem 1.29 (Lumer–Phillips). *If A generates a C^0 semigroup of contractions, then*

(i) $\overline{D(A)} = X$;

(ii) *A is dissipative. More precisely, for any $x \in D(A)$, we have*

$$\mathrm{Re}\langle x^*, Ax \rangle \leqslant 0, \forall\, x^* \in \mathrm{i}(x);$$

(iii) $\varrho(A) \supset \varrho^+$.

Conversely, if

(i') $\overline{D(A)} = X$;

(ii') *A is dissipative;*

(iii') $\varrho(A) \cap \varrho^+ \neq \emptyset$,

then A generates a C^0 semigroup of contractions.

Proof. Because of the Hille–Yosida Theorem 1.21, the operator A generates a C^0-semigroup of contractions if and only if the following conditions are satisfied:

(a) A is closed;

(b) $\overline{D(A)} = X$;

(c) $\varrho(A) \supset \mathbb{R}^+$;

(d) $\|R_\lambda\| \leqslant \dfrac{1}{\lambda} \qquad \forall\, \lambda > 0$.

Let us suppose that A generates the C^0 semigroup of contractions $T(t)$. Since (a)–(d) hold true, conditions (i) and (iii) are certainly satisfied. In order to prove (ii), let x^* denote any element in $\mathrm{i}(x)$. We have

$$\langle x^*, T(t)x - x \rangle = \langle x^*, T(t)x \rangle - \|x\|^2$$

and since

$$|\langle x^*, T(t)x \rangle| \leqslant \|x^*\|\,\|T(t)x\| \leqslant \|x\|^2,$$

we find

$$\mathrm{Re}\langle x^*, T(t)x - x \rangle = \mathrm{Re}\langle x^*, T(t)x \rangle - \|x\|^2 \leqslant 0.$$

Supposing $x \in D(A)$, dividing by t and letting $t \to 0^+$, we get

$$\mathrm{Re}\langle x^*, Ax \rangle \leqslant 0$$

and (ii) is proved.

Conversely, let A be an operator satisfying (i′)–(iii′). Condition (b) is obviously true.

Condition (a) follows from the fact that $\varrho(A) \neq \emptyset$ (see Lemma 1.17).

Let now $\mu \in \varrho(A) \cap \mathbb{R}^+$ and $\alpha = \frac{1}{\mu}$; since

$$I - \alpha A = I - \frac{1}{\mu}A = \frac{1}{\mu}(\mu I - A)$$

the existence of $(\mu I - A)^{-1}$ implies that $(I - \alpha A)^{-1}$ does exist and

$$(I - \alpha A)^{-1} = \mu\,(\mu I - A)^{-1}.$$

Because of the dissipativity of A we have (see Corollary 1.25) $\|(I - \alpha A)^{-1}\| \leqslant 1$, i.e.,

$$\|(\mu I - A)^{-1}\| \leqslant \frac{1}{\mu}. \tag{1.30}$$

If we choose λ such that $|\lambda - \mu| < \mu$, we get

$$\|(\lambda - \mu)(\mu I - A)^{-1}\| \leqslant \frac{|\lambda - \mu|}{\mu} < 1$$

and then the operator $I + (\lambda - \mu)(\mu I - A)^{-1}$ is invertible

Lemma 1.28 assures that $\lambda \in \varrho(A)$. We have shown that $\mu \in \varrho(A) \cap \mathbb{R}^+$ implies that all the interval $(0, 2\mu)$ is contained in $\varrho(A)$. Replacing μ by $\frac{3}{2}\mu$ we find that $\varrho(A)$ contains also every $\lambda > 0$ such that

$$\left| \lambda - \frac{3}{2}\mu \right| < \frac{3}{2}\mu,$$

i.e., $(0, 3\mu) \subset \varrho(A)$. By iterating the argument it follows that $\mathbb{R}^+ \subset \varrho(A)$ and (c) is proved.

Assertion (d) follows from (1.30), taking into account that $\mathbb{R}^+ \subset \varrho(A)$. $\qquad \square$

The Lumer–Phillips Theorem can be stated in an equivalent form by using the concept of m-dissipativity.

Theorem 1.30 (Lumer–Phillips). *The operator A generates a C^0 semigroup of contractions if, and only if, A is m-dissipative and $\overline{D(A)} = X$.*

Proof. The proof follows immediately from Theorem 1.29 and the definition of the m-dissipative operator. $\qquad \square$

The following theorem provides a useful sufficient condition for the generation of a semigroup of contractions.

Theorem 1.31. *Let A be a closed operator with $\overline{D(A)} = X$. If A and A^* are dissipative, then A generates a C^0 semigroup of contractions.*

Proof. Because of the Lumer–Phillips Theorem 1.30, we have to prove that A is m-dissipative. Since A is dissipative by hypothesis, all we have to show is that there exists $\lambda > 0$ such that $\mathscr{R}(\lambda I - A) = X$ (see Corollary 1.26).

Let λ be a positive number. In view of Corollary 1.27, $\mathscr{R}(\lambda I - A)$ is closed. If $\mathscr{R}(\lambda I - A) \neq X$, we can find $\varphi \in X^*$ such that $\varphi \neq 0$ and

$$\langle \varphi, \lambda x - Ax \rangle = 0, \qquad \forall\, x \in D(\lambda I - A) = D(A). \tag{1.31}$$

Condition (1.31) can be written as

$$\langle \lambda\varphi - A^*\varphi, x \rangle = 0, \qquad \forall\, x \in D(A).$$

From the density of $D(A)$ it follows that $\lambda\varphi - A^*\varphi = 0$.

On the other hand, in view of the dissipativity of A^* and Corollary 1.25, we have $\|\varphi\| \leqslant \|\lambda\varphi - A^*\varphi\|$. Then $\varphi = 0$ and this is absurd. $\qquad \square$

1.5   Time-dependent semi-bounded operators

So far we have considered dissipativity for operators which do not depend on t. In this section we are going to give some properties concerning ordinary differential equations in Banach spaces with variable dissipative coefficient operators.

Let B stand for a real Banach space. We denote by f_x a linear functional such that

$$\langle f_x, x \rangle = \|x\| \quad \text{and} \quad \|f_x\| = 1.$$

By using the notations of Section 1.1 we have $f_x = x^*/\|x^*\|$, where x^* belongs to the duality set of x.

The operator A acting in B with the domain $D(A)$ is semi-bounded above if there exists a real constant ω such that for every $x \in D(A)$,

$$\langle f_x, Ax \rangle \leqslant \omega \|x\|. \tag{1.32}$$

The dissipativity we have considered in the previous sections corresponds to the case $\omega = 0$.

If the norm in B is differentiable in the sense of Gâteaux, i.e., if for all elements x and h of B there exists the limit

$$\lim_{t \to 0} \frac{1}{t}(\|x + th\| - \|x\|),$$

then this limit is equal to $\langle f_x, h \rangle$. In fact, by definition of the functional f_x, one has, for $t > 0$,

$$\frac{1}{t}(\|x + th\| - \|x\|) \geqslant \frac{1}{t}[f_x(x + th) - \|x\|] = \langle f_x, h \rangle. \tag{1.33}$$

On the other hand

$$\frac{1}{t}(\|x\| - \|x - th\|) \leqslant \frac{1}{t}[\|x\| - f_x(x - th)] = \langle f_x, h \rangle. \tag{1.34}$$

Passing to the limit in (1.33) and (1.34), we obtain

$$\frac{d}{dt}\|x + th\|\bigg|_{t=0} = \langle f_x, h \rangle.$$

Hence condition (1.32) is equivalent to

$$\langle Ax, \Gamma x \rangle \leqslant \omega \|x\|$$

where Γ stands for the Gâteaux gradient of the norm in B.

If, in particular, $B = L^p(d\mu)$ $(1 < p < \infty)$, where μ is a measure, then

$$\Gamma x = \|x\|_{L^p(d\mu)}^{1-p} |x|^{p-2} x.$$

1.5.1 "Parabolic" equation

Let B denote an arbitrary Banach space. Consider the following equation, which generalizes the classical heat equation

$$\frac{dx}{dt} = A(t)x + \varphi(t), \quad t \in (0, T), \quad \varphi(t) \in B. \tag{1.35}$$

Here $A(t)$ is an operator acting in B (not necessarily linear), whose domain $D(A(t))$ does not depend on t.

The function $t \to x(t)$ in (1.35) is strongly continuous on $[0, T]$, belongs to $D(A(t))$ and has a weak first derivative on $(0, T)$. Let the function

$$[0, T] \ni t \to \|x(t)\|$$

be absolutely continuous on $[0, T]$. We assume that there exists an integrable function ω defined on $[0, T]$ subject to

$$\langle f_x, A(t)x \rangle \leqslant \omega(t)\|x\| \tag{1.36}$$

for all $x \in D(A(t))$.

By a solution of (1.35) we mean a function: $t \to x(t)$ fulfilling the previous conditions and satisfying equation (1.35).

Theorem 1.32. *For any solution of* (1.35) *the estimate*

$$\|x(t)\| \leqslant \|x(0)\|e^{\int_0^t \omega(\tau)d\tau} + \int_0^t \|\varphi(\tau)\|e^{\int_\sigma^\tau \omega(\sigma)d\sigma}d\tau \tag{1.37}$$

holds.

Proof. We note that for $t \in (0, T)$ we have:

$$\frac{d}{dt}\|x(t)\| \leqslant \left\langle f_{x(t)}, \frac{dx(t)}{dt} \right\rangle.$$

In fact, by definition of f_x,

$$\left\langle f_{x(t)}, \frac{dx(t)}{dt} \right\rangle = \lim_{\tau \to 0^+} \frac{1}{\tau}\{f_{x(t)}[x(t)] - f_{x(t)}[x(t - \tau)]\}$$

$$\geqslant \lim_{\tau \to 0^+} \frac{1}{\tau}\{\|x(t)\| - \|x(t - \tau)\|\} = \frac{d}{dt}\|x(t)\|.$$

By (1.35)

$$\left\langle f_{x(t)}, \frac{dx(t)}{dt} \right\rangle = f_{x(t)}[A(t)x(t)] + f_{x(t)}[\varphi(t)].$$

Now (1.36) implies

$$\frac{d}{dt}\|x(t)\| \leqslant \omega(t)\|x(t)\| + \|\varphi(t)\|$$

which leads directly to (1.37). □

Remark 1.33. If $\varphi(t) = 0$ and $\omega(t) \leqslant 0$ on $[0, T)$, then the solution of (1.35) satisfies the maximum principle

$$\|x(t)\| \leqslant \|x(0)\|.$$

1.5.2 "Elliptic" equation

Let us consider the equation

$$\frac{d^2 x}{dt^2} + A(t)x = 0, \quad t \in (0, T), \tag{1.38}$$

where $A(t)$ is the same operator as in Subsection 1.5.1.

By a solution of (1.38) we mean a function $[0, T] \ni t \to x(t) \in D(A(t))$, continuous on $[0, T]$ and having a weak second derivative on $(0, T)$. Assume that the first derivative of $\|x(t)\|$ exists and that it is absolutely continuous on $[0, T]$.

Theorem 1.34. *Let $t \to x(t)$ denote a solution of equation (1.38). If $\omega(t) \leqslant 0$, the maximum value of $\|x(t)\|$ on $[0, T]$ is attained at one of the endpoints of $[0, T]$.*

Proof. We have

$$\left\langle f_{x(t)}, \frac{d^2 x(t)}{dt^2} \right\rangle = \lim_{\tau \to 0^+} \frac{1}{\tau^2} \left\{ f_{x(t)}[x(t+\tau)] + f_{x(t)}[x(t-\tau)] - 2f_{x(t)}[x(t)] \right\}$$

$$\leqslant \lim_{\tau \to 0^+} \frac{1}{\tau^2} \left\{ \|x(t+\tau)\| + \|x(t-\tau)\| - 2\|x(t)\| \right\} = \frac{d^2}{dt^2}\|x(t)\|$$

and, keeping in mind (1.38) and (1.36),

$$\left\langle f_{x(t)}, \frac{d^2 x(t)}{dt^2} \right\rangle = -f_{x(t)}[A(t)x(t)] \geqslant -\omega(t)\|x(t)\|.$$

Therefore the function $t \to \|x(t)\|$ satisfies the differential inequality

$$\frac{d^2}{dt^2}\|x(t)\| + \omega(t)\|x(t)\| \geqslant 0.$$

Since the function $\|x(t)\|$ is convex, its maximum is attained either at $t = 0$ or $t = T$. The proof is complete. □

Remark 1.35. From the comparison theorem for ordinary differential equations, we have that

$$\|x(t)\| \leqslant w(t)$$

where w is a solution of the following two points problem:

$$\begin{cases} w''(t) + \omega(t)w(t) = 0 & \text{in } (0, T) \\ w(0) = \|x(0)\|, \ w(T) = \|x(T)\|. \end{cases}$$

For example, let $\omega(t) = \text{const} \leqslant 0$; then

$$\|x(t)\| \leqslant \frac{\sinh(\sqrt{|\omega|}(T-t))}{\sinh(\sqrt{|\omega|}T)}\|x(0)\| + \frac{\sinh(\sqrt{|\omega|}t)}{\sinh(\sqrt{|\omega|}T)}\|x(T)\|.$$

Therefore either

$$\|x(t)\| \leqslant \exp\left(-\sqrt{|\omega|}t\right)\|x(0)\| \quad \text{for } t \geqslant 0 \quad \text{or} \quad \liminf_{t \to +\infty} \frac{\|x(t)\|}{\exp(\sqrt{|\omega|}t)} > 0,$$

which can be interpreted as a variant of the Phragmén–Lindelöf principle.

1.6 Comments to Chapter 1

The concept of a semigroup goes back to J.A. Séguier, who introduced and named it in 1904. Later several applications to partial differential equations were discovered and the first systematic and comprehensive treatment of this theory can be found in Hille's book [37] (see also the revised and extended edition, Hille and Phillips [38]).

Detailed historical information on the subject can be found in Reed and Simon [89]. A huge list of references (updated to 1985) is in Goldstein [32].

Nowadays the theory of semigroups of operators is very well developed and there are several books describing it in great detail. We mention Davies [18, 19], Engel and Nagel [25], Fattorini [26], Goldstein [31], Kresin and Maz'ya [51], Ouhabaz [86], Pazy [87], Robinson [90] *et al.*

Much of the material of this chapter is taken from Goldstein [31] and Pazy [87].

Theorem 1.21 is due to Hille [37] and Yosida [99], who independently proved it in the late 1940s.

Theorem 1.29 was proved by Lumer and Phillips in [61]. In [77] Maz'ya and Sobolevskiĭ obtained independently of Lumer and Phillips a similar criterion under the assumption that the norm of the Banach space is Gâteaux-differentiable. Their result looks as follows

Theorem 1.36. *The closed and densely defined operator $A + \lambda I$ has a bounded inverse for all $\lambda \geqslant 0$ and satisfies the inequality*

$$\|[A + \lambda I]^{-1}\| \leqslant [\operatorname{Re}\lambda + \lambda_0]^{-1}$$

($\lambda_0 > 0$) if and only if, for any $v \in D(A)$ and $f \in D(A^*)$,

$$\operatorname{Re}\langle \Gamma v, Av \rangle \geqslant \lambda_0 \|v\|,$$
$$\operatorname{Re}\langle \Gamma^* f, A^* f \rangle \geqslant \lambda_0 \|f\|.$$

Here Γ^* stands for the Gâteaux gradient of the norm in B^*.

We remark that [77] was sent to the journal in 1960, before the Lumer–Phillips paper of 1961 [61] appeared.

Chapter 2

L^p-dissipativity of Scalar Second-order Operators with Complex Coefficients

In this chapter we start dealing with concrete problems. Let us consider a scalar second-order operator with complex coefficients

$$Au = \operatorname{div}(\mathscr{A} \nabla u) + \mathbf{b}\nabla u + \operatorname{div}(\mathbf{c}u) + au . \tag{2.1}$$

Section 2.1 is devoted to auxiliary material.

In Section 2.2 we deal with the operator $\operatorname{div}(\mathscr{A} \nabla)$, where $\mathscr{A}$ is a matrix whose entries are complex measures and whose imaginary part is symmetric. We prove that the relevant sesquilinear form $\mathscr{L}$ is L^p-dissipative if and only if the following algebraic condition is satisfied:

$$|p - 2| \, |\langle \operatorname{Im} \mathscr{A} \, \xi, \xi \rangle| \leqslant 2\sqrt{p - 1} \, \langle \operatorname{Re} \mathscr{A} \, \xi, \xi \rangle , \tag{2.2}$$

for any $\xi \in \mathbb{R}^n$. Here $|\cdot|$ denotes the total variation and the inequality has to be understood in the sense of measures.

We note that this criterion does not hold if either the operator A contains lower-order terms or $\operatorname{Im} \mathscr{A}$ is not symmetric. In these cases it is not possible to give algebraic characterizations of L^p-dissipativity. However, we find necessary and sufficient conditions for operator (2.1) with complex *constant* coefficients. This is the subject of Section 2.3.

Section 2.4 is concerned with the relations between the L^p-dissipativity of $\mathscr{L}$ and the L^p-dissipativity of the partial differential operator (2.1). We show also that if either the operator A contains lower-order terms or $\operatorname{Im} \mathscr{A}$ is not symmetric, then the algebraic condition (2.2) is necessary and sufficient for the so-called quasi-dissipativity.

The chapter is finished with Section 2.5, where we discuss a certain "quasi-commutativity" property of the composition operator on one hand and the Poisson operator P of the Dirichlet problem for the equation $\operatorname{div}(\mathscr{A} \nabla u) = 0$ on the other hand. This topic is somewhat different from our principal theme, but it is close to it in spirit.

In the special case of a Laplace operator our general result gives a series of sharp inequalities of type

$$\int_\Omega |\nabla(Ph)^{\alpha+1}|^2 dx \leqslant \frac{(\alpha+1)^2}{2\alpha+1} \int_\Omega |\nabla P(h^{\alpha+1})|^2 dx$$

for $\alpha > -1/2$ and for an arbitrary nonnegative function h defined on $\partial\Omega$.

2.1 Results on general operators with lower-order terms

2.1.1 Main Lemma

By $C_0(\Omega)$ we denote the space of complex-valued continuous functions having compact support in Ω. Let $C_0^1(\Omega)$ consist of all the functions in $C_0(\Omega)$ having continuous partial derivatives of the first order.

In what follows, $\mathscr{A}$ is an $n \times n$ matrix function with complex-valued entries $a^{hk} \in (C_0(\Omega))^*$, $\mathscr{A}^t$ is its transposed matrix and $\mathscr{A}^*$ is its adjoint matrix, i.e., $\mathscr{A}^* = \overline{\mathscr{A}}^t$.

Let $\mathbf{b} = (b_1, \ldots, b_n)$ and $\mathbf{c} = (c_1, \ldots, c_n)$ stand for complex-valued vectors with $b_j, c_j \in (C_0(\Omega))^*$. By a we mean a complex-valued scalar distribution in $(C_0^1(\Omega))^*$.

We denote by $\mathscr{L}(u, v)$ the sesquilinear form

$$\mathscr{L}(u, v) = -\int_\Omega (\langle \mathscr{A} \nabla u, \nabla v \rangle - \langle \mathbf{b}\nabla u, v \rangle + \langle u, \overline{\mathbf{c}}\nabla v \rangle - a\langle u, v \rangle)$$

defined on $C_0^1(\Omega) \times C_0^1(\Omega)$.

The integrals appearing in this definition have to be understood in a proper way. The entries a^{hk} being measures, the meaning of the first term is

$$\int_\Omega \langle \mathscr{A} \nabla u, \nabla v \rangle = \int_\Omega \partial_k u\, \partial_h \overline{v}\, da^{hk}.$$

Similar meanings have the terms involving $\mathbf{b}$ and $\mathbf{c}$. Finally, the last term is the action of the distribution $a \in (C_0^1(\Omega))^*$ on the function $\langle u, v \rangle$ belonging to $C_0^1(\Omega)$.

The form $\mathscr{L}$ is related to the operator

$$Au = \mathrm{div}(\mathscr{A} \nabla u) + \mathbf{b}\nabla u + \mathrm{div}(\mathbf{c}u) + au \qquad (2.3)$$

where div denotes the divergence operator.

The operator A acts from $C_0^1(\Omega)$ to $(C_0^1(\Omega))^*$ through the relation

$$\mathscr{L}(u, v) = \int_\Omega \langle Au, v \rangle$$

for any $u, v \in C_0^1(\Omega)$.

We start with the dissipativity of the form $\mathscr{L}$, according to Definition 1.4. The form $\mathscr{L}$ is L^p-*dissipative* if, for all $u \in C_0^1(\Omega)$,

$$\mathrm{Re}\,\mathscr{L}(u, |u|^{p-2}u) \leqslant 0 \qquad \text{if } p \geqslant 2, \tag{2.4}$$

$$\mathrm{Re}\,\mathscr{L}(|u|^{p'-2}u, u) \leqslant 0 \qquad \text{if } 1 < p < 2. \tag{2.5}$$

As in Example 1.6, the necessity of differentiating the case $1 < p < 2$ from $p \geqslant 2$ is due to the fact that $|u|^{q-2}u \in C_0^1(\Omega)$ for $q \geqslant 2$ and $u \in C_0^1(\Omega)$.

The following lemma will play a key role.

Lemma 2.1. *The form $\mathscr{L}$ is L^p-dissipative if and only if, for all $v \in C_0^1(\Omega)$,*

$$\mathrm{Re} \int_\Omega \left[\langle \mathscr{A}\,\nabla v, \nabla v \rangle - (1 - 2/p)\langle (\mathscr{A} - \mathscr{A}^*)\nabla(|v|), |v|^{-1}\overline{v}\nabla v \rangle \right.$$
$$\left. - (1 - 2/p)^2 \langle \mathscr{A}\,\nabla(|v|), \nabla(|v|) \rangle \right] + \int_\Omega \langle \mathrm{Im}(\mathbf{b} + \mathbf{c}), \mathrm{Im}(\overline{v}\nabla v) \rangle \tag{2.6}$$
$$+ \int_\Omega \mathrm{Re}(\mathrm{div}(\mathbf{b}/p - \mathbf{c}/p') - a)|v|^2 \geqslant 0.$$

Here and in the sequel the integrand is extended by zero on the set where v vanishes.

Proof. The proof of this lemma is quite technical.

Suppose that $p \geqslant 2$ and that (2.6) holds. Take $u \in C_0^1(\Omega)$ and set

$$v = |u|^{(p-2)/2}u. \tag{2.7}$$

The function v belongs to $C_0^1(\Omega)$ and $|v| = |u|^{p/2}$, i.e., $|u| = |v|^{2/p}$. From (2.7) it follows also that $u = |v|^{(2-p)/p}v$, $|u|^{p-2}u = |v|^{2(p-2)/p}|v|^{(2-p)/p}v = |v|^{(p-2)/p}v$.

A direct calculation shows that

$$\langle \mathscr{A}\,\nabla u, \nabla(|u|^{p-2}u) \rangle = \langle \mathscr{A}\,\nabla(|v|^{\frac{2-p}{p}}v), \nabla(|v|^{\frac{p-2}{p}}v) \rangle$$
$$= \langle \mathscr{A}\,(\nabla v - (1 - 2/p)|v|^{-1}v\nabla|v|), \nabla v + (1 - 2/p)|v|^{-1}v\nabla|v| \rangle$$
$$= \langle \mathscr{A}\,\nabla v, \nabla v \rangle - (1 - 2/p)\left(\langle |v|^{-1}v\,\mathscr{A}\,\nabla|v|, \nabla v \rangle - \langle \mathscr{A}\,\nabla v, |v|^{-1}v\nabla|v| \rangle \right)$$
$$- (1 - 2/p)^2 \langle \mathscr{A}\,\nabla|v|, \nabla|v| \rangle.$$

Since

$$\mathrm{Re}(\langle v\,\mathscr{A}\,\nabla|v|, \nabla v \rangle - \langle \mathscr{A}\,\nabla v, v\nabla|v| \rangle)$$
$$= \mathrm{Re}(v\langle \mathscr{A}\,\nabla|v|, \nabla v \rangle - \overline{v\,\langle \mathscr{A}^*\,\nabla|v|, \nabla v \rangle}) = \mathrm{Re}(\langle v(\mathscr{A} - \mathscr{A}^*)\nabla|v|, \nabla v \rangle)$$

we have

$$\mathrm{Re}\langle \mathscr{A} \nabla u, \nabla(|u|^{p-2}u)\rangle$$
$$= \mathrm{Re}\left[\langle \mathscr{A} \nabla v, \nabla v\rangle - (1 - 2/p)\langle(\mathscr{A} - \mathscr{A}^*)\nabla(|v|), |v|^{-1}\overline{v}\nabla v\rangle\right.$$
$$\left. - (1 - 2/p)^2\langle \mathscr{A} \nabla(|v|), \nabla(|v|)\rangle\right].$$

Moreover, we have

$$\langle \mathbf{b}\nabla u, |u|^{p-2}u\rangle = -(1 - 2/p)\,|v|\,\mathbf{b}\nabla|v| + \overline{v}\,\mathbf{b}\nabla v$$

and then

$$\mathrm{Re}\langle \mathbf{b}\nabla u, |u|^{p-2}u\rangle = 2\,\mathrm{Re}(\mathbf{b}/p)\,\mathrm{Re}(\overline{v}\nabla v) - (\mathrm{Im}\,\mathbf{b})\,\mathrm{Im}(\overline{v}\nabla v)$$
$$= \mathrm{Re}(\mathbf{b}/p)\nabla(|v|^2) - (\mathrm{Im}\,\mathbf{b})\,\mathrm{Im}(\overline{v}\nabla v).$$

An integration by parts gives

$$\int_\Omega \mathrm{Re}\langle \mathbf{b}\nabla u, |u|^{p-2}u\rangle = -\int_\Omega \mathrm{Re}(\nabla^t(\mathbf{b}/p))|v|^2 - \int_\Omega \langle \mathrm{Im}\,\mathbf{b}, \mathrm{Im}(\overline{v}\nabla v)\rangle. \tag{2.8}$$

In the same way we find

$$\mathrm{Re}\langle u, \overline{\mathbf{c}}\nabla(|u|^{p-2}u)\rangle = \mathrm{Re}\left((1 - 2/p)\,|v|\mathbf{c}\nabla|v| + v\,\mathbf{c}\nabla\overline{v}\right)$$
$$= 2\,\mathrm{Re}(\mathbf{c}/p')\,\mathrm{Re}(\overline{v}\nabla v) + (\mathrm{Im}\,\mathbf{c})\,\mathrm{Im}(\overline{v}\nabla v)$$
$$= \mathrm{Re}(\mathbf{c}/p')\nabla(|v|^2) + (\mathrm{Im}\,\mathbf{c})\,\mathrm{Im}(\overline{v}\nabla v)$$

and then

$$\int_\Omega \mathrm{Re}\langle u, \overline{\mathbf{c}}\nabla(|u|^{p-2}u)\rangle = -\int_\Omega \mathrm{Re}(\nabla^t(\mathbf{c}/p')|v|^2 + \int_\Omega \langle \mathrm{Im}\,\mathbf{c}, \mathrm{Im}(\overline{v}\nabla v)\rangle. \tag{2.9}$$

Finally, since we have also

$$\mathrm{Re}(a\langle u, |u|^{p-2}u\rangle = (\mathrm{Re}\,a)|u|^p = (\mathrm{Re}\,a)|v|^2,$$

the left-hand side in (2.6) is equal to $\mathrm{Re}\,\mathscr{L}(u, |u|^{p-2}u)$ and (2.4) follows from (2.6).

Let us suppose that $1 < p < 2$. Now (2.5) can be written as

$$\mathrm{Re}\int_\Omega (\langle \mathscr{A}^* \nabla u, \nabla(|u|^{p'-2}u)\rangle + \langle \overline{\mathbf{c}}\nabla u, |u|^{p'-2}u\rangle - \langle u, \mathbf{b}\nabla(|u|^{p'-2}u)\rangle$$
$$-a\langle u, |u|^{p'-2}u\rangle) \geqslant 0. \tag{2.10}$$

We know that this is true if

$$
\begin{aligned}
\text{Re} \int_\Omega &\Big[\langle \mathscr{A}^* \nabla v, \nabla v \rangle - (1 - 2/p') \langle (\mathscr{A}^* - \mathscr{A}) \nabla(|v|), |v|^{-1} \overline{v} \nabla v \rangle \\
&- (1 - 2/p')^2 \langle \mathscr{A}^* \nabla(|v|), \nabla(|v|) \rangle \Big] \\
&+ \int_\Omega \langle \text{Im}(-\overline{\mathbf{c}} - \overline{\mathbf{b}}), \text{Im}(\overline{v} \nabla v) \rangle \\
&+ \int_\Omega \text{Re} \left[\text{div} \left((-\overline{\mathbf{c}})/p' - (-\overline{\mathbf{b}})/p \right) - a \right] |v|^2 \geqslant 0
\end{aligned}
\tag{2.11}
$$

for any $v \in C_0^1(\Omega)$. This condition is exactly (2.6) and the sufficiency is proved also for $1 < p < 2$.

Vice versa, let us suppose (2.4) holds and let $v \in C_0^1(\Omega)$. Since the function $u = |v|^{\frac{2-p}{p}} v$ does not need to belong to $C_0^1(\Omega)$, we cannot proceed as for the sufficiency. In order to overcome this difficulty, set

$$
g_\varepsilon = (|v|^2 + \varepsilon^2)^{\frac{1}{2}}, \quad u_\varepsilon = g_\varepsilon^{\frac{2}{p}-1} v. \tag{2.12}
$$

We have

$$
\begin{aligned}
\langle \mathscr{A} \nabla u_\varepsilon, \nabla(|u_\varepsilon|^{p-2} u_\varepsilon) \rangle \\
= |u_\varepsilon|^{p-2} \langle \mathscr{A} \nabla u_\varepsilon, \nabla u_\varepsilon \rangle + (p-2)|u_\varepsilon|^{p-3} \langle \mathscr{A} \nabla u_\varepsilon, u_\varepsilon \nabla |u_\varepsilon| \rangle.
\end{aligned}
$$

On the other hand, since $\partial_h g_\varepsilon = g_\varepsilon^{-1} |v| \partial_h |v|$, we can write

$$
\begin{aligned}
|u_\varepsilon|^{p-2} &\partial_h u_\varepsilon \partial_k \overline{u}_\varepsilon \\
= g_\varepsilon^{2-p} |v|^{p-2} &\Big[(1 - 2/p)^2 g_\varepsilon^{-2} |v|^2 \partial_h g_\varepsilon \partial_k g_\varepsilon \\
&- (1 - 2/p) g_\varepsilon^{-1} (v \partial_h g_\varepsilon \partial_k \overline{v} + \overline{v} \partial_h v \partial_k g_\varepsilon) + \partial_h v \partial_k \overline{v} \Big] \\
= (1 - 2/p)^2 &g_\varepsilon^{-(p+2)} |v|^{p+2} \partial_h |v| \partial_k |v| \\
&- (1 - 2/p) g_\varepsilon^{-p} |v|^{p-1} (v \partial_h |v| \partial_k \overline{v} + \overline{v} \partial_h v \partial_k |v|) + g_\varepsilon^{2-p} |v|^{p-2} \partial_h v \partial_k \overline{v}.
\end{aligned}
$$

This leads to

$$
\begin{aligned}
|u_\varepsilon|^{p-2} \langle \mathscr{A} \nabla u_\varepsilon, \nabla u_\varepsilon \rangle = (1 - 2/p)^2 g_\varepsilon^{-(p+2)} |v|^{p+2} \langle \mathscr{A} \nabla |v|, \nabla |v| \rangle \\
- (1 - 2/p) g_\varepsilon^{-p} |v|^{p-1} (\langle \mathscr{A} v \nabla |v|, \nabla v \rangle + \langle \mathscr{A} \nabla v, v \nabla |v| \rangle) \\
+ g_\varepsilon^{2-p} |v|^{p-2} \langle \mathscr{A} \nabla v, \nabla v \rangle.
\end{aligned}
$$

In the same way

$$
\begin{aligned}
|u_\varepsilon|^{p-3} \langle \mathscr{A} \nabla u_\varepsilon, u_\varepsilon \nabla |u_\varepsilon| \rangle \\
= \Big[(1 - 2/p)^2 g_\varepsilon^{-(p+2)} |v|^{p+2} - (1 - 2/p) g_\varepsilon^{-p} |v|^p \Big] \langle \mathscr{A} \nabla |v|, \nabla |v| \rangle \\
+ \Big[-(1 - 2/p) g_\varepsilon^{-p} |v|^{p-1} + g_\varepsilon^{-p+2} |v|^{p-3} \Big] \langle \mathscr{A} \nabla v, v \nabla |v| \rangle.
\end{aligned}
$$

Observing that g_ε tends to $|v|$ as $\varepsilon \to 0$ and that $g_\varepsilon^{-1}|v| \leqslant 1$, referring to Lebesgue's dominated convergence theorem we find

$$\lim_{\varepsilon \to 0} \int_\Omega \langle \mathscr{A} \nabla u_\varepsilon, \nabla(|u_\varepsilon|^{p-2} u_\varepsilon) \rangle$$

$$= \int_\Omega \langle \mathscr{A} \nabla v, \nabla v \rangle - (1 - 2/p) \int_\Omega \frac{1}{|v|} \left(\langle v \mathscr{A} \nabla|v|, \nabla v \rangle - \langle \mathscr{A} \nabla v, v \nabla|v| \rangle \right) \quad (2.13)$$

$$- (1 - 2/p)^2 \int_\Omega \langle \mathscr{A} \nabla|v|, \nabla|v| \rangle \ .$$

Similar computations show that

$$\langle \mathbf{b}\nabla u_\varepsilon, |u_\varepsilon|^{p-2} u_\varepsilon \rangle = -(1 - 2/p) g_\varepsilon^{-p} |v|^{p+1} \mathbf{b}\nabla|v| + g_\varepsilon^{2-p} |v|^{p-2} \overline{v} \mathbf{b}\nabla v,$$

$$\langle u_\varepsilon, \overline{\mathbf{c}}\nabla(|u_\varepsilon|^{p-2} u_\varepsilon) \rangle = g_\varepsilon^{2-p} |v|^{p-2} \mathbf{c} \Big[(1 - p)(1 - 2/p) g_\varepsilon^{-2} |v|^3 \nabla|v|$$

$$+ (p - 2)|v|\nabla|v| + v\nabla\overline{v} \Big],$$

$$a\langle u_\varepsilon, |u_\varepsilon|^{p-2} u_\varepsilon \rangle = a g_\varepsilon^{2-p} |v|^p$$

from which follows

$$\lim_{\varepsilon \to 0} \int_\Omega \langle \mathbf{b}\nabla u_\varepsilon, |u_\varepsilon|^{p-2} u_\varepsilon \rangle = \int_\Omega \left(-(1 - 2/p)|v|\, \mathbf{b}\nabla|v| + \overline{v}\, \mathbf{b}\nabla v \right), \quad (2.14)$$

$$\lim_{\varepsilon \to 0} \int_\Omega \langle u_\varepsilon, \overline{\mathbf{c}}\nabla(|u_\varepsilon|^{p-2} u_\varepsilon) \rangle = \int_\Omega \left((1 - 2/p)|v|\mathbf{c}\nabla|v| + v\, \mathbf{c}\nabla\overline{v} \right),$$

$$\lim_{\varepsilon \to 0} \int_\Omega a\langle u_\varepsilon, |u_\varepsilon|^{p-2} u_\varepsilon \rangle = \int_\Omega a|v|^2. \quad (2.15)$$

From (2.13)–(2.15) we obtain that

$$\lim_{\varepsilon \to 0} \operatorname{Re} \mathscr{L}(u_\varepsilon, |u_\varepsilon|^{p-2} u_\varepsilon)$$

exists and is equal to the left-hand side of (2.6). This shows that (2.4) implies (2.6) and the necessity is proved for $p \geqslant 2$.

Let us assume $1 < p < 2$. Since (2.5) can be written as (2.10), replacing $\mathscr{A}$, $\mathbf{b}$, $\overline{\mathbf{c}}$ by $\mathscr{A}^*$, $-\overline{\mathbf{c}}$, $-\mathbf{b}$ respectively in formulas (2.13)–(2.15) we find that

$$\lim_{\varepsilon \to 0} \operatorname{Re} \mathscr{L}(|u_\varepsilon|^{p'-2} u_\varepsilon, u_\varepsilon)$$

exists and is equal to the left-hand side of (2.11). Thus (2.5) implies (2.6).   □

2.1.2   Necessary condition

Here we show some consequences of Lemma 2.1.

Corollary 2.2. *If the form $\mathscr{L}$ is L^p-dissipative, we have*

$$\langle \operatorname{Re} \mathscr{A} \, \xi, \xi \rangle \geqslant 0 \tag{2.16}$$

for any $\xi \in \mathbb{R}^n$.

Proof. We remark that condition (2.16) has to be understood in the sense of measures, i.e., it means that, for any $\xi \in \mathbb{R}^n$,

$$\int_\Omega \langle \operatorname{Re} \mathscr{A} \, \xi, \xi \rangle \, v \geqslant 0$$

for any nonnegative $v \in C_0(\Omega)$. Given a function v, let us set

$$X = \operatorname{Re}(|v|^{-1} \overline{v} \, \nabla v), \quad Y = \operatorname{Im}(|v|^{-1} \overline{v} \, \nabla v),$$

on the set $\{x \in \Omega \mid v \neq 0\}$. We have

$$\operatorname{Re}\langle \mathscr{A} \, \nabla v, \nabla v \rangle = \operatorname{Re} \langle \mathscr{A}(|v|^{-1} \overline{v} \, \nabla v), |v|^{-1} \overline{v} \, \nabla v \rangle$$
$$= \langle \operatorname{Re} \mathscr{A} \, X, X \rangle + \langle \operatorname{Re} \mathscr{A} \, Y, Y \rangle + \langle \operatorname{Im}(\mathscr{A} - \mathscr{A}^t) X, Y \rangle,$$
$$\operatorname{Re}\langle (\mathscr{A} - \mathscr{A}^*) \nabla(|v|), \nabla v \rangle |v|^{-1} v = \operatorname{Re}\langle (\mathscr{A} - \mathscr{A}^*) X, X + iY \rangle$$
$$= \langle \operatorname{Im}(\mathscr{A} - \mathscr{A}^*) X, Y \rangle,$$
$$\operatorname{Re}\langle \mathscr{A} \, \nabla|v|, \nabla|v| \rangle = \langle \operatorname{Re} \mathscr{A} \, X, X \rangle.$$

Since $\mathscr{L}$ is L^p-dissipative, (2.6) holds. Hence, keeping in mind that the next integral is extended on the set where v does not vanish,

$$\int_\Omega \Big\{ \frac{4}{p\,p'} \langle \operatorname{Re} \mathscr{A} \, X, X \rangle + \langle \operatorname{Re} \mathscr{A} \, Y, Y \rangle$$
$$+ 2\langle (p^{-1} \operatorname{Im} \mathscr{A} + p'^{-1} \operatorname{Im} \mathscr{A}^*) X, Y \rangle + \langle \operatorname{Im}(\mathbf{b} + \mathbf{c}), Y \rangle |v| \tag{2.17}$$
$$+ \operatorname{Re}\left[\operatorname{div}(\mathbf{b}/p - \mathbf{c}/p') - a\right] |v|^2 \Big\} \geqslant 0.$$

We define the function

$$v(x) = \varrho(x) \, e^{i\varphi(x)}$$

where ϱ and φ are real functions with $\varrho \in C_0^1(\Omega)$ and $\varphi \in C^1(\Omega)$. Since

$$|v|^{-1} \overline{v} \, \nabla v = |\varrho|^{-1}(\varrho \, e^{-i\varphi} (\nabla\varrho + i\varrho\nabla\varphi) \, e^{i\varphi}) = |\varrho|^{-1} \varrho\nabla\varrho + i|\varrho|\nabla\varphi$$

on the set $\{x \in \Omega \mid \varrho(x) \neq 0\}$, it follows from (2.17) that

$$\frac{4}{p\,p'} \int_\Omega \langle \operatorname{Re} \mathscr{A} \, \nabla\varrho, \nabla\varrho \rangle + \int_\Omega \varrho^2 \langle \operatorname{Re} \mathscr{A} \, \nabla\varphi, \nabla\varphi \rangle$$
$$+ 2 \int_\Omega \varrho\langle (p^{-1} \operatorname{Im} \mathscr{A} + p'^{-1} \operatorname{Im} \mathscr{A}^*) \nabla\varrho, \nabla\varphi \rangle \tag{2.18}$$
$$+ \int_\Omega \varrho\langle \operatorname{Im}(\mathbf{b} + \mathbf{c}), \nabla\varphi \rangle + \int_\Omega \operatorname{Re}\left[\operatorname{div}(\mathbf{b}/p - \mathbf{c}/p') - a\right] \varrho^2 \geqslant 0$$

for any $\varrho \in C_0^1(\Omega)$, $\varphi \in C^1(\Omega)$.

We choose φ by the equality

$$\varphi = \frac{\mu}{2} \log(\varrho^2 + \varepsilon)$$

where $\mu \in \mathbb{R}$ and $\varepsilon > 0$. Then (2.18) takes the form

$$\frac{4}{p\,p'} \int_\Omega \langle \operatorname{Re} \mathscr{A} \nabla \varrho, \nabla \varrho \rangle + \mu^2 \int_\Omega \frac{\varrho^4}{(\varrho^2 + \varepsilon)^2} \langle \operatorname{Re} \mathscr{A} \nabla \varrho, \nabla \varrho \rangle$$

$$+ 2\mu \int_\Omega \frac{\varrho^2}{\varrho^2 + \varepsilon} \langle (p^{-1} \operatorname{Im} \mathscr{A} + p'^{-1} \operatorname{Im} \mathscr{A}^*) \nabla \varrho, \nabla \varrho \rangle \qquad (2.19)$$

$$+ \mu \int_\Omega \frac{\varrho^3}{\varrho^2 + \varepsilon} \langle \operatorname{Im}(\mathbf{b} + \mathbf{c}), \nabla \varrho \rangle + \int_\Omega \operatorname{Re} \left[\operatorname{div}(\mathbf{b}/p - \mathbf{c}/p') - a\right] \varrho^2 \geqslant 0$$

Letting $\varepsilon \to 0^+$ in (2.19) leads to

$$\frac{4}{p\,p'} \int_\Omega \langle \operatorname{Re} \mathscr{A} \nabla \varrho, \nabla \varrho \rangle + \mu^2 \int_\Omega \langle \operatorname{Re} \mathscr{A} \nabla \varrho, \nabla \varrho \rangle$$

$$+ 2\mu \int_\Omega \langle (p^{-1} \operatorname{Im} \mathscr{A} + p'^{-1} \operatorname{Im} \mathscr{A}^*) \nabla \varrho, \nabla \varrho \rangle \qquad (2.20)$$

$$+ \mu \int_\Omega \varrho \langle \operatorname{Im}(\mathbf{b} + \mathbf{c}), \nabla \varrho \rangle + \int_\Omega \operatorname{Re} \left[\operatorname{div}(\mathbf{b}/p - \mathbf{c}/p') - a\right] \varrho^2 \geqslant 0.$$

Since this holds for any $\mu \in \mathbb{R}$, we have

$$\int_\Omega \langle \operatorname{Re} \mathscr{A} \nabla \varrho, \nabla \varrho \rangle \geqslant 0 \qquad (2.21)$$

for any $\varrho \in C_0^1(\Omega)$.

Taking $\varrho(x) = \psi(x) \cos\langle \xi, x \rangle$ with a real $\psi \in C_0^1(\Omega)$ and $\xi \in \mathbb{R}^n$, we find

$$\int_\Omega \{\langle \operatorname{Re} \mathscr{A} \nabla \psi, \nabla \psi \rangle \cos^2\langle \xi, x \rangle - [\langle \operatorname{Re} \mathscr{A} \xi, \nabla \psi \rangle$$

$$+ \langle \operatorname{Re} \mathscr{A} \nabla \psi, \xi \rangle] \sin\langle \xi, x \rangle \cos\langle \xi, x \rangle + \langle \operatorname{Re} \mathscr{A} \xi, \xi \rangle \psi^2(x) \sin^2\langle \xi, x \rangle\} \geqslant 0.$$

On the other hand, taking $\varrho(x) = \psi(x) \sin\langle \xi, x \rangle$,

$$\int_\Omega \{\langle \operatorname{Re} \mathscr{A} \nabla \psi, \nabla \psi \rangle \sin^2\langle \xi, x \rangle + [\langle \operatorname{Re} \mathscr{A} \xi, \nabla \psi \rangle$$

$$+ \langle \operatorname{Re} \mathscr{A} \nabla \psi, \xi \rangle] \sin\langle \xi, x \rangle \cos\langle \xi, x \rangle + \langle \operatorname{Re} \mathscr{A} \xi, \xi \rangle \psi^2(x) \cos^2\langle \xi, x \rangle\} \geqslant 0.$$

The two inequalities we have obtained lead to

$$\int_\Omega \langle \operatorname{Re} \mathscr{A} \nabla \psi, \nabla \psi \rangle + \int_\Omega \langle \operatorname{Re} \mathscr{A} \xi, \xi \rangle \psi^2 \geqslant 0.$$

Because of the arbitrariness of ξ, we find

$$\int_\Omega \langle \operatorname{Re} \mathscr{A} \xi, \xi \rangle \psi^2 \geqslant 0.$$

On the other hand, any nonnegative function $v \in C_0(\Omega)$ can be approximated in the uniform norm in Ω by a sequence ψ_n^2, with $\psi_n \in C_0^\infty(\Omega)$, and then $\langle \mathrm{Re}\,\mathscr{A}\,\xi, \xi \rangle$ is a nonnegative measure.   □

It will be clear later (see Example 2.10) that (2.16) is not sufficient for the L^p-dissipativity.

2.1.3   Sufficient condition

The next corollary provides a sufficient condition. It shows that the L^p-dissipativity of A follows from the nonnegativity of a certain polynomial (whose coefficients are measures) in $2n$ real variables. This polynomial depends on the real parameters α, β, which can be arbitrarily fixed.

Corollary 2.3. *Let* α, β *be two real constants. If*

$$
\begin{aligned}
\frac{4}{pp'}\langle \mathrm{Re}\,\mathscr{A}\,\xi, \xi \rangle &+ \langle \mathrm{Re}\,\mathscr{A}\,\eta, \eta \rangle + 2\langle(p^{-1}\,\mathrm{Im}\,\mathscr{A} + p'^{-1}\,\mathrm{Im}\,\mathscr{A}^*)\xi, \eta \rangle \\
&+ \langle \mathrm{Im}(\mathbf{b} + \mathbf{c}), \eta \rangle - 2\langle \mathrm{Re}(\alpha\mathbf{b}/p - \beta\mathbf{c}/p'), \xi \rangle \\
&+ \mathrm{Re}\,[\mathrm{div}\,((1 - \alpha)\mathbf{b}/p - (1 - \beta)\mathbf{c}/p') - a] \geqslant 0
\end{aligned}
\tag{2.22}
$$

for any $\xi, \eta \in \mathbb{R}^n$, *the form* $\mathscr{L}$ *is* L^p-*dissipative.*

Proof. In the proof of Lemma 2.1 we have integrated by parts in (2.8) and (2.9). More generally, we have

$$
\begin{aligned}
2/p \int_\Omega &\langle \mathrm{Re}\,\mathbf{b}, \mathrm{Re}(\overline{v}\nabla v) \rangle \\
&= 2\alpha/p \int_\Omega \langle \mathrm{Re}\,\mathbf{b}, \mathrm{Re}(\overline{v}\nabla v) \rangle - (1 - \alpha)/p \int_\Omega \mathrm{Re}(\nabla^t\mathbf{b})|v|^2 \,; \\
2/p' \int_\Omega &\langle \mathrm{Re}\,\mathbf{c}, \mathrm{Re}(\overline{v}\nabla v) \rangle \\
&= 2\beta/p' \int_\Omega \langle \mathrm{Re}\,\mathbf{c}, \mathrm{Re}(\overline{v}\nabla v) \rangle - (1 - \beta)/p' \int_\Omega \mathrm{Re}(\nabla^t\mathbf{c})|v|^2 \,.
\end{aligned}
$$

This leads us to write conditions (2.6) in a slightly different form:

$$
\begin{aligned}
\mathrm{Re} \int_\Omega &\Big[\langle \mathscr{A}\,\nabla v, \nabla v \rangle - (1 - 2/p)\langle(\mathscr{A} - \mathscr{A}^*)\nabla(|v|), |v|^{-1}\overline{v}\nabla v \rangle \\
&- (1 - 2/p)^2 \langle \mathscr{A}\,\nabla(|v|), \nabla(|v|) \rangle \Big] + \int_\Omega \langle \mathrm{Im}(\mathbf{b} + \mathbf{c}), \mathrm{Im}(\overline{v}\nabla v) \rangle \\
&- 2 \int_\Omega \langle \mathrm{Re}(\alpha\mathbf{b}/p - \beta\mathbf{c}/p'), \mathrm{Re}(\overline{v}\nabla v) \rangle \\
&+ \int_\Omega \mathrm{Re}(\mathrm{div}((1 - \alpha)\mathbf{b}/p - (1 - \beta)\mathbf{c}/p') - a)|v|^2 \geqslant 0.
\end{aligned}
$$

By using the functions X and Y introduced in Corollary 2.2, the left-hand side of the last inequality can be written as

$$\int_\Omega Q(X, Y)$$

where Q denotes the polynomial (2.22). The result follows from Lemma 2.1.   □

Generally speaking, conditions (2.22) are not necessary for L^p-dissipativity. We show this by the following example, where Im $\mathscr{A}$ is not symmetric. Later we give another example showing that, even for symmetric matrices Im $\mathscr{A}$, conditions (2.22) are not necessary for L^p-dissipativity (see Example 2.17). Nevertheless in the next section we show that the conditions are necessary for the L^p-dissipativity, provided the operator A has no lower-order terms and the matrix Im $\mathscr{A}$ is symmetric (see Theorem 2.7 and Remark 2.8).

Example 2.4. Let $n = 2$ and

$$\mathscr{A} = \begin{pmatrix} 1 & i\gamma \\ -i\gamma & 1 \end{pmatrix}$$

where γ is a real constant, $\mathbf{b} = \mathbf{c} = a = 0$. In this case polynomial (2.22) is given by

$$(\eta_1 + \gamma\xi_2)^2 + (\eta_2 - \gamma\xi_1)^2 - (\gamma^2 - 4/(pp'))|\xi|^2.$$

Taking $\gamma^2 > 4/(pp')$, condition (2.22) is not satisfied, while we have the L^p-dissipativity, because the corresponding operator A is nothing but the Laplacian.

2.1.4   Consequences of the main lemma

The next corollary is an interpolation result

Corollary 2.5. *If the form $\mathscr{L}$ is both L^p- and $L^{p'}$-dissipative, it is also L^r-dissipative for any r between p and p', i.e., for any r given by*

$$1/r = t/p + (1-t)/p' \qquad (0 \leqslant t \leqslant 1). \tag{2.23}$$

Proof. From the proof of Corollary 2.2 we know that (2.17) holds. In the same way, we find

$$\int_\Omega \left\{ \frac{4}{p'p} \langle \mathrm{Re}\,\mathscr{A}\, X, X \rangle + \langle \mathrm{Re}\,\mathscr{A}\, Y, Y \rangle \right.$$
$$- 2\langle (p'^{-1}\,\mathrm{Im}\,\mathscr{A} + p^{-1}\,\mathrm{Im}\,\mathscr{A}^*)X, Y \rangle + \langle \mathrm{Im}(\mathbf{b}+\mathbf{c}), Y \rangle |v| \tag{2.24}$$
$$\left. + \mathrm{Re}\left[\mathrm{div}\,(\mathbf{b}/p' - \mathbf{c}/p) - a\right]|v|^2 \right\} \geqslant 0.$$

We multiply (2.17) by t, (2.24) by $(1-t)$ and sum up. Since

$$t/p' + (1-t)/p = 1/r' \quad \text{and} \quad rr' \leqslant pp',$$

we find, keeping in mind Corollary 2.2,

$$\int_\Omega \left\{ \frac{4}{r\,r'} \langle \operatorname{Re} \mathscr{A} X, X \rangle + \langle \operatorname{Re} \mathscr{A} Y, Y \rangle \right.$$
$$- 2\langle (r^{-1} \operatorname{Im} \mathscr{A} + r'^{-1} \operatorname{Im} \mathscr{A}^*) X, Y \rangle + \langle \operatorname{Im}(\mathbf{b} + \mathbf{c}), Y \rangle |v|$$
$$\left. + \operatorname{Re}\left[\operatorname{div}(\mathbf{b}/r - \mathbf{c}/r') - a\right] |v|^2 \right\} \geqslant 0$$

and $\mathscr{L}$ is L^r-dissipative by Lemma 2.1.   □

Corollary 2.6. *Suppose that either*

$$\operatorname{Im} \mathscr{A} = 0, \qquad \operatorname{Re} \operatorname{div} \mathbf{b} = \operatorname{Re} \operatorname{div} \mathbf{c} = 0 \qquad (2.25)$$

or

$$\operatorname{Im} \mathscr{A} = \operatorname{Im} \mathscr{A}^t, \quad \operatorname{Im}(\mathbf{b} + \mathbf{c}) = 0, \quad \operatorname{Re} \operatorname{div} \mathbf{b} = \operatorname{Re} \operatorname{div} \mathbf{c} = 0. \qquad (2.26)$$

If $\mathscr{L}$ is L^p-dissipative, it is also L^r-dissipative for any r given by (2.23).

Proof. Assume that (2.25) holds. With the notation introduced in Corollary 2.2, inequality (2.6) reads as

$$\int_\Omega \left(\frac{4}{p\,p'} \langle \operatorname{Re} \mathscr{A} X, X \rangle + \langle \operatorname{Re} \mathscr{A} Y, Y \rangle + \langle \operatorname{Im}(\mathbf{b} + \mathbf{c}), Y \rangle |v| - \operatorname{Re} a |v|^2 \right) \geqslant 0.$$

Since the left-hand side does not change after replacing p by p', Lemma 2.1 gives the result.

Let (2.26) hold. Using the formula

$$p^{-1} \operatorname{Im} \mathscr{A} + p'^{-1} \operatorname{Im} \mathscr{A}^* = p^{-1} \operatorname{Im} \mathscr{A} - p'^{-1} \operatorname{Im} \mathscr{A}^t = -(1 - 2/p) \operatorname{Im} \mathscr{A}, \qquad (2.27)$$

we obtain

$$\int_\Omega \left(\frac{4}{p\,p'} \langle \operatorname{Re} \mathscr{A} x, x \rangle + \langle \operatorname{Re} \mathscr{A} Y, Y \rangle - 2(1 - 2/p)\langle \operatorname{Im} \mathscr{A} X, Y \rangle - \operatorname{Re} a |v|^2 \right) \geqslant 0.$$

Replacing v by $\bar{v}$, we find

$$\int_\Omega \left(\frac{4}{p\,p'} \langle \operatorname{Re} \mathscr{A} x, x \rangle + \langle \operatorname{Re} \mathscr{A} Y, Y \rangle + 2(1 - 2/p)\langle \operatorname{Im} \mathscr{A} X, Y \rangle - \operatorname{Re} a |v|^2 \right) \geqslant 0$$

and we have the $L^{p'}$-dissipativity by $1 - 2/p = -1 + 2/p'$. The reference to Corollary 2.5 completes the proof.   □

2.2   The operator : $u \to \operatorname{div}(\mathscr{A}\,\nabla u)$. The main theorem

In this section we consider operator (2.3) without lower-order terms:

$$Au = \operatorname{div}(\mathscr{A}\,\nabla u) \tag{2.28}$$

with the coefficients $a^{hk} \in (C_0(\Omega))^*$. The following theorem contains an algebraic necessary and sufficient condition for the L^p-dissipativity.

Theorem 2.7. *Let the matrix* $\operatorname{Im}\mathscr{A}$ *be symmetric, i.e.,* $\operatorname{Im}\mathscr{A}^t = \operatorname{Im}\mathscr{A}$. *The form*

$$\mathscr{L}(u,v) = \int_\Omega \langle \mathscr{A}\,\nabla u, \nabla v \rangle$$

is L^p-*dissipative if and only if*

$$|p-2|\,|\langle \operatorname{Im}\mathscr{A}\,\xi,\xi\rangle| \leqslant 2\sqrt{p-1}\,\langle \operatorname{Re}\mathscr{A}\,\xi,\xi\rangle \tag{2.29}$$

for any $\xi \in \mathbb{R}^n$, *where* $|\cdot|$ *denotes the total variation.*

Proof. Sufficiency. In view of Corollary 2.3 the form $\mathscr{L}$ is L^p-dissipative if

$$\frac{4}{p\,p'}\langle \operatorname{Re}\mathscr{A}\,\xi,\xi\rangle + \langle \operatorname{Re}\mathscr{A}\,\eta,\eta\rangle - 2(1-2/p)\langle \operatorname{Im}\mathscr{A}\,\xi,\eta\rangle \geqslant 0 \tag{2.30}$$

for any $\xi, \eta \in \mathbb{R}^n$.

By putting

$$\lambda = \frac{2\sqrt{p-1}}{p}\,\xi$$

we write (2.30) in the form

$$\langle \operatorname{Re}\mathscr{A}\,\lambda,\lambda\rangle + \langle \operatorname{Re}\mathscr{A}\,\eta,\eta\rangle - \frac{p-2}{\sqrt{p-1}}\langle \operatorname{Im}\mathscr{A}\,\lambda,\eta\rangle \geqslant 0.$$

Then (2.30) is equivalent to

$$\mathscr{S}(\xi,\eta) := \langle \operatorname{Re}\mathscr{A}\,\xi,\xi\rangle + \langle \operatorname{Re}\mathscr{A}\,\eta,\eta\rangle - \frac{p-2}{\sqrt{p-1}}\langle \operatorname{Im}\mathscr{A}\,\xi,\eta\rangle \geqslant 0$$

for any $\xi, \eta \in \mathbb{R}^n$.

For any nonnegative $\varphi \in C_0(\Omega)$, define

$$\lambda_\varphi = \min_{|\xi|^2 + |\eta|^2 = 1} \int_\Omega \mathscr{S}(\xi,\eta)\,\varphi.$$

Let us fix ξ_0, η_0 such that $|\xi_0|^2 + |\eta_0|^2 = 1$ and

$$\lambda_\varphi = \int_\Omega \mathscr{S}(\xi_0,\eta_0)\,\varphi.$$

We have the algebraic system

$$
\begin{cases}
\displaystyle \int_{\Omega} \left(2\,\mathrm{Re}\,\mathscr{A}\,\xi_0 - \frac{p-2}{2\sqrt{p-1}}\,\mathrm{Im}(\mathscr{A} - \mathscr{A}^*)\eta_0 \right) \varphi = 2\,\lambda_{\varphi}\,\xi_0 \\[4mm]
\displaystyle \int_{\Omega} \left(2\,\mathrm{Re}\,\mathscr{A}\,\eta_0 - \frac{p-2}{2\sqrt{p-1}}\,\mathrm{Im}(\mathscr{A} - \mathscr{A}^*)\xi_0 \right) \varphi = 2\,\lambda_{\varphi}\,\eta_0\,.
\end{cases}
$$

This implies

$$
\int_{\Omega} \left(2\,\mathrm{Re}\,\mathscr{A}(\xi_0 - \eta_0) + \frac{p-2}{2\sqrt{p-1}}\,\mathrm{Im}(\mathscr{A} - \mathscr{A}^*)(\xi_0 - \eta_0) \right) \varphi = 2\,\lambda_{\varphi}\,(\xi_0 - \eta_0)
$$

and therefore

$$
\int_{\Omega} \left(2\langle \mathrm{Re}\,\mathscr{A}(\xi_0 - \eta_0), \xi_0 - \eta_0 \rangle + \frac{p-2}{\sqrt{p-1}}\langle \mathrm{Im}\,\mathscr{A}(\xi_0 - \eta_0), \xi_0 - \eta_0 \rangle \right) \varphi
$$
$$
= 2\,\lambda_{\varphi}\,|\xi_0 - \eta_0|^2.
$$

The left-hand side is nonnegative because of (2.29). Hence, if $\lambda_{\varphi} < 0$, we find $\xi_0 = \eta_0$. On the other hand we have

$$
\lambda_{\varphi} = \int_{\Omega} \mathscr{S}(\xi_0, \xi_0)\,\varphi = \int_{\Omega} \left(2\langle \mathrm{Re}\,\mathscr{A}\,\xi_0, \xi_0 \rangle - \frac{p-2}{\sqrt{p-1}}\langle \mathrm{Im}\,\mathscr{A}\,\xi_0, \xi_0 \rangle \right) \varphi \geq 0.
$$

This shows that $\lambda_{\varphi} \geq 0$ for any nonnegative φ and the sufficiency is proved.

Necessity. We know from the proof of Corollary 2.2 that if $\mathscr{L}$ is L^p-dissipative, then (2.20) holds for any $\varrho \in C_0^1(\Omega)$, $\mu \in \mathbb{R}$. In the present case, keeping in mind (2.27), (2.20) can be written as

$$
\int_{\Omega} \langle \mathscr{B}\,\nabla\varrho, \nabla\varrho \rangle \geq 0,
$$

where

$$
\mathscr{B} = \frac{4}{p\,p'}\,\mathrm{Re}\,\mathscr{A} + \mu^2\,\mathrm{Re}\,\mathscr{A} - 2\,\mu\,(1 - 2/p)\,\mathrm{Im}\,\mathscr{A}\,.
$$

In the proof of Corollary 2.2, we have also seen that from (2.21) for any $\varrho \in C_0^1(\Omega)$, (2.16) follows. In the same way, the last relation implies $\langle \mathscr{B}\,\xi, \xi \rangle \geq 0$, i.e.,

$$
\frac{4}{p\,p'}\,\langle \mathrm{Re}\,\mathscr{A}\,\xi, \xi \rangle + \mu^2\langle \mathrm{Re}\,\mathscr{A}\,\xi, \xi \rangle - 2\,\mu\,(1 - 2/p)\langle \mathrm{Im}\,\mathscr{A}\,\xi, \xi \rangle \geq 0
$$

for any $\xi \in \mathbb{R}^n$, $\mu \in \mathbb{R}$.

Because of the arbitrariness of μ we have

$$
\int_{\Omega} \langle \mathrm{Re}\,\mathscr{A}\,\xi, \xi \rangle\,\varphi \geq 0
$$

$$
(1 - 2/p)^2 \left(\int_{\Omega} \langle \mathrm{Im}\,\mathscr{A}\,\xi, \xi \rangle\,\varphi \right)^2 \leq \frac{4}{p\,p'} \left(\int_{\Omega} \langle \mathrm{Re}\,\mathscr{A}\,\xi, \xi \rangle\,\varphi \right)^2,
$$

i.e.,

$$|p - 2| \left| \int_\Omega \langle \operatorname{Im} \mathscr{A} \, \xi, \xi \rangle \, \varphi \right| \leqslant 2 \, \sqrt{p - 1} \, \int_\Omega \langle \operatorname{Re} \mathscr{A} \, \xi, \xi \rangle \, \varphi$$

for any $\xi \in \mathbb{R}^n$ and for any nonnegative $\varphi \in C_0(\Omega)$.

We have

$$|p - 2| \left| \int_\Omega \langle \operatorname{Im} \mathscr{A} \, \xi, \xi \rangle \varphi \right| \leqslant 2 \sqrt{p - 1} \int_\Omega \langle \operatorname{Re} \mathscr{A} \, \xi, \xi \rangle |\varphi|$$

for any $\varphi \in C_0(\Omega)$ and this implies (2.29), because

$$|p - 2| \int_\Omega |\langle \operatorname{Im} \mathscr{A} \, \xi, \xi \rangle| \, g = |p - 2| \sup_{\substack{\varphi \in C_0(\Omega) \\ |\varphi| \leqslant g}} \left| \int_\Omega \langle \operatorname{Im} \mathscr{A} \, \xi, \xi \rangle \varphi \right|$$

$$\leqslant 2 \sqrt{p - 1} \sup_{\substack{\varphi \in C_0(\Omega) \\ |\varphi| \leqslant g}} \int_\Omega \langle \operatorname{Re} \mathscr{A} \, \xi, \xi \rangle |\varphi| \leqslant 2 \sqrt{p - 1} \int_\Omega \langle \operatorname{Re} \mathscr{A} \, \xi, \xi \rangle g$$

for any nonnegative $g \in C_0(\Omega)$. $\qquad\qquad\qquad\qquad\qquad\qquad\qquad\qquad\qquad\qquad\quad$ $\square$

Remark 2.8. From the proof of Theorem 2.7 we see that condition (2.29) holds if and only if

$$\frac{4}{p\,p'} \langle \operatorname{Re} \mathscr{A} \, \xi, \xi \rangle + \langle \operatorname{Re} \mathscr{A} \, \eta, \eta \rangle - 2(1 - 2/p)\langle \operatorname{Im} \mathscr{A} \, \xi, \eta \rangle \geqslant 0$$

for any $\xi, \eta \in \mathbb{R}^n$. This means that conditions (2.22) are necessary and sufficient for the operators considered in Theorem 2.7.

Remark 2.9. Let us assume that either A has lower-order terms or they are absent and $\operatorname{Im} \mathscr{A}$ is not symmetric. Using the same arguments as in Theorem 2.7, one could prove that (2.29) is still a necessary condition for A to be L^p-dissipative. However, in general, it is not sufficient. This is shown by the next example (see also Theorem 2.15 below for the particular case of constant coefficients).

Example 2.10. Let $n = 2$ and let Ω be a bounded domain. Denote by σ a not identically vanishing real function in $C_0^2(\Omega)$ and let $\lambda \in \mathbb{R}$. Consider operator (2.28) with

$$\mathscr{A} = \begin{pmatrix} 1 & i\lambda\partial_1(\sigma^2) \\ -i\lambda\partial_1(\sigma^2) & 1 \end{pmatrix},$$

i.e.,

$$Au = \partial_1(\partial_1 u + i\lambda\partial_1(\sigma^2) \, \partial_2 u) + \partial_2(-i\lambda\partial_1(\sigma^2) \, \partial_1 u + \partial_2 u),$$

where $\partial_i = \partial/\partial x_i$ $(i = 1, 2)$.

By definition, we have L^2-dissipativity if and only if

$$\operatorname{Re} \int_\Omega ((\partial_1 u + i\lambda\partial_1(\sigma^2) \, \partial_2 u)\partial_1 \overline{u} + (-i\lambda\partial_1(\sigma^2) \, \partial_1 u + \partial_2 u)\partial_2 \overline{u}) \, dx \geqslant 0$$

for any $u \in C_0^1(\Omega)$, i.e., if and only if

$$\int_\Omega |\nabla u|^2 dx - 2\lambda \int_\Omega \partial_1(\sigma^2) \operatorname{Im}(\partial_1 \overline{u} \, \partial_2 u) \, dx \geqslant 0$$

for any $u \in C_0^1(\Omega)$. Taking $u = \sigma \exp(itx_2)$ ($t \in \mathbb{R}$), we obtain, in particular,

$$t^2 \int_\Omega \sigma^2 dx - t\lambda \int_\Omega (\partial_1(\sigma^2))^2 dx + \int_\Omega |\nabla \sigma|^2 dx \geqslant 0. \tag{2.31}$$

Since

$$\int_\Omega (\partial_1(\sigma^2))^2 dx > 0,$$

we can choose $\lambda \in \mathbb{R}$ so that (2.31) is impossible for all $t \in \mathbb{R}$. Thus A is not L^2-dissipative, although (2.29) is satisfied.

Since A can be written as

$$Au = \Delta u - i\lambda(\partial_{21}(\sigma^2) \, \partial_1 u - \partial_{11}(\sigma^2) \, \partial_2 u),$$

the same example shows that (2.29) is not sufficient for the L^2-dissipativity in the presence of lower-order terms, even if $\operatorname{Im} \mathscr{A}$ is symmetric.

Remark 2.11. It is nice to remark that from (2.29) we can immediately deduce the following facts: let A be the differential operator (2.28) satisfying the hypothesis of Theorem 2.7. Let us suppose that A is a degenerate elliptic operator (i.e., it satisfies (2.16)). Then

(i) the corresponding form $\mathscr{L}$ if L^2-dissipative;
(ii) if the operator A has real coefficients ($\operatorname{Im} \mathscr{A} = 0$), then the corresponding form $\mathscr{L}$ is L^p-dissipative for any p.

Remark 2.12. In view of Theorem 2.7, it is now clear why condition (2.16) cannot be sufficient for the L^p-dissipativity when $p \neq 2$.

2.3 Operators with lower-order terms

We know from Remark 2.9 that, if the partial differential operator A contains lower-order terms, the algebraic condition (2.29) is not necessary and sufficient for the L^p-dissipativity. One could ask if there are other algebraic necessary and sufficient conditions for these more general operators.

Generally speaking, this is not possible. We can convince ourselves of that by means of the following examples.

Example 2.13. Let A be the operator

$$Au = \Delta u + a(x)u$$

in a bounded domain $\Omega \subset \mathbb{R}^n$, where $a(x)$ is a real smooth function. Denote by λ_1 the first eigenvalue of the Dirichlet problem for a Laplace equation in Ω. A sufficient condition for A to be L^2-dissipative is $\operatorname{Re} a \leqslant \lambda_1$ and, in general, one cannot give an explicit value of λ_1.

Example 2.14. Let A be the operator

$$Au = \nabla(\mathscr{A}\, \nabla u) + \mu u$$

in a domain $\Omega \subset \mathbb{R}^n$, where $\mathscr{A}$ is a matrix with real continuous entries and μ is a nonnegative measure. Lemma 2.1 shows that A is L^p-dissipative if and only if

$$\int_\Omega |w|^2 d\mu \leqslant \frac{4}{pp'} \int_\Omega \langle \mathscr{A}\, \nabla w, \nabla w\rangle\, dx, \qquad \forall\, w \in C_0^\infty(\Omega). \tag{2.32}$$

If (2.32) holds, then

$$\frac{\mu(F)}{\operatorname{cap}_\Omega(F)} \leqslant \frac{4}{pp'} \tag{2.33}$$

for any compact set $F \subset \Omega$, where $\operatorname{cap}_\Omega(F)$ is the relative capacity of F,

$$\operatorname{cap}_\Omega(F) = \inf\left\{ \int_\Omega \langle \mathscr{A}\, \nabla u, \nabla u\rangle\, dx \;:\; u \in C_0^\infty(\Omega),\; u \geqslant 1 \text{ on } F\right\}.$$

In fact, if $u \in C_0^\infty(\Omega)$, with $u \geqslant 1$ on F, (2.32) implies that

$$\mu(F) \leqslant \int_F u^2 d\mu \leqslant \int_\Omega u^2 d\mu \leqslant \frac{4}{pp'} \int_\Omega \langle \mathscr{A}\, \nabla u, \nabla u\rangle\, dx$$

and then

$$\mu(F) \leqslant \frac{4}{pp'} \inf_{\substack{u \in C_0^\infty(\Omega) \\ u \geqslant 1 \text{ on } F}} \int_\Omega \langle \mathscr{A}\, \nabla u, \nabla u\rangle\, dx,$$

i.e., (2.33).

On the other hand, if

$$\frac{\mu(F)}{\operatorname{cap}_\Omega(F)} \leqslant \frac{1}{pp'} \tag{2.34}$$

for any compact set $F \subset \Omega$, then (2.32) holds. This result is due to V. Maz'ya (see [73, Th. 2.3.3]; see also [66, 67, 72]). One can show that the necessary condition (2.33) is not sufficient and the sufficient condition (2.34) is not necessary.

In connection with the L^p-dissipativity of the operator A see also comments to Chapter 7.

However, if the operator has constant coefficients, then one can still give necessary and sufficient conditions. This is the subject of the following subsection.

2.3.1 Operators with constant complex coefficients

In this section we characterize the L^p-dissipativity for the operator (2.3) with constant complex coefficients. Without loss of generality we can write A as

$$Au = \nabla^t(\mathscr{A} \nabla u) + \mathbf{b}\nabla u + au, \tag{2.35}$$

assuming that the matrix $\mathscr{A}$ is symmetric.

Theorem 2.15. *Let Ω be an open set in $\mathbb{R}^n$ which contains balls of arbitrarily large radius. The operator A is L^p-dissipative if and only if there exists a real constant vector V such that*

$$2 \operatorname{Re} \mathscr{A} V + \operatorname{Im} \mathbf{b} = 0, \tag{2.36}$$

$$\operatorname{Re} a + \langle \operatorname{Re} \mathscr{A} V, V \rangle \leqslant 0 \tag{2.37}$$

and the inequality

$$|p - 2| \, |\langle \operatorname{Im} \mathscr{A} \, \xi, \xi \rangle| \leqslant 2\sqrt{p - 1} \, \langle \operatorname{Re} \mathscr{A} \, \xi, \xi \rangle \tag{2.38}$$

holds for any $\xi \in \mathbb{R}^n$.

Proof. First, let us prove the theorem for the special case $\mathbf{b} = 0$, i.e., for the operator

$$A = \nabla^t(\mathscr{A} \nabla u) + au.$$

If A is L^p-dissipative, (2.6) holds for any $v \in C_0^1(\Omega)$. We find, by repeating the arguments used in the proof of Theorem 2.7, that

$$\begin{aligned}
\frac{4}{p\,p'} & \int_\Omega \langle \operatorname{Re} \mathscr{A} \, \nabla \varrho, \nabla \varrho \rangle \, dx + \mu^2 \int_\Omega \langle \operatorname{Re} \mathscr{A} \, \nabla \varrho, \nabla \varrho \rangle \, dx \\
& - 2\,\mu\,(1 - 2/p) \int_\Omega \langle \operatorname{Im} \mathscr{A} \, \nabla \varrho, \nabla \varrho \rangle \, dx - (\operatorname{Re} a) \int_\Omega \varrho^2 dx \geqslant 0
\end{aligned} \tag{2.39}$$

for any $\varrho \in C_0^\infty(\Omega)$ and for any $\mu \in \mathbb{R}$. As in the proof of Theorem 2.7 this implies (2.38). On the other hand, we can find a sequence of balls contained in Ω with centers x_m and radii m. Set

$$\varrho_m(x) = m^{-n/2}\sigma\left((x - x_m)/m\right),$$

where $\sigma \in C_0^\infty(\mathbb{R}^n)$, $\operatorname{spt} \sigma \subset B_1(0)$ and

$$\int_{B_1(0)} \sigma^2(x) \, dx = 1.$$

Putting in (2.39) $\mu = 1$ and $\varrho = \varrho_m$, we obtain

$$\begin{aligned}
\frac{4}{p\,p'} & \int_{B_1(0)} \langle \operatorname{Re} \mathscr{A} \, \nabla \sigma, \nabla \sigma \rangle \, dy + \int_{B_1(0)} \langle \operatorname{Re} \mathscr{A} \, \nabla \sigma, \nabla \sigma \rangle \, dy \\
& - 2\,(1 - 2/p) \int_{B_1(0)} \langle \operatorname{Im} \mathscr{A} \, \nabla \sigma, \nabla \sigma \rangle \, dy - m^2(\operatorname{Re} a) \geqslant 0
\end{aligned}$$

for any $m \in \mathbb{N}$. This implies $\operatorname{Re} a \leqslant 0$. Note that in this case the algebraic system (2.36) has always the trivial solution and that for any eigensolution V (if they exist) we have $\langle \operatorname{Re} \mathscr{A} V, V \rangle = 0$. Then (2.37) is satisfied.

Conversely, if (2.38) is satisfied, we have (see Remark 2.8)

$$\frac{4}{p\,p'} \langle \operatorname{Re} \mathscr{A}\, \xi, \xi \rangle + \langle \operatorname{Re} \mathscr{A}\, \eta, \eta \rangle - 2\,(1 - 2/p) \langle \operatorname{Im} \mathscr{A}\, \xi, \xi \rangle \geqslant 0$$

for any $\xi, \eta \in \mathbb{R}^n$. If also (2.37) is satisfied (i.e., if $\operatorname{Re} a \leqslant 0$), A is L^p-dissipative in view of Corollary 2.3.

Let us consider the operator in the general form (2.35). If A is L^p-dissipative, we find, by repeating the arguments employed in the proof of Theorem 2.7, that

$$\frac{4}{p\,p'} \int_\Omega \langle \operatorname{Re} \mathscr{A}\, \nabla \varrho, \nabla \varrho \rangle \, dx + \int_\Omega \varrho^2 \langle \operatorname{Re} \mathscr{A}\, \nabla \varphi, \nabla \varphi \rangle \, dx$$
$$- 2\,(1 - 2/p) \int_\Omega \varrho \, \langle \operatorname{Im} \mathscr{A}\, \nabla \varrho, \nabla \varphi \rangle \, dx$$
$$+ \int_\Omega \varrho^2 \langle \operatorname{Im} \mathbf{b}, \nabla \varphi \rangle \, dx - \operatorname{Re} a \int_\Omega \varrho^2 dx \geqslant 0$$

for any $\varrho \in C_0^1(\Omega)$, $\varphi \in C^1(\Omega)$. By fixing ϱ and choosing $\varphi = t\langle \eta, x \rangle$ ($t \in \mathbb{R}$, $\eta \in \mathbb{R}^n$) we get

$$\frac{4}{p\,p'} \int_\Omega \langle \operatorname{Re} \mathscr{A}\, \nabla \varrho, \nabla \varrho \rangle \, dx + (t^2 \langle \operatorname{Re} \mathscr{A}\, \eta, \eta \rangle + t \, \langle \operatorname{Im} \mathbf{b}, \eta \rangle - \operatorname{Re} a) \int_\Omega \varrho^2 \, dx \geqslant 0$$

for any $t \in \mathbb{R}$. This leads to

$$|\langle \operatorname{Im} \mathbf{b}, \eta \rangle|^2 \leqslant K \, \langle \operatorname{Re} \mathscr{A}\, \eta, \eta \rangle$$

for any $\eta \in \mathbb{R}^n$ and this inequality shows that system (2.36) is solvable. Let V be a solution of this system and let

$$z = e^{-i\langle V, x \rangle} u.$$

One checks directly that

$$Au = (\nabla^t(\mathscr{A}\, \nabla z) + \langle \mathbf{c}, \nabla z \rangle + \alpha z) e^{i\langle V, x \rangle}$$

where

$$\mathbf{c} = 2i\, \mathscr{A}\, V + \mathbf{b}, \quad \alpha = a + i\langle \mathbf{b}, V \rangle - \langle \mathscr{A}\, V, V \rangle.$$

Since we have

$$\int_\Omega \langle Au, u \rangle |u|^{p-2} dx = \int_\Omega \langle \nabla^t(\mathscr{A}\, \nabla z) + \langle \mathbf{c}, \nabla z \rangle + \alpha z, z \rangle |z|^{p-2} dx \,,$$

the L^p-dissipativity of A is equivalent to the L^p-dissipativity of the operator

$$\nabla^t(\mathscr{A}\,\nabla z) + \langle \mathbf{c}, \nabla z\rangle + \alpha z\,.$$

On the other hand Lemma 2.1 shows that, as far as the first-order terms are concerned, the $\mathrm{Re}\,\mathbf{b}$ does not play any role. Since $\mathrm{Im}\,\mathbf{c} = \mathbf{0}$ because of (2.36), the L^p-dissipativity of A is equivalent to the L^p-dissipativity of the operator

$$\nabla^t(\mathscr{A}\,\nabla z) + \alpha z\,. \tag{2.40}$$

By what we have already proved above, the last operator is L^p-dissipative if and only if (2.38) is satisfied and $\mathrm{Re}\,\alpha \leqslant 0$. From (2.36) it follows that $\mathrm{Re}\,\alpha$ is equal to the left-hand side of (2.37).

Conversely, if there exists a solution V of (2.36), (2.37), and if (2.38) is satisfied, operator (2.40) is L^p-dissipative. Since this is equivalent to the L^p-dissipativity of A, the proof is complete. $\qquad\square$

Corollary 2.16. *Let Ω be an open set in $\mathbb{R}^n$ which contains balls of arbitrarily large radius. Let us suppose that the matrix $\mathrm{Re}\,\mathscr{A}$ is not singular. The operator A is L^p-dissipative if and only if (2.38) holds and*

$$4\,\mathrm{Re}\,a \leqslant -\langle(\mathrm{Re}\,\mathscr{A})^{-1}\,\mathrm{Im}\,\mathbf{b}, \mathrm{Im}\,\mathbf{b}\rangle. \tag{2.41}$$

Proof. If $\mathrm{Re}\,\mathscr{A}$ is not singular, the only vector V satisfying (2.36) is

$$V = -(1/2)(\mathrm{Re}\,\mathscr{A})^{-1}\,\mathrm{Im}\,\mathbf{b}$$

and (2.37) is satisfied if and only if (2.41) holds. The result follows from Theorem 2.15. $\qquad\square$

Example 2.17. Let $n = 1$ and $\Omega = \mathbb{R}^1$. Consider the operator

$$\left(1 + 2\,\frac{\sqrt{p-1}}{p-2}\,i\right) u'' + 2iu' - u,$$

where $p \neq 2$ is fixed. Conditions (2.38) and (2.41) are satisfied and this operator is L^p-dissipative, in view of Corollary 2.16.

On the other hand, the polynomial considered in Corollary 2.3 (with $\alpha = \beta = 0$) is

$$Q(\xi, \eta) = \left(2\,\frac{\sqrt{p-1}}{p}\,\xi - \eta\right)^2 + 2\eta + 1$$

which is not nonnegative for any $\xi, \eta \in \mathbb{R}$. This shows that, in general, condition (2.22) is not necessary for the L^p-dissipativity, even if the matrix $\mathrm{Im}\,\mathscr{A}$ is symmetric.

2.4 Equivalence between the dissipativity of the operator and the dissipativity of the associated form

In this section we study the relations between the dissipativity of the operator A and the dissipativity of the form $\mathscr{L}$ (see Definition 1.4). We also deal with the question whether A generates a contractive or quasi-contractive semigroup.

In all this section A is the operator

$$Au = \operatorname{div}(\mathscr{A}\,\nabla u) + \mathbf{b}\nabla u + a\,u \tag{2.42}$$

with smooth coefficients: $a^{hk}, b^h \in C^1(\overline{\Omega})$, $a \in C^0(\overline{\Omega})$ where Ω is a bounded domain in $\mathbb{R}^n$. We consider A as an operator defined on the set

$$D(A) = W^{2,p}(\Omega) \cap \mathring{W}^{1,p}(\Omega). \tag{2.43}$$

We suppose that the boundary $\partial\Omega$ satisfies the following smoothness assumptions:

- if $p > n$ the boundary $\partial\Omega$ belongs to $W^{2-1/p}$;
- if $p \leqslant n$ we have

$$\int_0^1 \left| \frac{\omega(t)}{t} \right|^p dt < \infty,$$

where $\omega(t)$ is the modulus of continuity of the normal to $\partial\Omega$.

These assumptions ensure the invertibility of operator A (see Maz'ya and Shaposhnikova [76, Ch. 14] for this and other conditions).

2.4.1 L^p-dissipativity of $\mathscr{L}$ and L^p-dissipativity of A

We recall that the operator A is L^p-dissipative if

$$\operatorname{Re} \int_\Omega \langle Au, u \rangle |u|^{p-2} dx \leqslant 0 \tag{2.44}$$

for any $u \in D(A)$.

The aim of this subsection is to show that the L^p-dissipativity of A is equivalent to the L^p-dissipativity of the sesquilinear form

$$\mathscr{L}(u,v) = -\int_\Omega \left(\langle \mathscr{A}\,\nabla u, \nabla v \rangle - \langle \mathbf{b}\nabla u, v \rangle - a\langle u, v \rangle \right).$$

In order to obtain that, we need some lemmas.

Lemma 2.18. *The form $\mathscr{L}$ is L^p-dissipative if and only if*

$$
\operatorname{Re} \int_\Omega \Big[\langle \mathscr{A}\,\nabla v, \nabla v \rangle - (1 - 2/p)\langle (\mathscr{A} - \mathscr{A}^*)\nabla(|v|), |v|^{-1}\overline{v}\nabla v \rangle
$$
$$
- (1 - 2/p)^2 \langle \mathscr{A}\,\nabla(|v|), \nabla(|v|) \rangle \Big] dx \tag{2.45}
$$
$$
+ \int_\Omega \langle \operatorname{Im} \mathbf{b}, \operatorname{Im}(\overline{v}\nabla v) \rangle dx + \int_\Omega \operatorname{Re}(\nabla^t(\mathbf{b}/p) - a)|v|^2 dx \geqslant 0
$$

for any $v \in H_0^1(\Omega)$.

Proof. Sufficiency. We know from Lemma 2.1 that $\mathscr{L}$ is L^p-dissipative if and only if (2.45) holds for any $v \in C_0^1(\Omega)$. Since $C_0^1(\Omega) \subset H_0^1(\Omega)$, the sufficiency follows.

Necessity. Given $v \in H_0^1(\Omega)$, we can find a sequence $\{v_n\} \subset C_0^1(\Omega)$ such that $v_n \to v$ in $H_0^1(\Omega)$. Let us show that

$$
\chi_{E_n}|v_n|^{-1}\overline{v}_n \nabla v_n \to \chi_E |v|^{-1}\overline{v}\nabla v \quad \text{in } L^2(\Omega) \tag{2.46}
$$

where $E_n = \{x \in \Omega \mid v_n(x) \neq 0\}$, $E = \{x \in \Omega \mid v(x) \neq 0\}$. We may assume $v_n(x) \to v(x)$, $\nabla v_n(x) \to \nabla v(x)$ almost everywhere in Ω. We see that

$$
\chi_{E_n}|v_n|^{-1}\overline{v}_n \nabla v_n \to \chi_E |v|^{-1}\overline{v}\nabla v \tag{2.47}
$$

almost everywhere on the set $E \cup \{x \in \Omega \setminus E \mid \nabla v(x) = 0\}$. Since the set $\{x \in \Omega \setminus E \mid \nabla v(x) \neq 0\}$ has zero measure, we can say that (2.47) holds almost everywhere in Ω.

Moreover, since

$$
\int_G |\chi_{E_n}|v_n|^{-1}\overline{v}_n \nabla v_n|^2 dx \leqslant \int_G |\nabla v_n|^2 dx
$$

for any measurable set $G \subset \Omega$ and $\{\nabla v_n\}$ is convergent in $L^2(\Omega)$, the sequence $\{|\chi_{E_n}|v_n|^{-1}\overline{v}_n \nabla v_n - \chi_E |v|^{-1}\overline{v}\nabla v|^2\}$ has uniformly absolutely continuous integrals. Now we may appeal to Vitali's Theorem (see, e.g., Rudin [91, p. 133]) to obtain (2.46).

From this it follows that (2.45) for any $v \in H_0^1(\Omega)$ implies (2.45) for any $v \in C_0^1(\Omega)$. Lemma 2.1 shows that $\mathscr{L}$ is L^p-dissipative. $\qquad\square$

Lemma 2.19. *The form $\mathscr{L}$ is L^p-dissipative if and only if*

$$
\operatorname{Re} \int_\Omega (\langle \mathscr{A}\,\nabla u, \nabla(|u|^{p-2}u) \rangle - \langle \mathbf{b}\nabla u, |u|^{p-2}u \rangle - a\,|u|^p) dx \geqslant 0 \tag{2.48}
$$

for any $u \in \Xi$, where Ξ denotes the space $\{u \in C^2(\overline{\Omega}) \mid u|_{\partial\Omega} = 0\}$.

Proof. Necessity. Since $\mathscr{L}$ is L^p-dissipative, (2.45) holds for any $v \in H_0^1(\Omega)$. Let $u \in \Xi$. We introduce the function

$$\varrho_\varepsilon(s) = \begin{cases} \varepsilon^{\frac{p-2}{2}} & \text{if } 0 \leqslant s \leqslant \varepsilon \\ s^{\frac{p-2}{2}} & \text{if } s > \varepsilon. \end{cases}$$

Setting

$$v_\varepsilon = \varrho_\varepsilon(|u|)\, u$$

a direct computation shows that $u = \sigma_\varepsilon(|v_\varepsilon|)\, v_\varepsilon$ and $\varrho_\varepsilon^2(|u|)\, u = [\sigma_\varepsilon(|v_\varepsilon|)]^{-1}\, v_\varepsilon$, where

$$\sigma_\varepsilon(s) = \begin{cases} \varepsilon^{\frac{2-p}{2}} & \text{if } 0 \leqslant s \leqslant \varepsilon^{\frac{p}{2}} \\ s^{\frac{2-p}{p}} & \text{if } s > \varepsilon^{\frac{p}{2}}. \end{cases}$$

Therefore

$$\langle \mathscr{A}\, \nabla u, \nabla[\varrho_\varepsilon^2(|u|)\, u] \rangle = \langle \mathscr{A}\, \nabla[\sigma_\varepsilon(|v_\varepsilon|)\, v_\varepsilon], \nabla[(\sigma_\varepsilon(|v_\varepsilon|))^{-1} v_\varepsilon] \rangle$$

$$= \langle \mathscr{A}\, [\sigma_\varepsilon(|v_\varepsilon|)\, \nabla v_\varepsilon + \sigma_\varepsilon'(|v_\varepsilon|)\, v_\varepsilon\, \nabla |v_\varepsilon|], \sigma_\varepsilon(|v_\varepsilon|)^{-1} \nabla v_\varepsilon$$

$$- \sigma_\varepsilon'(|v_\varepsilon|)\sigma_\varepsilon^{-2}(|v_\varepsilon|)v_\varepsilon\, \nabla |v_\varepsilon| \rangle$$

$$= \langle \mathscr{A}\, \nabla v_\varepsilon, \nabla v_\varepsilon \rangle + \sigma_\varepsilon'(|v_\varepsilon|)\sigma_\varepsilon(|v_\varepsilon|)^{-1} \left(\langle v_\varepsilon\, \mathscr{A}\, \nabla |v_\varepsilon|, \nabla v_\varepsilon \rangle - \langle \mathscr{A}\, \nabla v_\varepsilon, v_\varepsilon\, \nabla |v_\varepsilon| \rangle \right)$$

$$- \sigma_\varepsilon'(|v_\varepsilon|)^2 \sigma_\varepsilon(|v_\varepsilon|)^{-2} \langle v_\varepsilon\, \mathscr{A}\, \nabla |v_\varepsilon|, v_\varepsilon \nabla |v_\varepsilon| \rangle.$$

Since

$$\frac{\sigma_\varepsilon'(|v_\varepsilon|)}{\sigma_\varepsilon(|v_\varepsilon|)} = \begin{cases} 0 & \text{if } 0 < |u| < \varepsilon \\ -(1 - 2/p)\, |v_\varepsilon|^{-1} & \text{if } |u| > \varepsilon \end{cases}$$

we may write

$$\int_\Omega \langle \mathscr{A}\, \nabla u, \nabla[\varrho_\varepsilon^2(|u|)\, u] \rangle\, dx = \int_\Omega \langle \mathscr{A}\, \nabla v_\varepsilon, \nabla v_\varepsilon \rangle\, dx$$

$$- (1 - 2/p) \int_{E_\varepsilon} \frac{1}{|v_\varepsilon|} \left(\langle v_\varepsilon\, \mathscr{A}\, \nabla |v_\varepsilon|, \nabla v_\varepsilon \rangle - \langle \mathscr{A}\, \nabla v_\varepsilon, v_\varepsilon\, \nabla |v_\varepsilon| \rangle \right)\, dx$$

$$- (1 - 2/p)^2 \int_{E_\varepsilon} \langle \mathscr{A}\, \nabla |v_\varepsilon|, \partial_h \nabla |v_\varepsilon| \rangle\, dx$$

where $E_\varepsilon = \{x \in \Omega \mid |u(x)| > \varepsilon\}$. Then

$$\int_\Omega \langle \mathscr{A}\, \nabla u, \nabla[\varrho_\varepsilon^2(|u|)\, u] \rangle\, dx = \int_\Omega \langle \mathscr{A}\, \nabla v_\varepsilon, \nabla v_\varepsilon \rangle\, dx$$

$$- (1 - 2/p) \int_\Omega \frac{1}{|v_\varepsilon|} \left(\langle v_\varepsilon\, \mathscr{A}\, \nabla |v_\varepsilon|, \nabla v_\varepsilon \rangle - \langle \mathscr{A}\, \nabla v_\varepsilon, v_\varepsilon\, \nabla |v_\varepsilon| \rangle \right)\, dx$$

$$- (1 - 2/p)^2 \int_\Omega \langle \mathscr{A}\, \nabla |v_\varepsilon|, \nabla |v_\varepsilon| \rangle\, dx + R(\varepsilon)$$

where

$$R(\varepsilon) = (1 - 2/p) \int_{\Omega \setminus E_\varepsilon} \frac{1}{|v_\varepsilon|} \left(v_\varepsilon \langle \mathscr{A} \nabla v_\varepsilon |, \nabla v_\varepsilon \rangle - \langle \mathscr{A} \nabla v_\varepsilon, v_\varepsilon \nabla |v_\varepsilon| \rangle \right) dx$$

$$- (1 - 2/p)^2 \int_{\Omega \setminus E_\varepsilon} \langle \mathscr{A} \nabla |v_\varepsilon|, \nabla |v_\varepsilon| \rangle \, dx.$$

It is proved in Langer [54] that if $u \in C^2(\overline{\Omega})$ and $u|_{\partial \Omega} = 0$, then

$$\lim_{\varepsilon \to 0} \varepsilon^r \int_{\Omega \setminus E_\varepsilon} |\nabla u|^2 dx = 0 \tag{2.49}$$

for any $r > -1$. Since

$$|\nabla |v_\varepsilon| \, | = \left| \operatorname{Re} \left(\frac{\overline{v}_\varepsilon \nabla v_\varepsilon}{|v_\varepsilon|} \chi_{E_0} \right) \right| \leqslant |\nabla v_\varepsilon| = \varepsilon^{\frac{p-2}{2}} |\nabla u|$$

in $E_0 \setminus E_\varepsilon$, we obtain

$$\left| \int_{\Omega \setminus E_\varepsilon} \langle \mathscr{A} \nabla |v_\varepsilon|, \nabla |v_\varepsilon| \rangle \, dx \right| \leqslant K \, \varepsilon^{p-2} \int_{\Omega \setminus E_\varepsilon} |\nabla u|^2 dx \to 0$$

as $\varepsilon \to 0$. We have also

$$|v_\varepsilon|^{-1} |\langle v_\varepsilon \mathscr{A} \nabla |v_\varepsilon|, \nabla v_\varepsilon \rangle - \langle \mathscr{A} \nabla v_\varepsilon, v_\varepsilon \nabla |v_\varepsilon| \rangle| \leqslant K \, \varepsilon^{p-2} |\nabla u|^2$$

and thus $R(\varepsilon) = o(1)$ as $\varepsilon \to 0$.

We have proved that

$$\operatorname{Re} \int_\Omega \langle \mathscr{A} \nabla u, \nabla [\varrho_\varepsilon^2(|u|) \, u] \rangle \, dx$$

$$= \operatorname{Re} \left[\int_\Omega \langle \mathscr{A} \nabla v_\varepsilon, \nabla v_\varepsilon \rangle \, dx - (1 - 2/p) \int_\Omega \langle (\mathscr{A} - \mathscr{A}^*) \nabla |v_\varepsilon|, |v_\varepsilon|^{-1} \overline{v}_\varepsilon \nabla v_\varepsilon \rangle dx \right.$$

$$\left. - (1 - 2/p)^2 \int_\Omega \langle \mathscr{A} \nabla |v_\varepsilon|, \nabla |v_\varepsilon| \rangle \, dx \right] + o(1). \tag{2.50}$$

By means of similar computations, we find by the identity

$$\int_\Omega \langle \mathbf{b} \nabla u, |u|^{p-2} u \rangle dx$$

$$= \int_{\Omega \setminus E_\varepsilon} \langle \mathbf{b} \nabla u, |u|^{p-2} u \rangle dx - (1 - 2/p) \int_{E_\varepsilon} \langle \mathbf{b}, |v_\varepsilon| \nabla(|v_\varepsilon|) \rangle dx + \int_{E_\varepsilon} \langle \mathbf{b} \nabla v_\varepsilon, v_\varepsilon \rangle dx$$

that

$$\operatorname{Re} \int_\Omega \langle \mathbf{b} \nabla u, |u|^{p-2} u \rangle dx$$

$$= \int_\Omega \langle \operatorname{Re}(\mathbf{b}/p), \nabla(|v_\varepsilon|^2) \rangle dx - \int_\Omega \langle \operatorname{Im} \mathbf{b}, \operatorname{Im}(\overline{v}_\varepsilon \nabla v) \rangle dx + o(1). \tag{2.51}$$

Moreover

$$\int_\Omega |u|^p dx = \int_{E_\varepsilon} |u|^p dx + \int_{\Omega \setminus E_\varepsilon} |u|^p dx$$

$$= \int_{E_\varepsilon} |v_\varepsilon|^2 dx + \int_{\Omega \setminus E_\varepsilon} |u|^p dx = \int_\Omega |v_\varepsilon|^2 dx + o(1). \tag{2.52}$$

Equalities (2.50), (2.51) and (2.52) lead to

$$\mathrm{Re} \int_\Omega (\langle \mathscr{A} \nabla u, \nabla[\varrho_\varepsilon^2(|u|)\, u]\rangle - \langle \mathbf{b}\nabla u, |u|^{p-2}u\rangle - a|u|^p) dx$$

$$= \mathrm{Re} \Big[\int_\Omega \langle \mathscr{A} \nabla v_\varepsilon, \nabla v_\varepsilon\rangle\, dx - (1 - 2/p) \int_\Omega \langle (\mathscr{A} - \mathscr{A}^*)\, \nabla |v_\varepsilon|, \nabla v_\varepsilon\rangle) v_\varepsilon |v_\varepsilon|^{-1} dx$$

$$- (1 - 2/p)^2 \int_\Omega \langle \mathscr{A} \nabla |v_\varepsilon|, \nabla |v_\varepsilon|\rangle\, dx \Big] + \int_\Omega \mathrm{Re}(\nabla^t (\mathbf{b}/p)|v_\varepsilon|^2 dx$$

$$+ \int_\Omega \langle \mathrm{Im}\, \mathbf{b}, \mathrm{Im}(\overline{v}_\varepsilon \nabla v)\rangle dx - \int_\Omega \mathrm{Re}\, a\, |v_\varepsilon|^2 dx + o(1). \tag{2.53}$$

As far as the left-hand side of (2.53) is concerned, we have

$$\int_\Omega \langle \mathscr{A} \nabla u, \nabla[\varrho_\varepsilon^2(|u|)\, u]\rangle\, dx$$

$$= \varepsilon^{p-2} \int_{\Omega \setminus E_\varepsilon} \langle \mathscr{A} \nabla u, \nabla u\rangle\, dx + \int_{E_\varepsilon} \langle \mathscr{A} \nabla u, \nabla(|u|^{p-2}u)\rangle\, dx$$

and then

$$\lim_{\varepsilon \to 0} \mathrm{Re} \int_\Omega (\langle \mathscr{A} \nabla u, \nabla[\varrho_\varepsilon^2(|u|)\, u]\rangle - \langle \mathbf{b}\nabla u, |u|^{p-2}u\rangle - a|u|^p) dx$$

$$= \int_\Omega \langle \nabla u, \nabla(|u|^{p-2}u)\rangle - \langle \mathbf{b}\nabla u, |u|^{p-2}u\rangle - a|u|^p) dx.$$

Letting $\varepsilon \to 0$ in (2.53), we complete the proof of the necessity.

Sufficiency. Suppose that (2.48) holds. Let $v \in \Xi$ and let u_ε be defined by (2.12). We have $u_\varepsilon \in \Xi$ and arguing as in the necessity part of Lemma 2.1, we find (2.13), (2.14) and (2.15). These limit relations lead to (2.45) for any $v \in \Xi$ and thus (2.45) is true for any $v \in H_0^1(\Omega)$ (see the proof of Lemma 2.18). In view of Lemma 2.18, the form $\mathscr{L}$ is L^p-dissipative.   □

Theorem 2.20. *The operator A is L^p-dissipative if and only if the form $\mathscr{L}$ is L^p-dissipative.*

Proof. Necessity. Let $u \in \Xi$ and $g_\varepsilon = (|u|^2 + \varepsilon^2)^{\frac{1}{2}}$. Since $g_\varepsilon^{p-2}\overline{u} \in \Xi$ we have

$$- \int_\Omega \langle \nabla^t (\mathscr{A} \nabla u), u\rangle g_\varepsilon^{p-2} dx = \int_\Omega \langle \mathscr{A} \nabla u, \nabla(g_\varepsilon^{p-2}u)\rangle dx$$

and since

$$\partial_h(g_\varepsilon^{p-2}\overline{u}) = (p-2)g_\varepsilon^{p-4}\operatorname{Re}(\langle \partial_h u, u\rangle)\,\overline{u} + g_\varepsilon^{p-2}\partial_h\overline{u}$$

we have also

$$\partial_h(g_\varepsilon^{p-2}\overline{u})$$
$$= \begin{cases} (p-2)|u|^{p-4}\operatorname{Re}(\langle \partial_h u, u\rangle)\,\overline{u} + |u|^{p-2}\partial_h\overline{u} = \partial_h(|u|^{p-2}\overline{u}) & \text{if } x \in F_0 \\ \varepsilon^{p-2}\partial_h\overline{u} & \text{if } x \in \Omega \setminus F_0. \end{cases}$$

We find, keeping in mind (2.49), that

$$\lim_{\varepsilon \to 0} \int_\Omega \langle \mathscr{A}\,\nabla u, \nabla(g_\varepsilon^{p-2}u)\rangle dx = \int_\Omega \langle \mathscr{A}\,\nabla u, \nabla(|u|^{p-2}u)\rangle dx\,.$$

On the other hand, using Lemma 4.35, we see that

$$\lim_{\varepsilon \to 0} \int_\Omega \langle \nabla^t(\mathscr{A}\,\nabla u), u\rangle g_\varepsilon^{p-2} dx = \int_\Omega \langle \nabla^t(\mathscr{A}\,\nabla u), u\rangle |u|^{p-2} dx.$$

Then

$$-\int_\Omega \langle \nabla^t(\mathscr{A}\,\nabla u), u\rangle |u|^{p-2} dx = \int_\Omega \langle \mathscr{A}\,\nabla u, \nabla(|u|^{p-2}u)\rangle dx \qquad (2.54)$$

for any $u \in \Xi$. Hence

$$-\int_\Omega \langle Au, u\rangle |u|^{p-2} dx = \int_\Omega (\langle \mathscr{A}\,\nabla u, \nabla(|u|^{p-2}u)\rangle) - \langle \mathbf{b}\nabla u, |u|^{p-2}u\rangle - a\,|u|^p)dx\,.$$

Therefore (2.48) holds. We can conclude now that the form $\mathscr{L}$ is L^p-dissipative, because of Lemma 2.19.

Sufficiency. Given $u \in D(A)$, we can find a sequence $\{u_n\} \subset \Xi$ such that $u_n \to u$ in $W^{2,p}(\Omega)$. Keeping in mind (2.54), we have

$$-\int_\Omega \langle Au, u\rangle |u|^{p-2} dx = -\lim_{n \to \infty} \int_\Omega \langle Au_n, u_n\rangle |u_n|^{p-2} dx$$

$$= \lim_{n \to \infty} \int_\Omega \langle \mathscr{A}\,\nabla u_n, \nabla(|u_n|^{p-2}u_n)\rangle - \langle \mathbf{b}\nabla u_n, |u_n|^{p-2}u_n\rangle - a\,|u_n|^p)dx.$$

Since $\mathscr{L}$ is L^p-dissipative, (2.48) holds for any $u \in \Xi$ and (2.44) is true for any $u \in D(A)$. $\qquad \square$

2.4.2   Intervals of dissipativity

The next result permits us to determine the best interval of p's for which the operator

$$Au = \nabla^t(\mathscr{A}\,\nabla u) \tag{2.55}$$

is L^p-dissipative. We set

$$\lambda = \inf_{(\xi,x)\in\mathcal{M}} \frac{\langle \mathrm{Re}\,\mathscr{A}(x)\xi,\xi\rangle}{|\langle \mathrm{Im}\,\mathscr{A}(x)\xi,\xi\rangle|}$$

where $\mathcal{M}$ is the set of (ξ,x) with $\xi \in \mathbb{R}^n$, $x \in \Omega$ such that $\langle \mathrm{Im}\,\mathscr{A}(x)\xi,\xi\rangle \neq 0$.

Corollary 2.21. *Let A be the operator* (2.55). *Let us suppose that the matrix* $\mathrm{Im}\,\mathscr{A}$ *is symmetric and that*

$$\langle \mathrm{Re}\,\mathscr{A}(x)\xi,\xi\rangle \geqslant 0 \tag{2.56}$$

for any $x \in \Omega$, $\xi \in \mathbb{R}^n$. If $\mathrm{Im}\,\mathscr{A}(x) = 0$ for any $x \in \Omega$, A is L^p-dissipative for any $p > 1$. If $\mathrm{Im}\,\mathscr{A}$ does not vanish identically on Ω, A is L^p-dissipative if and only if

$$2 + 2\lambda(\lambda - \sqrt{\lambda^2 + 1}) \leqslant p \leqslant 2 + 2\lambda(\lambda + \sqrt{\lambda^2 + 1}). \tag{2.57}$$

Proof. When $\mathrm{Im}\,\mathscr{A}(x) = 0$ for any $x \in \Omega$, the statement follows from Theorem 2.7. Let us assume that $\mathrm{Im}\,\mathscr{A}$ does not vanish identically; note that this implies $\mathcal{M} \neq \emptyset$.

Necessity. If the operator (2.55) is L^p-dissipative, Theorem 2.7 shows that

$$|p - 2|\,|\langle \mathrm{Im}\,\mathscr{A}(x)\xi,\xi\rangle| \leqslant 2\sqrt{p-1}\,\langle \mathrm{Re}\,\mathscr{A}(x)\xi,\xi\rangle \tag{2.58}$$

for any $x \in \Omega$, $\xi \in \mathbb{R}^n$. In particular we have

$$\frac{|p-2|}{2\sqrt{p-1}} \leqslant \frac{\langle \mathrm{Re}\,\mathscr{A}(x)\xi,\xi\rangle}{|\langle \mathrm{Im}\,\mathscr{A}(x)\xi,\xi\rangle|}$$

for any $(\xi,x) \in \mathcal{M}$ and then

$$\frac{|p-2|}{2\sqrt{p-1}} \leqslant \lambda.$$

This inequality is equivalent to (2.57).

Sufficiency. If (2.57) holds, we have $(p-2)^2 \leqslant 4(p-1)\lambda^2$. Note that $p > 1$, because $2 + 2\lambda(\lambda - \sqrt{\lambda^2 + 1}) > 1$.

Since $\lambda \geqslant 0$ in view of (2.56), we find $|p - 2| \leqslant 2\sqrt{p-1}\lambda$ and (2.58) is true for any $(\xi,x) \in \mathcal{M}$. On the other hand, if $x \in \Omega$ and $\xi \in \mathbb{R}^n$ with $(\xi,x) \notin \mathcal{M}$, (2.58) is trivially satisfied and then it holds for any $x \in \Omega$, $\xi \in \mathbb{R}^n$. Theorem 2.7 gives the result.   □

The next corollary provides a characterization of operators which are L^p-dissipative only for $p = 2$.

Corollary 2.22. *Let A be as in Corollary 2.21. The operator A is L^p-dissipative only for $p = 2$ if and only if $\operatorname{Im} \mathscr{A}$ does not vanish identically and $\lambda = 0$.*

Proof. Inequalities (2.57) are satisfied only for $p = 2$ if and only if $\lambda(\lambda - \sqrt{\lambda^2 - 1}) = \lambda(\lambda + \sqrt{\lambda^2 + 1})$ and this happens if and only if $\lambda = 0$. Thus the result is a consequence of Corollary 2.21. □

2.4.3 Contractive semigroups generated by the operator : $u \to \operatorname{div}(\mathscr{A} \nabla u)$

Let A be the operator $\operatorname{div}(\mathscr{A} \nabla u)$ with smooth coefficients. In this subsection we want to investigate when A generates a contraction semigroup.

In the next theorem we suppose that A is strongly elliptic, i.e.,

$$\langle \operatorname{Re} \mathscr{A}(x)\xi, \xi \rangle > 0$$

for any $x \in \overline{\Omega}$, $\xi \in \mathbb{R}^n \setminus \{0\}$.

Theorem 2.23. *Let A be the strongly elliptic operator (2.55) with $\operatorname{Im}\mathscr{A} = \operatorname{Im}\mathscr{A}^t$. The operator A generates a contraction semigroup on L^p if and only if*

$$|p - 2| \, |\langle \operatorname{Im} \mathscr{A}(x)\xi, \xi \rangle| \leqslant 2\sqrt{p-1} \, \langle \operatorname{Re} \mathscr{A}(x)\xi, \xi \rangle \tag{2.59}$$

for any $x \in \Omega$, $\xi \in \mathbb{R}^n$.

Proof. Sufficiency. It is a classical result that the operator A defined on (2.43) and acting in $L^p(\Omega)$ is a densely defined closed operator (see Agmon, Douglis and Nirenberg [1], Maz'ya and Shaposhnikova [76, Ch. 14]).

From Theorem 2.7 we know that the form $\mathscr{L}$ is L^p-dissipative and Theorem 2.20 shows that A is L^p-dissipative. Finally the formal adjoint operator

$$A^* u = \nabla^t (\mathscr{A}^* \nabla u)$$

with $D(A^*) = W^{2,p'}(\Omega) \cap \mathring{W}^{1,p'}(\Omega)$, is the adjoint operator of A and since $\operatorname{Im} \mathscr{A}^* = \operatorname{Im}(\mathscr{A}^*)^t$ and (2.59) can be written as

$$|p' - 2| \, |\langle \operatorname{Im} \mathscr{A}^*(x)\xi, \xi \rangle| \leqslant 2\sqrt{p'-1} \, \langle \operatorname{Re} \mathscr{A}^*(x)\xi, \xi \rangle, \tag{2.60}$$

we have also the $L^{p'}$-dissipativity of A^*.

The result is a consequence of Theorem 1.31.

Necessity. If A generates a contraction semigroup on L^p, it is L^p-dissipative. Therefore (2.59) holds because of Theorem 2.7. □

2.4.4 Quasi-dissipativity and quasi-contractivity

We know that, in case either A has lower-order terms or they are absent and $\operatorname{Im} \mathscr{A}$ is not symmetric, condition (2.59) is not sufficient for the L^p-dissipativity. As we shall see now, it turns out that, for these more general operators, (2.59) is necessary and sufficient for the quasi-dissipativity of A (see Definition 1.7), i.e., the dissipativity of $A - \omega I$ for a suitable $\omega \geqslant 0$. In other words, A is L^p-quasi-dissipative if there exists $\omega \geqslant 0$ such that

$$\operatorname{Re} \int_\Omega \langle Au, u \rangle |u|^{p-2} dx \leqslant \omega \, \|u\|_p^p$$

for any $u \in D(A)$.

As a consequence, condition (2.59) is necessary and sufficient for the quasi-contractivity of the semigroup generated by A (see Theorem 2.27 below).

Lemma 2.24. *The operator* (2.42) *is L^p-quasi-dissipative if and only if there exists $\omega \geqslant 0$ such that*

$$\operatorname{Re} \int_\Omega \Big[\langle \mathscr{A} \nabla v, \nabla v \rangle - (1 - 2/p)\langle (\mathscr{A} - \mathscr{A}^*)\nabla(|v|), |v|^{-1}\overline{v}\nabla v \rangle$$

$$- (1 - 2/p)^2 \langle \mathscr{A} \nabla(|v|), \nabla(|v|) \rangle \Big] dx + \int_\Omega \langle \operatorname{Im} \mathbf{b}, \operatorname{Im}(\overline{v}\nabla v) \rangle \, dx \qquad (2.61)$$

$$+ \int_\Omega \operatorname{Re}(\operatorname{div}(\mathbf{b}/p) - a)|v|^2 dx \geqslant -\omega \int_\Omega |v|^2 dx$$

for any $v \in H_0^1(\Omega)$.

Proof. The result follows from Lemma 2.18. □

Theorem 2.25. *The strongly elliptic operator* (2.42) *is L^p-quasi-dissipative if and only if*

$$|p - 2| \, |\langle \operatorname{Im} \mathscr{A}(x)\xi, \xi \rangle| \leqslant 2\sqrt{p - 1} \, \langle \operatorname{Re} \mathscr{A}(x)\xi, \xi \rangle \qquad (2.62)$$

for any $x \in \Omega$, $\xi \in \mathbb{R}^n$.

Proof. Necessity. By using the functions X, Y introduced in Corollary 2.2, we write condition (2.61) in the form

$$\int_\Omega \Big\{ \frac{4}{p\,p'} \langle \operatorname{Re} \mathscr{A} X, X \rangle + \langle \operatorname{Re} \mathscr{A} Y, Y \rangle + 2\langle (p^{-1}\operatorname{Im} \mathscr{A} + p'^{-1}\operatorname{Im} \mathscr{A}^*)X, Y \rangle$$

$$+ \langle \operatorname{Im} \mathbf{b}, Y \rangle |v| + \operatorname{Re}[\operatorname{div}(\mathbf{b}/p) - a + \omega]|v|^2 \Big\} dx \geqslant 0 \,.$$

As in the proof of Corollary 2.2, this inequality implies

$$\frac{4}{p\,p'}\int_\Omega \langle \mathrm{Re}\,\mathscr{A}\,\nabla\varrho, \nabla\varrho\rangle dx + \mu^2 \int_\Omega \langle \mathrm{Re}\,\mathscr{A}\,\nabla\varrho, \nabla\varrho\rangle dx$$

$$+ 2\mu \int_\Omega \langle (p^{-1}\,\mathrm{Im}\,\mathscr{A} + p'^{-1}\,\mathrm{Im}\,\mathscr{A}^*)\nabla\varrho, \nabla\varrho\rangle dx$$

$$+ \mu \int_\Omega \varrho\langle \mathrm{Im}\,\mathbf{b}, \nabla\varrho\rangle dx + \int_\Omega \mathrm{Re}\,[\mathrm{div}\,(\mathbf{b}/p) - a + \omega]\,\varrho^2 dx \geqslant 0$$

for any $\varrho \in C_0^1(\Omega)$, $\mu \in \mathbb{R}$. Since

$$\langle \mathrm{Im}\,\mathscr{A}^*\,\nabla\varrho, \nabla\varrho\rangle = -\langle \mathrm{Im}\,\mathscr{A}^t\,\nabla\varrho, \nabla\varrho\rangle = -\langle \mathrm{Im}\,\mathscr{A}\,\nabla\varrho, \nabla\varrho\rangle$$

we have

$$\frac{4}{p\,p'}\int_\Omega \langle \mathrm{Re}\,\mathscr{A}\,\nabla\varrho, \nabla\varrho\rangle dx + \mu^2 \int_\Omega \langle \mathrm{Re}\,\mathscr{A}\,\nabla\varrho, \nabla\varrho\rangle dx$$

$$- 2(1 - 2/p)\mu \int_\Omega \langle \mathrm{Im}\,\mathscr{A}\,\nabla\varrho, \nabla\varrho\rangle dx$$

$$+ \mu \int_\Omega \varrho\langle \mathrm{Im}\,\mathbf{b}, \nabla\varrho\rangle dx + \int_\Omega \mathrm{Re}\,[\mathrm{div}\,(\mathbf{b}/p) - a + \omega]\,\varrho^2 dx \geqslant 0$$

for any $\varrho \in C_0^1(\Omega)$, $\mu \in \mathbb{R}$.

Taking $\varrho(x) = \psi(x)\cos\langle\xi, x\rangle$ and $\varrho(x) = \psi(x)\sin\langle\xi, x\rangle$ with $\psi \in C_0^1(\Omega)$ and arguing as in the proof of Corollary 2.2, we find

$$\int_\Omega \langle \mathscr{B}\,\nabla\psi, \nabla\psi\rangle dx + \int_\Omega \langle \mathscr{B}\,\xi, \xi\rangle \psi^2 dx$$

$$+ \mu \int_\Omega \langle \mathrm{Im}\,\mathbf{b}, \nabla\psi\rangle \psi\, dx + \int_\Omega \mathrm{Re}\,[\mathrm{div}\,(\mathbf{b}/p) - a + \omega]\,\psi^2 dx \geqslant 0,$$

where $\mu \in \mathbb{R}$ and

$$\mathscr{B} = \frac{4}{p\,p'}\,\mathrm{Re}\,\mathscr{A} + \mu^2\,\mathrm{Re}\,\mathscr{A} - 2(1 - 2/p)\mu\,\mathrm{Im}\,\mathscr{A} .$$

Because of the arbitrariness of ξ we see that

$$\int_\Omega \langle \mathscr{B}\,\xi, \xi\rangle \psi^2 dx \geqslant 0$$

for any $\psi \in C_0^1(\Omega)$. Hence $\langle \mathscr{B}\,\xi, \xi\rangle \geqslant 0$, i.e.,

$$\frac{4}{p\,p'}\langle \mathrm{Re}\,\mathscr{A}\,\xi, \xi\rangle + \mu^2\langle \mathrm{Re}\,\mathscr{A}\,\xi, \xi\rangle - 2(1 - 2/p)\mu\langle \mathrm{Im}\,\mathscr{A}\,\xi, \xi\rangle \geqslant 0$$

for any $x \in \Omega$, $\xi \in \mathbb{R}^n$, $\mu \in \mathbb{R}$. Inequality (2.62) follows from the arbitrariness of μ.

Sufficiency. Assume first that $\operatorname{Im}\mathscr{A}$ is symmetric. By repeating the first part of the proof of sufficiency of Theorem 2.7, we find that (2.62) implies

$$\frac{4}{pp'}\langle\operatorname{Re}\mathscr{A}\,\xi,\xi\rangle + \langle\operatorname{Re}\mathscr{A}\,\eta,\eta\rangle - 2(1-p/2)\langle\operatorname{Im}\mathscr{A}\,\xi,\eta\rangle \geqslant 0 \qquad (2.63)$$

for any $x\in\Omega$, $\xi,\eta\in\mathbb{R}^n$.

In order to prove (2.61), it is not restrictive to suppose

$$\operatorname{Re}(\operatorname{div}(\mathbf{b}/p) - a) = 0.$$

Since A is strongly elliptic, there exists a non-singular real matrix $\mathscr{C}\in C^1(\overline{\Omega})$ such that

$$\langle\operatorname{Re}\mathscr{A}\,\eta,\eta\rangle = \langle\mathscr{C}\,\eta,\mathscr{C}\,\eta\rangle$$

for any $\eta\in\mathbb{R}^n$. Setting

$$\mathscr{S} = (1-2/p)(\mathscr{C}^t)^{-1}\operatorname{Im}\mathscr{A},$$

we have

$$|\mathscr{C}\,\eta - \mathscr{S}\,\xi|^2 = \langle\operatorname{Re}\mathscr{A}\,\eta,\eta\rangle - 2(1-p/2)\langle\operatorname{Im}\mathscr{A}\,\xi,\eta\rangle + |\mathscr{S}\,\xi|^2.$$

This leads to the identity

$$\frac{4}{pp'}\langle\operatorname{Re}\mathscr{A}\,\xi,\xi\rangle + \langle\operatorname{Re}\mathscr{A}\,\eta,\eta\rangle - 2(1-p/2)\langle\operatorname{Im}\mathscr{A}\,\xi,\eta\rangle$$
$$= |\mathscr{C}\,\eta - \mathscr{S}\,\xi|^2 + \frac{4}{pp'}\langle\operatorname{Re}\mathscr{A}\,\xi,\xi\rangle - |\mathscr{S}\,\xi|^2 \qquad (2.64)$$

for any $\xi,\eta\in\mathbb{R}^n$. In view of (2.63), putting $\eta = \mathscr{C}^{-1}\mathscr{S}\,\xi$ in (2.64), we obtain

$$\frac{4}{pp'}\langle\operatorname{Re}\mathscr{A}\,\xi,\xi\rangle - |\mathscr{S}\,\xi|^2 \geqslant 0 \qquad (2.65)$$

for any $\xi\in\mathbb{R}^n$.

On the other hand, we may write

$$\langle\operatorname{Im}\mathbf{b}, Y\rangle = \langle(\mathscr{C}^{-1})^t\operatorname{Im}\mathbf{b}, \mathscr{C}\,Y\rangle$$
$$= \langle(\mathscr{C}^{-1})^t\operatorname{Im}\mathbf{b}, \mathscr{C}\,Y - \mathscr{S}\,X\rangle + \langle(\mathscr{C}^{-1})^t\operatorname{Im}\mathbf{b}, \mathscr{S}\,X\rangle.$$

By the Cauchy inequality

$$\int_\Omega\langle(\mathscr{C}^{-1})^t\operatorname{Im}\mathbf{b}, \mathscr{C}\,Y - \mathscr{S}\,X\rangle|v|\,dx$$
$$\geqslant -\int_\Omega|\mathscr{C}\,Y - \mathscr{S}\,X|^2 dx - \frac{1}{4}\int_\Omega|(\mathscr{C}^{-1})^t\operatorname{Im}\mathbf{b}|^2|v|^2 dx$$

and, integrating by parts,

$$\int_\Omega \langle (\mathscr{C}^{-1})^t \operatorname{Im} \mathbf{b}, \mathscr{S} X \rangle |v| \, dx = \frac{1}{2} \int_\Omega \langle (\mathscr{C}^{-1} \mathscr{S})^t \operatorname{Im} \mathbf{b}, \nabla(|v|^2) \rangle \, dx$$

$$= -\frac{1}{2} \int_\Omega \nabla^t ((\mathscr{C}^{-1} \mathscr{S})^t \operatorname{Im} \mathbf{b}) \, |v|^2 \, dx.$$

This implies that there exists $\omega \geq 0$ such that

$$\int_\Omega \langle \operatorname{Im} \mathbf{b}, Y \rangle |v| \, dx \geq -\int_\Omega |\mathscr{C} Y - \mathscr{S} X|^2 dx - \omega \int_\Omega |v|^2 dx$$

and then, in view of (2.64),

$$\int_\Omega \left\{ \frac{4}{pp'} \langle \operatorname{Re} \mathscr{A} X, X \rangle + \langle \operatorname{Re} \mathscr{A} Y, Y \rangle + 2(1 - p/2)\langle \operatorname{Im} \mathscr{A} X, Y \rangle + \langle \operatorname{Im} \mathbf{b}, Y \rangle |v| \right\} dx$$

$$\geq \int_\Omega \left(\frac{4}{pp'} \langle \operatorname{Re} \mathscr{A} X, X \rangle - |\mathscr{S} X|^2 \right) dx - \omega \int_\Omega |v|^2 dx.$$

Inequality (2.65) gives the result.

We have proved the sufficiency under the assumption $\operatorname{Im} \mathscr{A}^t = \operatorname{Im} \mathscr{A}$. In the general case, the operator A can be written in the form

$$Au = \nabla^t ((\mathscr{A} + \mathscr{A}^t)\nabla u)/2 + \mathbf{c}\nabla u + au$$

where

$$\mathbf{c} = \nabla^t (\mathscr{A} - \mathscr{A}^t)/2 + \mathbf{b}.$$

Since $(\mathscr{A} + \mathscr{A}^t)$ is symmetric, we know that A is L^p-quasi-dissipative if and only if

$$|p - 2| \, |\langle \operatorname{Im}(\mathscr{A} + \mathscr{A}^t)\xi, \xi \rangle| \leq 2\sqrt{p - 1} \, \langle \operatorname{Re}(\mathscr{A} + \mathscr{A}^t)\xi, \xi \rangle$$

for any $\xi \in \mathbb{R}^n$, which is exactly condition (2.62). $\qquad \square$

With Theorem 2.25 in hand, we may obtain the following corollary.

Corollary 2.26. *Let A be the strongly elliptic operator (2.42). If $\operatorname{Im} \mathscr{A}(x) = 0$ for any $x \in \Omega$, A is L^p-quasi-dissipative for any $p > 1$. If $\operatorname{Im} \mathscr{A}$ does not vanish identically on Ω, A is L^p-quasi-dissipative if and only if (2.57) holds.*

Proof. The proof is similar to that of Corollary 2.21, the role of Theorem 2.7 being played by Theorem 2.25. $\qquad \square$

The next theorem gives a criterion for the L^p-quasi-contractivity of the semigroup generated by A (i.e., the L^p-contractivity of the semigroup generated by $A - \omega I$).

Theorem 2.27. *Let A be the strongly elliptic operator (2.42). The operator A generates a quasi-contraction semigroup on L^p if and only if (2.59) holds for any $x \in \Omega$, $\xi \in \mathbb{R}^n$.*

Proof. Sufficiency. Let us consider A as an operator defined on (2.43) and acting in $L^p(\Omega)$. As in the proof of Theorem 2.23, one can see that A is a densely defined closed operator and that the formal adjoint coincides with the adjoint A^*. Theorem 2.25 shows that A is L^p-quasi-dissipative. On the other hand, condition (2.60) holds and then A^* is $L^{p'}$-quasi-dissipative. As in Theorem 2.23, this implies that A generates a quasi-contraction semigroup on L^p.

Necessity. If A generates a quasi-contraction semigroup on L^p, A is L^p-quasi-dissipative and (2.59) holds.   □

2.5   A quasi-commutative property of the Poisson and composition operators

In this section we deal with the composition of a function of one variable and a solution of an elliptic equation.

To be more precise, we consider an elliptic second-order formally self-adjoint differential operator L in a bounded domain Ω. We denote by Ph the L-harmonic function with the Dirichlet data h on $\partial\Omega$. The Dirichlet integral corresponding to the operator L will be denoted by $D[u]$. We also introduce a real-valued function Φ on the line $\mathbb{R}$ and denote by $\Phi \circ u$ the composition of Φ and u.

We want to show that the Dirichlet integrals of the functions $\Phi \circ Ph$ and $P(\Phi \circ h)$ are comparable. First of all, clearly, the inequality

$$D[P(\Phi \circ h)[\leqslant D[\Phi \circ Ph]$$

is valid. Hence we only need to check the opposite estimate

$$D[\Phi \circ Ph] \leqslant C\, D[P(\Phi \circ h)]. \tag{2.66}$$

We find a condition on Φ which is both necessary and sufficient for (2.66).

Moreover, we prove that the two Dirichlet integrals are comparable if and only if the derivative $\Psi = \Phi'$ satisfies the reverse Cauchy inequality

$$\frac{1}{b-a}\int_a^b \Psi^2(t)\, dt \leqslant C\left(\frac{1}{b-a}\int_a^b \Psi(t)\, dt\right)^2 \tag{2.67}$$

for any interval $(a,b) \subset \mathbb{R}$.

We add that the constants C appearing in (2.66) and (2.67) are the same.

At the end of the section, this result is illustrated for harmonic functions and for $\Psi(t) = |t|^\alpha$ with $\alpha > -1/2$. In particular, we obtain the sharp inequalities

$$\int_\Omega |\nabla(|Ph|\,Ph)|^2 dx \leqslant \frac{3}{2} \int_\Omega |\nabla P(|h|\,h)|^2 dx\,,$$

$$\int_\Omega |\nabla(Ph)^3|^2 dx \leqslant \frac{9}{4} \int_\Omega |\nabla P(h^3)|^2 dx \tag{2.68}$$

for any $h \in W^{1/2,2}(\partial\Omega)$ and

$$\int_\Omega |\nabla(Ph)^2|^2 dx \leqslant \frac{4}{3} \int_\Omega |\nabla P(h^2)|^2 dx\,,$$

$$\int_\Omega |\nabla(Ph)^3|^2 dx \leqslant \frac{9}{5} \int_\Omega |\nabla P(h^3)|^2 dx \tag{2.69}$$

for any nonnegative $h \in W^{1/2,2}(\partial\Omega)$. Here, P is the harmonic Poisson operator.

To avoid technical complications connected with non-smoothness of the boundary, we only deal with domains bounded by surfaces of class C^∞, although, in principle, this restriction can be significantly weakened.

Here we follow our article Cialdea–Maz'ya [13].

2.5.1 Preliminaries

Here all functions are assumed to take real values and the notation ∂_i stands for $\partial/\partial x_i$.

Let L be the second-order differential operator

$$Lu = -\partial_i(a_{ij}(x)\,\partial_j u)$$

defined in a bounded domain $\Omega \subset \mathbb{R}^n$.

The coefficients a_{ij} are measurable and bounded. The operator L is uniformly elliptic, i.e., there exists $\lambda > 0$ such that

$$a_{ij}(x)\xi_i\xi_j \geqslant \lambda\,|\xi|^2 \tag{2.70}$$

for all $\xi \in \mathbb{R}^n$ and for almost every $x \in \Omega$.

Let Ψ be a function defined on $\mathbb{R}$ such that, for any $N \in \mathbb{N}$, the functions

$$\Psi_N(t) = \begin{cases} \Psi(t), & |\Psi(t)| \leqslant N, \\ N\operatorname{sign}(\Psi(t)), & |\Psi(t)| > N, \end{cases} \tag{2.71}$$

are continuous. We suppose that there exists a constant C such that, for any finite interval $\sigma \subset \mathbb{R}$, we have

$$\overline{\Psi^2} \leqslant C(\overline{\Psi})^2, \tag{2.72}$$

where $\overline{u}$ denotes the mean value of u on σ.

Also let

$$\Phi(t) = \int_0^t \Psi(\tau)\,d\tau, \quad t \in \mathbb{R}.$$

Let $W^{1/2,2}(\partial\Omega)$ be the trace space for the Sobolev space $W^{1,2}(\Omega)$, and let P denote the Poisson operator, i.e., the solution operator:

$$W^{1/2,2}(\partial\Omega) \ni h \to u \in W^{1,2}(\Omega)$$

for the Dirichlet problem

$$\begin{cases} Lu = 0 & \text{in } \Omega, \\ \operatorname{tr} u = h & \text{on } \partial\Omega, \end{cases} \tag{2.73}$$

where $\operatorname{tr} u$ is the trace of a function $u \in W^{1,2}(\Omega)$ on $\partial\Omega$.

We introduce the Dirichlet integral

$$D[u] = \int_\Omega a_{ij}\partial_i u\,\partial_j u\,dx$$

and the bilinear form

$$D[u, v] = \int_\Omega a_{ij}\partial_i u\,\partial_j v\,dx.$$

In the sequel, we consider $D[P(\Phi \circ h)]$ and $D[\Phi \circ Ph]$ for $h \in W^{1/2,2}(\partial\Omega)$. Since $h \in W^{1/2,2}(\partial\Omega)$ implies neither $\Phi \circ h \in W^{1/2,2}(\partial\Omega)$ nor $\Phi \circ Ph \in W^{1,2}(\Omega)$, we have to specify what $D[P(\Phi \circ h)]$ and $D[\Phi \circ Ph]$ mean.

We define

$$D[P(\Phi \circ h)] = \liminf_{k \to \infty} D[P(\Phi_k \circ h)], \tag{2.74}$$

where

$$\Phi_k(t) = \int_0^t \Psi_k(\tau)\,d\tau \tag{2.75}$$

and Ψ_k is given by (2.71). Note that if h belongs to $W^{1/2,2}(\partial\Omega)$, $D[P(\Phi_k \circ h)]$ makes sense. In fact,

$$|\Phi_k \circ h| \leqslant k\,|h|, \quad |\Phi_k \circ h(x) - \Phi_k \circ h(y)| \leqslant k\,|h(x) - h(y)|$$

imply that $\Phi_k \circ h$ belongs to $W^{1/2,2}(\partial\Omega)$.

In order to accept the definition (2.74), we have to show that if the left-hand side of (2.74) makes sense because $\Phi \circ h$ belongs to $W^{1/2,2}(\partial\Omega)$, then (2.74) holds. In fact, we have the following assertion.

Lemma 2.28. *Let* $h \in W^{1/2,2}(\partial\Omega)$ *be such that also* $\Phi \circ h$ *belongs to* $W^{1/2,2}(\partial\Omega)$. *Then*

$$D[P(\Phi \circ h)] = \lim_{k \to \infty} D[P(\Phi_k \circ h)]. \tag{2.76}$$

Proof. It is obvious that

$$\Phi_k \circ h(x) - \Phi \circ h(x) = \int_0^{h(x)} [\Psi_k(t) - \Psi(t)] \, dt \to 0 \quad \text{a.e.}$$

and

$$|[\Phi_k \circ h(x) - \Phi \circ h(x)] - [\Phi_k \circ h(y) - \Phi \circ h(y)]| \leqslant 2 |\Phi \circ h(x) - \Phi \circ h(y)|.$$

By the Lebesgue dominated convergence theorem, these inequalities imply $\Phi_k \circ h \to \Phi \circ h$ in $W^{1/2,2}(\partial\Omega)$. Therefore, $P(\Phi_k \circ h) \to P(\Phi \circ h)$ in $W^{1,2}(\Omega)$ and (2.76) holds. $\qquad\square$

As far as $D[\Phi \circ Ph]$ is concerned, we remark that

$$D[\Phi_k \circ Ph] = \int_\Omega (\Psi_k(Ph))^2 a_{ij} \partial_i(Ph) \partial_j(Ph) \, dx$$

tends to

$$\int_\Omega (\Psi(Ph))^2 a_{ij} \partial_i(Ph) \partial_j(Ph) \, dx$$

because of the monotone convergence theorem. Therefore, we set

$$D[\Phi \circ Ph] = \lim_{k \to \infty} D[\Phi_k \circ Ph].$$

Note that neither $D[\Phi \circ Ph]$ nor $D[P(\Phi \circ h)]$ needs to be finite.

Let $G(x,y)$ be the Green function of the Dirichlet problem (2.73), and let $\partial/\partial\nu$ be the co-normal operator

$$\frac{\partial}{\partial\nu} = a_{ij} \cos(n, x_j) \, \partial_i,$$

where n is the exterior unit normal.

Lemma 2.29. *Let the coefficients* a_{ij} *of the operator* L *belong to* $C^\infty(\overline\Omega)$. *There exist two positive constants* c_1 *and* c_2 *such that*

$$\frac{c_1}{|x-y|^n} \leqslant \frac{\partial^2 G(x,y)}{\partial\nu_x \partial\nu_y} \leqslant \frac{c_2}{|x-y|^n} \tag{2.77}$$

for any $x, y \in \partial\Omega$, $x \neq y$.

Proof. Let us fix a point x_0 on $\partial\Omega$. We consider a neighborhood of x_0 and introduce local coordinates $y = (y', y_n)$ in such a way that x_0 corresponds to $y = 0$, $y_n = 0$ is the tangent hyperplane and locally Ω is contained in the half-space $y_n > 0$. We may suppose that this change of variables is such that $a_{ij}(x_0) = \delta_{ij}$.

It is known (cf. Maz'ya–Plamenevskii [75]) that the Poisson kernel

$$(\partial/\partial\nu_y)G(x, y)$$

in a neighborhood of x_0 is given by

$$2\,\omega_n^{-1}\,y_n|y|^{-n} + \mathcal{O}\left(|y|^{2-n-\varepsilon}\right),$$

where ω_n is the measure of the unit sphere in $\mathbb{R}^n$ and $\varepsilon > 0$. Moreover the derivative of the Poisson kernel with respect to y_n is equal to

$$2\,\omega_n^{-1}\left(|y|^2 - y_n^2\right)|y|^{-n-2} + \mathcal{O}\left(|y|^{1-n-\varepsilon}\right)$$

which becomes

$$2\,\omega_n^{-1}\,|y'|^{-n} + \mathcal{O}\left(|y'|^{1-n-\varepsilon}\right) \tag{2.78}$$

for $y_n = 0$. Formula (2.78) and the arbitrariness of x_0 imply (2.77). □

2.5.2   The main result

Theorem 2.30. *If $\Psi : \mathbb{R} \to \mathbb{R}_+$ satisfies the condition (2.72), then*

$$D[\Phi \circ Ph] \leqslant C\,D[P(\Phi \circ h)] \tag{2.79}$$

for any $h \in W^{1/2,2}(\partial\Omega)$, where C is the constant in (2.72).

Proof. We suppose temporarily that $a_{ij} \in C^\infty$.

Let u be a solution of the equation $Lu = 0$, $u \in C^\infty(\overline{\Omega})$. We show that

$$D[u] = \frac{1}{2}\int\limits_{\partial\Omega}\int\limits_{\partial\Omega}(\operatorname{tr} u(x) - \operatorname{tr} u(y))^2\frac{\partial^2 G(x, y)}{\partial\nu_x\partial\nu_y}\,d\sigma_x d\sigma_y. \tag{2.80}$$

In fact, since

$$L_x[(u(x) - u(y))^2] = 2\,a_{hk}\partial_h u\,\partial_k u\,,$$

the integration by parts in (2.80) gives

$$\int\limits_{\partial\Omega}\int\limits_{\partial\Omega}(\operatorname{tr} u(x) - \operatorname{tr} u(y))^2\frac{\partial^2 G(x, y)}{\partial\nu_x\partial\nu_y}\,d\sigma_x d\sigma_y = 2\int\limits_{\Omega}a_{hk}\partial_h u\,\partial_k u\,dx\int\limits_{\partial\Omega}\frac{\partial G(x, y)}{\partial\nu_y}\,d\sigma_y$$

$$= 2D[u]\,,$$

and (2.80) is proved.

Let $u \in W^{1,2}(\Omega)$ be a solution of $Lu = 0$ in Ω, and let $\{u_k\}$ be a sequence of $C^\infty(\overline{\Omega})$ functions which tends to u in $W^{1,2}(\Omega)$. Since $\operatorname{tr} u_k \to \operatorname{tr} u$ in $W^{1/2,2}(\partial\Omega)$, we see that $P(\operatorname{tr} u_k)$ tends to $P(\operatorname{tr} u) = u$ in $W^{1,2}(\Omega)$ and, therefore, $D[P(\operatorname{tr} u_k)] \to D[u]$. This implies that (2.80) holds for any u in $W^{1,2}(\Omega)$ with $Lu = 0$ in Ω.

Let now u and v belong to $W^{1,2}(\Omega)$, and let $Lu = Lv = 0$ in Ω. Since

$$D[u,v] = 4^{-1}(D[u+v] - D[u-v]),$$

we can write

$$D[u,v] = \frac{1}{2} \int_{\partial\Omega} \int_{\partial\Omega} (\operatorname{tr} u(x) - \operatorname{tr} u(y))(\operatorname{tr} v(x) - \operatorname{tr} v(y)) \frac{\partial^2 G(x,y)}{\partial \nu_x \partial \nu_y} d\sigma_x d\sigma_y. \quad (2.81)$$

Note also that, if $h \in W^{1/2,2}(\partial\Omega)$ and $g \in W^{1,2}(\Omega)$, we have

$$D[Ph, P(\operatorname{tr} g)] = D[Ph, g]. \quad (2.82)$$

Suppose now that $h \in W^{1/2,2}(\partial\Omega)$ is such that $\Phi \circ h \in W^{1/2,2}(\partial\Omega)$. We have

$$D[\Phi \circ Ph] = \int_\Omega a_{ij} \partial_i (\Phi \circ Ph) \partial_j (\Phi \circ Ph) \, dx = \int_\Omega a_{ij} (\Psi(Ph))^2 \partial_i(Ph) \partial_j(Ph) \, dx.$$

The last integral can be written as

$$\int_\Omega a_{ij} \partial_i(Ph) \partial_j \left(\int_0^{Ph} \Psi^2(\tau) \, d\tau \right) dx,$$

and we have proved that

$$D[\Phi \circ Ph] = D\left(Ph, \int_0^{Ph} \Psi^2(\tau) \, d\tau \right).$$

From (2.81) and (2.82) we get

$$D[\Phi \circ Ph] = \frac{1}{2} \int_{\partial\Omega} \int_{\partial\Omega} (h(y) - h(x)) \int_{h(x)}^{h(y)} \Psi^2(\tau) \, d\tau \frac{\partial^2 G(x,y)}{\partial \nu_x \partial \nu_y} d\sigma_x d\sigma_y. \quad (2.83)$$

In view of (2.77), $\partial^2 G(x,y)/\partial \nu_x \partial \nu_y$ is positive, and the condition (2.72) leads to

$$D[\Phi \circ Ph] \leqslant \frac{C}{2} \int_{\partial\Omega} \int_{\partial\Omega} \left(\int_{h(x)}^{h(y)} \Psi(\tau) \, d\tau \right)^2 \frac{\partial^2 G(x,y)}{\partial \nu_x \partial \nu_y} d\sigma_x d\sigma_y. \quad (2.84)$$

The inequality (2.79) is proved since the right-hand side of (2.84) is nothing but $C\, D[P(\Phi \circ h)]$ (cf. (2.80)).

For any $h \in W^{1/2,2}(\partial\Omega)$, the inequality (2.79) follows from

$$D[\Phi \circ Ph] = \lim_{k\to\infty} D[\Phi_k \circ Ph] \leqslant C \liminf_{k\to\infty} D[P(\Phi_k \circ h)].$$

Let us suppose now that a_{ij} only belong to $L^\infty(\Omega)$. There exist $a_{ij}^{(k)} \in C^\infty(\mathbb{R}^n)$ such that $a_{ij}^{(k)} \to a_{ij}$ in measure as $k \to \infty$. We can assume that

$$\|a_{ij}^{(k)}\|_{L^\infty(\Omega)} \leqslant K$$

and that the operators $L^{(k)}u = -\partial_i(a_{ij}^{(k)}\partial_j u)$ satisfy the ellipticity condition (2.70) with the same constant λ.

Let u be a solution of the Dirichlet problem (2.73), and let u_k satisfy

$$\begin{cases} L^{(k)}u_k = 0 & \text{in } \Omega, \\ \operatorname{tr} u_k = h & \text{on } \partial\Omega. \end{cases}$$

Denote by A and $A^{(k)}$ the matrices $\{a_{ij}\}$ and $\{a_{ij}^{(k)}\}$ respectively. Since we can write

$$\operatorname{div} A\nabla(u - u_k) = -\operatorname{div} A\nabla u_k = -\operatorname{div}(A - A_k)\nabla u_k,$$

we find that

$$D[u - u_k] \leqslant \|(A - A_k)\nabla u_k\| \, \|\nabla(u - u_k)\|.$$

Then there exists a constant K such that

$$\|\nabla(u - u_k)\| \leqslant K \, \|(A - A_k)\nabla u_k\|. \tag{2.85}$$

Denoting by D_k the quadratic form

$$D_k[u] = \int_\Omega a_{ij}^{(k)} \partial_i u \, \partial_j u \, dx,$$

we have

$$D_k[u_k] = \min_{\substack{u \in W^{1,2}(\Omega) \\ \operatorname{tr} u = h}} D_k[u] \quad \text{and} \quad \lambda \|\nabla u_k\|^2 \leqslant D_k[u_k].$$

This shows that the sequence $\|\nabla u_k\|$ is bounded and the right-hand side of (2.85) tends to 0 as $k \to \infty$.

We are now in a position to prove (2.79). Clearly, it is enough to show that

$$D[\Phi_m \circ Ph] \leqslant C\, D[P(\Phi_m \circ h)], \tag{2.86}$$

where Φ_m is given by (2.75) for any $h \in W^{1/2,2}(\partial\Omega)$ such that $\Phi \circ h$ belongs to the same space.

Because of what we have proved when the coefficients are smooth, we may write

$$D_k[\Phi_m \circ P_k h] \leqslant C\,D_k[P_k(\Phi_m \circ h)]\,,$$

where P_k denotes the Poisson operator for $L^{(k)}$.

Formula (2.85) shows that $P_k(\Phi_m \circ h)$ tends to $P(\Phi_m \circ h)$ (as $k \to \infty$) in $W^{1,2}(\Omega)$ and thus

$$\lim_{k\to\infty} D_k[P_k(\Phi_m \circ h)] = D[P(\Phi_m \circ h)]. \tag{2.87}$$

On the other hand, we have

$$\nabla\Phi_m(P_k h) - \nabla\Phi_m(Ph) = \Psi_m(P_k h)(\nabla P_k h - \nabla Ph) + (\Psi_m(P_k h) - \Psi_m(Ph))\,\nabla Ph.$$

By the continuity of Ψ_m, we find

$$\|\nabla\Phi_m(P_k h) - \nabla\Phi_m(Ph)\|_{L^2(\Omega)} \to 0.$$

This implies that $D_k[\Phi_m \circ P_k h] \to D[\Phi_m \circ Ph]$, which together with (2.87), leads to (2.86). $\qquad\square$

Under the assumption that the coefficients of the operator are smooth, we can prove the inverse of Theorem 2.30.

Theorem 2.31. *Let the coefficients of the operator L belong to $C^\infty(\overline{\Omega})$. If (2.79) holds for any $h \in W^{1/2,2}(\partial\Omega)$, then (2.72) is true with the same constant C.*

Proof. Let Γ be a subdomain of $\partial\Omega$ with smooth nonempty boundary. We choose a sufficiently small $\varepsilon > 0$ and denote by $[\Gamma]_\varepsilon$ the ε-neighborhood of Γ. We set $\gamma_\varepsilon = [\Gamma]_\varepsilon \setminus \Gamma$ and denote by $\delta(x)$ the distance from the point x to Γ.

Let a and b be different real numbers, and let h be the function defined on $\partial\Omega$ by

$$h(x) = \begin{cases} a, & x \in \Gamma, \\ a + (b - a)\,\varepsilon^{-1}\delta(x), & x \in \gamma_\varepsilon, \\ b, & x \in \partial\Omega \setminus [\Gamma]_\varepsilon. \end{cases}$$

We know from (2.83) that $D[\Phi \circ Ph]$ is equal to

$$\int_{\partial\Omega} d\sigma_x \int_{\partial\Omega} Q(x,y)\,d\sigma_y,$$

where

$$Q(x,y) = \frac{1}{2}\,(h(y) - h(x))\left(\int_{h(x)}^{h(y)} \Psi^2(\tau)\,d\tau\right)\frac{\partial^2 G(x,y)}{\partial\nu_x\partial\nu_y}.$$

We can write

$$\int_{\partial\Omega} d\sigma_x \int_{\partial\Omega} Q(x,y)\, d\sigma_y = I_1 + I_2 + I_3,$$

where

$$I_1 = \int_{\gamma_\varepsilon} d\sigma_x \int_{\gamma_\varepsilon} Q(x,y)\, d\sigma_y\,,$$

$$I_2 = \int_{\partial\Omega\backslash\gamma_\varepsilon} d\sigma_x \int_{\gamma_\varepsilon} Q(x,y)\, d\sigma_y + \int_{\gamma_\varepsilon} d\sigma_x \int_{\partial\Omega\backslash\gamma_\varepsilon} Q(x,y)\, d\sigma_y\,,$$

$$I_3 = \int_{\Gamma} d\sigma_x \int_{\partial\Omega\backslash[\Gamma]_\varepsilon} Q(x,y)\, d\sigma_y + \int_{\partial\Omega\backslash[\Gamma]_\varepsilon} d\sigma_x \int_{\Gamma} Q(x,y)\, d\sigma_y\,.$$

The right-hand estimate in (2.77) leads to

$$I_1 \leqslant c \left(\frac{b-a}{\varepsilon} \right)^2 \int_{\gamma_\varepsilon} d\sigma_x \int_{\gamma_\varepsilon} (\delta(x) - \delta(y))^2 |x-y|^{-n} d\sigma_y. \qquad (2.88)$$

The integral

$$\int_{\gamma_\varepsilon} (\delta(x) - \delta(y))^2 |x-y|^{-n} d\sigma_y$$

with $x \in \gamma_\varepsilon$ is majorized by

$$c \int_0^\varepsilon (t - \delta(x))^2 dt \int_{\mathbb{R}^{n-2}} (|\eta| + (t - \delta(x)))^{-n} d\eta = \mathcal{O}(\varepsilon)\,.$$

This estimate and (2.88) imply $I_1 = \mathcal{O}(1)$ as $\varepsilon \to 0^+$.
Since

$$\left| \int_{\Gamma} d\sigma_x \int_{\gamma_\varepsilon} Q(x,y)\, d\sigma_y \right| \leqslant c \left(\frac{b-a}{\varepsilon} \right)^2 \int_{\Gamma} d\sigma_x \int_{\gamma_\varepsilon} \delta^2(y) |x-y|^{-n} d\sigma_y$$

and the integral

$$\int_{\gamma_\varepsilon} \delta^2(y) d\sigma_y \int_{\Gamma} |x-y|^{-n} d\sigma_x$$

does not exceed

$$c \int_0^\varepsilon t^2 dt \int_{\mathbb{R}^{n-2}} (|\eta| + t)^{-n+1} d\eta = \mathcal{O}(\varepsilon^2)\,,$$

we find

$$\int_{\Gamma} d\sigma_x \int_{\gamma_\varepsilon} Q(x,y)\, d\sigma_y = \mathcal{O}(1). \tag{2.89}$$

Analogously,

$$\left| \int_{\partial\Omega\setminus[\Gamma]_\varepsilon} d\sigma_x \int_{\gamma_\varepsilon} Q(x,y)\, d\sigma_y \right| \leqslant c \left(\frac{b-a}{\varepsilon}\right)^2 \int_{\partial\Omega\setminus[\Gamma]_\varepsilon} d\sigma_x \int_{\gamma_\varepsilon} (\varepsilon - \delta(y))^2 |x-y|^{-n} d\sigma_y$$

$$= \mathcal{O}(1). \tag{2.90}$$

Exchanging the roles of x and y in the previous argument, we arrive at the estimate

$$\int_{\gamma_\varepsilon} d\sigma_x \int_{\partial\Omega\setminus\gamma_\varepsilon} Q(x,y)\, d\sigma_y = \mathcal{O}(1).$$

Combining this with (2.89) and (2.90), we see that $I_2 = \mathcal{O}(1)$.

Let us consider I_3. Since the two terms in the definition of I_3 are equal, we have

$$I_3 = 2\,(b-a) \int_a^b \Psi^2(\tau)\, d\tau \int_\Gamma d\sigma_x \int_{\partial\Omega\setminus[\Gamma]_\varepsilon} \frac{\partial^2 G(x,y)}{\partial\nu_x\partial\nu_y}\, d\sigma_y\,.$$

By the left inequality in (2.77),

$$I_3 \geqslant 2\,c_1(b-a) \int_a^b \Psi^2(\tau)\, d\tau \int_\Gamma d\sigma_x \int_{\partial\Omega\setminus[\Gamma]_\varepsilon} |x-y|^{-n} d\sigma_y\,.$$

There exists $\varrho > 0$ such that the integral over $\Gamma \times (\partial\Omega \setminus [\Gamma]_\varepsilon)$ of $|x-y|^{-n}$ admits the lower estimate by

$$c \int_{-\varrho}^0 dt \int_\varepsilon^\varrho ds \int_{|\tau|\leqslant\varrho} d\tau \int_{|\eta|\leqslant\varrho} (|\eta - \tau| + |s - t|)^{-n} d\eta$$

which implies

$$\int_\Gamma d\sigma_x \int_{\partial\Omega\setminus[\Gamma]_\varepsilon} \frac{\partial^2 G(x,y)}{\partial\nu_x\partial\nu_y}\, d\sigma_y \geqslant c \log(1/\varepsilon). \tag{2.91}$$

Now we deal with $D[P(\Phi \circ h)]$. This can be written as $J_1 + J_2 + J_3$, where J_s are defined as I_s $(s = 1, 2, 3)$ with the only difference that $Q(x,y)$ is replaced by

$$\frac{1}{2}\left(\int_{h(x)}^{h(y)} \Psi(\tau)\, d\tau \right)^2 \frac{\partial^2 G(x,y)}{\partial\nu_x\partial\nu_y}\,.$$

As before, $J_1 = \mathcal{O}(1)$, $J_2 = \mathcal{O}(1)$, and

$$J_3 = 2 \left(\int_a^b \Psi(\tau) d\tau \right)^2 \int_\Gamma d\sigma_x \int_{\partial\Omega\backslash[\Gamma]_\varepsilon} \frac{\partial^2 G(x,y)}{\partial\nu_x \partial\nu_y} d\sigma_y .$$

The inequality (2.79) for the function h can be written as

$$I_3 + \mathcal{O}(1) \leqslant C \left(J_3 + \mathcal{O}(1) \right),$$

i.e.,

$$\mathcal{O}(1) + (b-a) \int_a^b \Psi^2(\tau) \, d\tau \int_\Gamma d\sigma_x \int_{\partial\Omega\backslash[\Gamma]_\varepsilon} \frac{\partial^2 G(x,y)}{\partial\nu_x \partial\nu_y} d\sigma_y$$

$$\leqslant C \left(\mathcal{O}(1) + \left(\int_a^b \Psi(\tau) d\tau \right)^2 \int_\Gamma d\sigma_x \int_{\partial\Omega\backslash[\Gamma]_\varepsilon} \frac{\partial^2 G(x,y)}{\partial\nu_x \partial\nu_y} d\sigma_y \right).$$

Dividing both sides by

$$\int_\Gamma d\sigma_x \int_{\partial\Omega\backslash[\Gamma]_\varepsilon} \frac{\partial^2 G(x,y)}{\partial\nu_x \partial\nu_y} d\sigma_y$$

and letting $\varepsilon \to 0^+$, we arrive at (2.67), referring to (2.91). The proof is complete.
□

Remark. Inspection of the proofs of Theorems 2.30 and 2.31 shows that if the coefficients are smooth and Dirichlet data on $\partial\Omega$ are nonnegative, we have also the following result: *The inequality*

$$D[\Phi \circ Ph] \leqslant C_+ \, D[P(\Phi \circ h)]$$

holds for any nonnegative $h \in W^{1/2,2}(\partial\Omega)$ *if and only if*

$$\overline{\Psi^2} \leqslant C_+ (\overline{\Psi})^2$$

for all finite intervals $\sigma \subset \mathbb{R}_+$.

Example. As a simple application of this theorem, we consider the case of the Laplace operator and the function $\Psi(t) = |t|^\alpha$ ($\alpha > -1/2$). Let C_α and $C_{\alpha,+}$ be the following constants:

$$C_\alpha = \frac{(\alpha+1)^2}{2\alpha+1} \sup_{t\in\mathbb{R}} \frac{(1-t)(1-t^{2\alpha+1})}{(1-|t|^\alpha t)^2}, \qquad (2.92)$$

$$C_{\alpha,+} = \frac{(\alpha+1)^2}{2\alpha+1} \sup_{t\in\mathbb{R}_+} \frac{(1-t)(1-t^{2\alpha+1})}{(1-t^{\alpha+1})^2} . \qquad (2.93)$$

Theorem 2.30 shows that

$$\int_\Omega |\nabla(|Ph|^\alpha Ph)|^2 dx \leqslant C_\alpha \int_\Omega |\nabla P(|h|^\alpha h)|^2 dx, \qquad (2.94)$$

for any $h \in W^{1/2,2}(\partial\Omega)$, where P denotes the harmonic Poisson operator for Ω.
In view of the remark above, we have also

$$\int_\Omega |\nabla(Ph)^{\alpha+1}|^2 dx \leqslant C_{\alpha,+} \int_\Omega |\nabla P(h^{\alpha+1})|^2 dx \qquad (2.95)$$

for any nonnegative $h \in W^{1/2,2}(\partial\Omega)$. By Theorem 2.31, the constants C_α and $C_{\alpha,+}$ are the best possible in (2.94) and (2.95).

Hence the problem of finding explicitly the best constant in such inequalities is reduced to the determination of the supremum in (2.92) and (2.93).

One can check that $C_{\alpha,+} = (\alpha+1)^2/(2\alpha+1)$. In the particular cases $\alpha = 1, 2$, C_1 and C_2 can also be determined explicitly. This leads to the inequalities (2.68)–(2.69).

2.6 Comments to Chapter 2

The results contained in Sections 2.1–2.4 and in Section 2.5 are borrowed from Cialdea and Maz'ya [11] and [13], respectively.

We remark that the proof of Corollary 2.16 was given in [11]. The same result, obtained with a different approach, can be found also in the earlier paper by Kresin and Maz'ya [50].

Amann [3] gives a sufficient condition for the contractivity in L^1 of the semigroup generated by a second-order scalar elliptic operator with real coefficients

$$-\partial_j(a_{jk}\partial_k u) + a_j\partial_j u + a_0 u, \qquad (2.96)$$

$a_{jk} = a_{kj} \in C^1(\overline{\Omega}), a_j, a_0 \in C(\overline{\Omega})$. He proves also that the related operator generates a contractive semigroup for all $p \in [1, +\infty)$ simultaneously if and only if $a_0 \geqslant 0$ and $a_0 - \partial_j a_j \geqslant O$. These results generalize earlier results due to Brezis and Strauss [8].

Several properties of the semigroups generated by a linear differential operator can be deduced from the so-called Gaussian upper bounds of the relevant heat kernels. The term "Gaussian" means that the estimates contain an exponential factor similar to the one related to the Laplace operator: $(4\pi t)^{-n/2}\exp(-|x-y|^2/(4t))$. Among the many paper dealing with this approach, we mention Daners [17], Davies [20, 21], Grigor'yan [33], Karrmann [45], Ouhabaz [84, 85] and the monographs by Davies [19] and Ouhabaz [86].

In [57] Liskevich studies the semigroup generated by a general second-order differential operator (for the operator $-\Delta + b(x) \cdot \nabla$ see also Kovalenko, Semenov [48]). In particular, he obtains sufficient conditions for the L^p quasi-contractivity of this semigroup. More general conditions have been obtained in Sobol, Vogt [93] and Liskevich, Sobol, Vogt [60].

The contractivity and the analyticity of the semigroup generated by elliptic operators on $\mathbb{R}^n$ with singular coefficients is the object of the paper [80] by Metafune, Pallara, Prüss and Schnaubelt.

Questions related to Markovian and sub-Markovian semigroups in L^p are studied in Liskevich, Perelmuter [58] and Liskevich, Semenov [59].

Strichartz [97, 98] has considered the heat semigroup related to the Hodge–de Rham Laplacian on differential forms defined on a Riemaniann manifold. He shows that, differently from the case of functions, the heat semigroup for differential forms is often L^p-contractive only for $p = 2$. In fact, Strichartz shows that on a compact Riemaniann manifold with non-trivial real cohomology in dimension k, if the curvature operator is positive definite at one point, then the heat semigroup for k-forms is contractive only for $p = 2$; if the curvature operator is everywhere nonnegative, the heat semigroup for k-forms is contractive for all $p \in [1, +\infty]$.

Hömberg, Krumbiegel and Rehberg [39] used some of the techniques introduced in Cialdea and Maz'ya [11] to show the dissipativity of a certain operator connected to the problem of the existence of an optimal control for the heat equation with dynamic boundary condition.

The results described in Section 2.1 allows Nittka [81] to consider the case of partial differential operators with complex coefficients.

Ostermann and Schratz [83] obtain the stability of a numerical procedure for solving a certain evolution problem. The necessary and sufficient condition (2.29) show that their result does not require the contractivity of the corresponding semigroup.

We remark in conclusion that this chapter and Chapters 3–6, where we deal with L^p-dissipativity, have no significant intersections with the existing books concerning dissipative differential operators and contractive semigroups: Davies [18, 19], Engel and Nagel [25] Fattorini [26], Goldstein [31], Hille [37], Kresin and Maz'ya [51], Ouhabaz [86], Pazy [87], Robinson [90] *et al.*

Chapter 3

Elasticity System

Let us consider the classical operator of linear elasticity

$$Eu = \Delta u + (1 - 2\nu)^{-1}\nabla \operatorname{div} u \qquad (3.1)$$

where ν is the Poisson ratio. Throughout this chapter, we assume that either $\nu > 1$ or $\nu < 1/2$. It is well known that E is strongly elliptic if and only if these inequalities hold (see, for instance, Gurtin [35, p. 86]).

For the planar elasticity we prove that Lamé operator is L^p-dissipative if and only if

$$\left(\frac{1}{2} - \frac{1}{p}\right)^2 \leqslant \frac{2(\nu - 1)(2\nu - 1)}{(3 - 4\nu)^2}. \qquad (3.2)$$

The result is followed by two corollaries concerning the comparison between the Lamé operator and the Laplacian from the point of view of the L^p-dissipativity.

In Section 3.3 we show that condition (3.2) is necessary for the L^p-dissipativity of operator (3.1), even when the Poisson ratio is not constant. In the same section we give a more strict explicit condition which is sufficient for the L^p-dissipativity of (3.1).

In Section 3.4 we give necessary and sufficient conditions for a weighted L^p-dissipativity, i.e., for the validity of the inequality

$$\int_\Omega (\Delta u + (1 - 2\nu)^{-1}\nabla \operatorname{div} u)\,|u|^{p-2}u\,\frac{dx}{|x|^\alpha} \leqslant 0$$

under the condition that the vector u is rotationally invariant, i.e., u depends only on $\varrho = |x|$ and u_ϱ is the only nonzero spherical component of u. Namely we show that this holds if and only if

$$-(p - 1)(n + p' - 2) \leqslant \alpha \leqslant n + p - 2$$

where $p' = p/(p - 1)$.

3.1 L^p-dissipativity of planar elasticity

The aim of this section is to give a necessary and sufficient condition for the L^p-dissipativity of operator (3.1) in the case $n = 2$.

We start with a result holding in any number of variables. Let Ω be an arbitrary domain in $\mathbb{R}^n$ and let us consider the bilinear form related to elasticity operator

$$-\int_\Omega (\langle \nabla u, \nabla v \rangle + (1 - 2\nu)^{-1} \operatorname{div} u \operatorname{div} v)\, dx, \qquad (3.3)$$

where $\langle \cdot, \cdot \rangle$ denotes the scalar product in $\mathbb{R}^n$.

According to Definition 1.4, the form (3.3) is L^p-dissipative in Ω if

$$-\int_\Omega \langle \nabla u, \nabla(|u|^{p-2}u) \rangle + (1 - 2\nu)^{-1} \operatorname{div} u \operatorname{div}(|u|^{p-2}u) \leqslant 0 \qquad \text{if } p \geqslant 2, \quad (3.4)$$

$$-\int_\Omega \langle \nabla u, \nabla(|u|^{p'-2}u) \rangle + (1 - 2\nu)^{-1} \operatorname{div} u \operatorname{div}(|u|^{p'-2}u) \leqslant 0 \qquad \text{if } p < 2, \quad (3.5)$$

for all $u \in (C_0^1(\Omega))^n$.

In the following lemma we demonstrate the equivalence between the L^p-dissipativity of the form (3.3) and the positivity of a certain form, quadratic in the first derivatives.

Lemma 3.1. *Let Ω be a domain of $\mathbb{R}^n$. The form (3.3) is L^p-dissipative if and only if*

$$\int_\Omega \left[-C_p |\nabla|v||^2 + \sum_{j=1}^2 |\nabla v_j|^2 - \gamma C_p |v|^{-2} |v_h \partial_h |v||^2 + \gamma |\operatorname{div} v|^2 \right] dx \geqslant 0 \qquad (3.6)$$

for any $v \in (C_0^1(\Omega))^n$, where

$$C_p = (1 - 2/p)^2, \qquad \gamma = (1 - 2\nu)^{-1}. \qquad (3.7)$$

Proof. Sufficiency. First suppose $p \geqslant 2$. Let $u \in (C_0^1(\Omega))^n$ and set $v = |u|^{p-2}u$. We have $v \in (C_0^1(\Omega))^2$ and $u = |v|^{(2-p)/p}v$. One checks directly that

$$\langle \nabla u, \nabla(|u|^{p-2}u) \rangle + (1 - 2\nu)^{-1} \operatorname{div} u \operatorname{div}(|u|^{p-2}u)$$
$$= \sum_j |\nabla v_j|^2 - C_p |\nabla|v||^2 - \gamma C_p ||v_h \partial_h |v||^2 + \gamma |\operatorname{div} v|^2.$$

The left-hand side of (3.4) being equal to the left-hand side of (3.6), inequality (3.4) is satisfied for any $u \in (C_0^1(\Omega))^n$.

If $1 < p < 2$ we find

$$\langle \nabla u, \nabla(|u|^{p'-2}u) \rangle + (1 - 2\nu)^{-1} \operatorname{div} u \operatorname{div}(|u|^{p'-2}u)$$
$$= \sum_j |\nabla v_j|^2 - C_{p'} |\nabla|v||^2 - \gamma C_{p'} ||v_h \partial_h |v||^2 + \gamma |\operatorname{div} v|^2.$$

and since $1 - 2/p' = -1 + 2/p$ (which implies $C_p = C_{p'}$), we get the result also in this case.

Necessity. Let $p \geqslant 2$ and set

$$g_\varepsilon = (|v|^2 + \varepsilon^2)^{1/2}, \quad u_\varepsilon = g_\varepsilon^{2/p-1} v,$$

where $v \in (C_0^1(\Omega))^n$. We have

$$\langle \nabla u_\varepsilon, \nabla(|u_\varepsilon|^{p-2} u_\varepsilon) \rangle = |u_\varepsilon|^{p-2} \langle \partial_h u_\varepsilon, \partial_h u_\varepsilon \rangle + (p-2)|u_\varepsilon|^{p-3} \langle \partial_h u_\varepsilon, u_\varepsilon \rangle \partial_h |u_\varepsilon|.$$

A direct computation shows that

$$\langle \nabla u_\varepsilon, \nabla(|u_\varepsilon|^{p-2} u_\varepsilon) \rangle = [(1-2/p)^2 g_\varepsilon^{-(p+2)} |v|^p - 2(1-2/p) g_\varepsilon^{-p} |v|^{p-2}] \sum_k |v_j \partial_k v_j|^2$$

$$+ g_\varepsilon^{2-p} |v|^{p-2} \langle \partial_h v, \partial_h v \rangle,$$

$$|u_\varepsilon|^{p-3} \langle \partial_h u_\varepsilon, u_\varepsilon \rangle \partial_h |u_\varepsilon| = \{(1-2/p)[(1-2/p) g_\varepsilon^{-(p+2)} |v|^p - g_\varepsilon^{-p} |v|^{p-2}]$$

$$+ [g_\varepsilon^{2-p} |v|^{p-4} - (1-2/p) g_\varepsilon^{-p} |v|^{p-2}]\} \sum_k |v_j \partial_k v_j|^2$$

on the set $E = \{x \in \Omega \mid |v(x)| > 0\}$. The inequality $g_\varepsilon^a \leqslant |v|^a$ for $a \leqslant 0$, shows that the right-hand sides are dominated by L^1 functions. Since $g_\varepsilon \to |v|$ pointwise as $\varepsilon \to 0^+$, we find

$$\lim_{\varepsilon \to 0^+} \langle \nabla u_\varepsilon, \nabla(|u_\varepsilon|^{p-2} u_\varepsilon) \rangle$$

$$= \langle \partial_h v, \partial_h v \rangle + [(1-2/p)^2 - 2(1-2/p) + 4(p-2)/p^2]|v|^{-2} \sum_k |v_j \partial_k v_j|^2$$

$$= -(1-2/p)^2 |\nabla|v||^2 + \sum_j |\nabla v_j|^2$$

and dominated convergence gives

$$\lim_{\varepsilon \to 0^+} \int_E \langle \nabla u_\varepsilon, \nabla(|u_\varepsilon|^{p-2} u_\varepsilon) \rangle dx = \int_E \left[-C_p |\nabla|v||^2 + \sum_j |\nabla v_j|^2 \right] dx. \qquad (3.8)$$

Similar arguments show that

$$\lim_{\varepsilon \to 0^+} \int_E \operatorname{div} u_\varepsilon \operatorname{div}(|u_\varepsilon|^{p-2} u_\varepsilon) dx = \int_E \left[-C_p |v|^{-2} |v_h \partial_h |v||^2 + |\operatorname{div} v|^2 \right] dx. \qquad (3.9)$$

Formulas (3.8) and (3.9) lead to

$$\lim_{\varepsilon \to 0^+} \int_\Omega (\langle \nabla u_\varepsilon, \nabla(|u_\varepsilon|^{p-2} u_\varepsilon) \rangle + \gamma \operatorname{div}(|u_\varepsilon|^{p-2} u_\varepsilon) dx \qquad (3.10)$$

$$= \int_\Omega \left(-C_p |\nabla|v||^2 + \sum_j |\nabla v_j|^2 - \gamma C_p |v|^{-2} |v_h \partial_h |v||^2 + \gamma |\operatorname{div} v|^2 \right) dx.$$

The function u_ε being in $(C_0^1(\Omega))^n$, the left-hand side is greater than or equal to zero and (3.6) follows.

If $1 < p < 2$, we can write, in view of (3.10),

$$\lim_{\varepsilon \to o^+} \int_\Omega (\langle \nabla u_\varepsilon, \nabla(|u_\varepsilon|^{p'-2}u_\varepsilon)\rangle + \gamma \operatorname{div}(|u_\varepsilon|^{p'-2}u_\varepsilon)dx$$

$$= \int_\Omega \left(-C_{p'}|\nabla|v||^2 + \sum_j |\nabla v_j|^2 - \gamma C_{p'}|v|^{-2}|v_h\partial_h|v||^2 + \gamma|\operatorname{div} v|^2 \right) dx.$$

Since $C_{p'} = C_p$, (3.5) implies (3.6). $\qquad\qquad\qquad\qquad\qquad\qquad\qquad\qquad\square$

The next lemma concerns the case $n = 2$ and provides a necessary algebraic condition for the L^p-dissipativity of the form (3.3).

Lemma 3.2. *Let Ω be a domain of $\mathbb{R}^2$. If the form (3.3) is L^p-dissipative, we have*

$$-C_p[|\xi|^2 + \gamma \langle \xi, \omega \rangle^2]\langle \lambda, \omega \rangle^2 + |\xi|^2|\lambda|^2 + \gamma \langle \xi, \lambda \rangle^2 \geqslant 0 \qquad (3.11)$$

for any $\xi, \lambda, \omega \in \mathbb{R}^2$, $|\omega| = 1$ (the constants C_p and γ being given by (3.7)).

Proof. Assume first that $\Omega = \mathbb{R}^2$. Let us fix $\omega \in \mathbb{R}^2$ with $|\omega| = 1$ and take $v(x) = w(x)\,\eta(\log|x|/\log R)$, where

$$w(x) = \mu\,\omega + \psi(x)$$

$\mu, R \in \mathbb{R}^+$, $\psi \in (C_0^\infty(\mathbb{R}^2))^2$, $\eta \in C^\infty(\mathbb{R}^2)$, $\eta(t) = 1$ if $t \leqslant 1/2$ and $\eta(t) = 0$ if $t \geqslant 1$.

On the set where $v \neq 0$ one has

$$\langle \nabla|v|, \nabla|v|\rangle = \langle \nabla|w|, \nabla|w|\rangle\,\eta^2(\log|x|/\log R)$$
$$+ 2(\log R)^{-1}|w|\,\langle \nabla|w|, x\rangle\,|x|^{-2}\eta(\log|x|/\log R)\,\eta'(\log|x|/\log R)$$
$$+ (\log R)^{-2}|w|^2|x|^{-2}\,(\eta'(\log|x|/\log R))^2.$$

Choose δ such that $\operatorname{spt}\psi \subset B_\delta(0)$ and $R > \delta^2$. If $|x| > \delta$ one has $w(x) = \mu\,\omega$ and then $\nabla|w| = 0$, while if $|x| < \delta$, then $\eta(\log|x|/\log R) = 1$, $\eta'(\log|x|/\log R) = 0$. Therefore

$$\int_{\mathbb{R}^2} \langle \nabla|v|, \nabla|v|\rangle\,dx$$

$$= \int_{B_\delta(0)} \langle \nabla|w|, \nabla|w|\rangle\,dx + \frac{1}{\log^2 R}\int_{B_R(0)\setminus B_{\sqrt R}(0)} \frac{|w|^2}{|x|^2}\,(\eta'(\log|x|/\log R))^2 dx.$$

Since

$$\lim_{R \to +\infty} \frac{1}{\log^2 R}\int_{B_R(0)\setminus B_{\sqrt R}(0)} \frac{dx}{|x|^2} = 0,$$

we find

$$\lim_{R \to +\infty} \int_{\mathbb{R}^2} \langle \nabla|v|, \nabla|v|\rangle\,dx = \int_{B_\delta(0)} \langle \nabla|w|, \nabla|w|\rangle\,dx.$$

By similar arguments we obtain

$$\lim_{R \to +\infty} \int_{\mathbb{R}^2} \left[-C_p |\nabla|v||^2 + \sum_{j=1}^{2} |\nabla v_j|^2 - \gamma\, C_p\, |v|^{-2} |v_h \partial_h |v||^2 + \gamma\, |\operatorname{div} v|^2 \right] dx$$

$$= \int_{B_\delta(0)} \left[-C_p |\nabla|w||^2 + \sum_{j=1}^{2} |\nabla w_j|^2 - \gamma\, C_p\, |w|^{-2} |w_h \partial_h |w||^2 + \gamma\, |\operatorname{div} w|^2 \right] dx\,.$$

In view of Lemma 3.1, (3.6) holds. Putting v in this formula and letting $R \to +\infty$, we find

$$\int_{B_\delta(0)} \left[-C_p |\nabla|w||^2 + \sum_{j=1}^{2} |\nabla w_j|^2 - \gamma\, C_p\, |w|^{-2} |w_h \partial_h |w||^2 + \gamma\, |\operatorname{div} w|^2 \right] dx \geq 0\,.$$

$$(3.12)$$

From the identities

$$\partial_h w = \partial_h \psi, \qquad \operatorname{div} w = \operatorname{div} \psi,$$

$$|\nabla|w||^2 = |\mu\,\omega + \psi|^{-2} \sum_{h=1}^{2} \langle \mu\,\omega + \psi, \partial_h \psi \rangle^2,$$

$$|w|^{-2} |w_h \partial_h w|^2 = |\mu\,\omega + \psi|^{-4} |(\mu\,\omega_h + \psi_h)\langle \mu\,\omega + \psi, \partial_h \psi \rangle|^2$$

we infer, letting $\mu \to +\infty$ in (3.12),

$$\int_{\mathbb{R}^2} \left[-C_p \sum_{h=1}^{2} \langle \omega, \partial_h \psi \rangle^2 + \sum_{j=1}^{2} |\nabla \psi_j|^2 - \gamma\, C_p |\omega_h \langle \omega, \partial_h \psi \rangle|^2 + \gamma\, |\operatorname{div} \psi|^2 \right] dx \geq 0\,.$$

$$(3.13)$$

Putting in (3.13)

$$\psi(x) = \lambda\, \varphi(x)\, \cos(\mu\langle \xi, x \rangle) \quad \text{and} \quad \psi(x) = \lambda\, \varphi(x)\, \sin(\mu\langle \xi, x \rangle)$$

where $\lambda \in \mathbb{R}^2$, $\varphi \in C_0^\infty(\mathbb{R}^2)$ and μ is a real parameter, we obtain (3.11) by standard arguments (see, e.g., Fichera [29, pp. 107–108]).

If $\Omega \neq \mathbb{R}^2$, fix $x_0 \in \Omega$ and $0 < \varepsilon < \operatorname{dist}(x_0, \partial\Omega)$. Given $\psi \in (C_0^1(\Omega))^2$, put the function

$$v(x) = \psi((x - x_0)/\varepsilon)$$

in (3.6). By a change of variables we find

$$\int_{\mathbb{R}^2} \left[-C_p |\nabla|\psi||^2 + \sum_{j=1}^{2} |\nabla \psi_j|^2 - \gamma\, C_p\, |\psi|^{-2} |\psi_h \partial_h |\psi||^2 + \gamma\, |\operatorname{div} \psi|^2 \right] dx \geq 0\,.$$

The arbitrariness of $\psi \in (C_0^1(\Omega))^2$ and what we have proved for $\mathbb{R}^2$ gives the result. $\qquad\square$

We are now in a position to give a necessary and sufficient condition for the L^p-dissipativity of planar elasticity.

Theorem 3.3. *Let* $n = 2$. *The form* (3.1) *is* L^p-*dissipative if and only if*

$$\left(\frac{1}{2} - \frac{1}{p}\right)^2 \leqslant \frac{2(\nu - 1)(2\nu - 1)}{(3 - 4\nu)^2}. \tag{3.14}$$

Proof. Necessity. In view of Lemma 3.2, the L^p-dissipativity of (3.1) implies the algebraic inequality (3.11) for any ξ, λ, $\omega \in \mathbb{R}^2$, $|\omega| = 1$.

Without loss of generality we may suppose $\xi = (1, 0)$ and (3.11) can be written as

$$-C_p(1 + \gamma\omega_1^2)(\lambda_j\omega_j)^2 + |\lambda|^2 + \gamma\lambda_1^2 \geqslant 0 \tag{3.15}$$

for any λ, $\omega \in \mathbb{R}^2$, $|\omega| = 1$.

Condition (3.15) holds if and only if

$$-C_p(1 + \gamma\omega_1^2)\omega_1^2 + 1 + \gamma \geqslant 0,$$

$$[C_p(1 + \gamma\omega_1^2)\omega_1\omega_2]^2 \leqslant [-C_p(1 + \gamma\omega_1^2)\omega_1^2 + 1 + \gamma][-C_p(1 + \gamma\omega_1^2)\omega_2^2 + 1]$$

for any $\omega \in \mathbb{R}^2$, $|\omega| = 1$.

In particular, the second condition has to be satisfied. This can be written in the form

$$1 + \gamma - C_p(1 + \gamma\omega_1^2)(1 + \gamma\omega_2^2) \geqslant 0 \tag{3.16}$$

for any $\omega \in \mathbb{R}^2$, $|\omega| = 1$. The minimum of the left-hand side of (3.16) on the unit sphere is given by

$$1 + \gamma - C_p(1 + \gamma/2)^2.$$

Hence (3.16) is satisfied if and only if $1 + \gamma - C_p(1 + \gamma/2)^2 \geqslant 0$. The last inequality means

$$\frac{2(1 - \nu)}{1 - 2\nu} - \left(\frac{p - 2}{p}\right)^2 \left(\frac{3 - 4\nu}{2(1 - 2\nu)}\right)^2 \geqslant 0,$$

i.e., (3.14). From the identity $4/(pp') = 1 - (1 - 2/p)^2$, it follows that (3.14) can be written also as

$$\frac{4}{pp'} \geqslant \frac{1}{(3 - 4\nu)^2}. \tag{3.17}$$

Sufficiency. In view of Lemma 3.1, E is L^p-dissipative if and only if (3.6) holds for any $v \in (C_0^1(\Omega))^2$. Choose $v \in (C_0^1(\Omega))^2$ and define

$$X_1 = |v|^{-1}(v_1\partial_1|v| + v_2\partial_2|v|), \qquad X_2 = |v|^{-1}(v_2\partial_1|v| - v_1\partial_2|v|)$$

$$Y_1 = |v|[\partial_1(|v|^{-1}v_1) + \partial_2(|v|^{-1}v_2)], \quad Y_2 = |v|[\partial_1(|v|^{-1}v_2) - \partial_2(|v|^{-1}v_1)]$$

on the set $E = \{x \in \Omega \mid v \neq 0\}$. From the identities

$$|\nabla|v||^2 = X_1^2 + X_2^2, \quad Y_1 = (\partial_1 v_1 + \partial_2 v_2) - X_1, \quad Y_2 = (\partial_1 v_2 - \partial_2 v_1) - X_2$$

it follows that

$$Y_1^2 + Y_2^2 = |\nabla|v||^2 + (\partial_1 v_1 + \partial_2 v_2)^2 + (\partial_1 v_2 - \partial_2 v_1)^2$$
$$- 2(\partial_1 v_1 + \partial_2 v_2)X_1 - 2(\partial_1 v_2 - \partial_2 v_1)X_2.$$

Keeping in mind that $\partial_h |v| = |v|^{-1} v_j \partial_h v_j$, one can check that

$$(\partial_1 v_1 + \partial_2 v_2)(v_1 \partial_1 |v| + v_2 \partial_2 |v|) + (\partial_1 v_2 - \partial_2 v_1)(v_2 \partial_1 |v| - v_1 \partial_2 |v|)$$
$$= |v| \, |\nabla|v||^2 + |v|(\partial_1 v_1 \partial_2 v_2 - \partial_2 v_1 \partial_1 v_2),$$

which implies

$$\sum_j |\nabla v_j|^2 = X_1^2 + X_2^2 + Y_1^2 + Y_2^2. \tag{3.18}$$

Thus (3.6) can be written as

$$\int_E \left[\frac{4}{pp'}(X_1^2 + X_2^2) + Y_1^2 + Y_2^2 - \gamma\, C_p X_1^2 + \gamma\,(X_1 + Y_1)^2 \right] dx \geqslant 0. \tag{3.19}$$

Let us prove that

$$\int_E X_1 Y_1 dx = - \int_E X_2 Y_2 dx. \tag{3.20}$$

Since $X_1 + Y_1 = \operatorname{div} v$ and $X_2 + Y_2 = \partial_1 v_2 - \partial_2 v_1$, keeping in mind (3.18), we may write

$$2 \int_E (X_1 Y_1 + X_2 Y_2) dx = \int_E [(X_1 + Y_1)^2 + (X_2 + Y_2)^2 - (X_1^2 + X_2^2 + Y_1^2 + Y_2^2)]\, dx$$
$$= \int_E [(\operatorname{div} v)^2 + (\partial_1 v_2 - \partial_2 v_1)^2 - \sum_j |\nabla v_j|^2]\, dx,$$

i.e.,

$$\int_E (X_1 Y_1 + X_2 Y_2) dx = \int_E (\partial_1 v_1 \partial_2 v_2 - \partial_1 v_2 \partial_2 v_1)\, dx.$$

The set $\{x \in \Omega \setminus E \mid \nabla v(x) \neq 0\}$ has zero measure and then

$$\int_E (X_1 Y_1 + X_2 Y_2) dx = \int_\Omega (\partial_1 v_1 \partial_2 v_2 - \partial_1 v_2 \partial_2 v_1)\, dx.$$

There exists a sequence $\{v^{(n)}\} \subset C_0^\infty(\Omega)$ such that $v^{(n)} \to v$, $\nabla v^{(n)} \to \nabla v$ uniformly in Ω and hence

$$\int_\Omega \partial_1 v_1 \partial_2 v_2 dx = \lim_{n \to \infty} \int_\Omega \partial_1 v_1^{(n)} \partial_2 v_2^{(n)} dx$$
$$= \lim_{n \to \infty} \int_\Omega \partial_1 v_2^{(n)} \partial_2 v_1^{(n)} dx = \int_\Omega \partial_1 v_2 \partial_2 v_1 dx$$

and (3.20) is proved. In view of this, (3.19) can be written as

$$\int_E \left(\frac{4}{p\,p'}(1+\gamma)X_1^2 + 2\vartheta\gamma\,X_1 Y_1 + (1+\gamma)Y_1^2 \right) dx$$
$$+ \int_E \left(\frac{4}{p\,p'}X_2^2 - 2(1-\vartheta)\gamma\,X_2 Y_2 + Y_2^2 \right) dx \geqslant 0$$

for any fixed $\vartheta \in \mathbb{R}$.

If we choose

$$\vartheta = \frac{2(1-\nu)}{3-4\nu}$$

we find

$$(1-\vartheta)\gamma = \frac{1}{3-4\nu}, \qquad \vartheta^2\gamma^2 = \frac{(1+\gamma)^2}{(3-4\nu)^2}.$$

Inequality (3.17) leads to

$$\vartheta^2\gamma^2 \leqslant \frac{4}{p\,p'}(1+\gamma)^2, \qquad (1-\vartheta)^2\gamma^2 \leqslant \frac{4}{p\,p'}.$$

Observing that (3.14) implies $1+\gamma = 2(1-\nu)(1-2\nu)^{-1} \geqslant 0$, we get

$$\frac{4}{p\,p'}(1+\gamma)x_1^2 + 2\vartheta\gamma\,x_1 y_1 + (1+\gamma)y_1^2 \geqslant 0,$$

$$\frac{4}{p\,p'}x_2^2 - 2(1-\vartheta)\gamma\,x_2 y_2 + y_2^2 \geqslant 0$$

for any $x_1, x_2, y_1, y_2 \in \mathbb{R}$. This shows that (3.19) holds. Then (3.6) is true for any $v \in (C_0^1(\Omega))^2$ and the proof is complete. $\qquad\square$

In the next theorem Ω is a bounded domain satisfying the same smoothness assumption as in Section 2.4 (see p. 48).

Theorem 3.4. *Let E be the two-dimensional elasticity operator* (3.1) *with domain* $(W^{2,p}(\Omega) \cap \mathring{W}^{1,p}(\Omega))^2$. *The operator E is L^p-dissipative if and only if condition* (3.14) *holds.*

Proof. By means of the same arguments as in Section 2.4.1, we have the equivalence between the L^p-dissipativity of the form (3.3) and the L^p-dissipativity of the elasticity operator (3.1). The result follows from Theorem 3.3. $\qquad\square$

3.2 Comparison between E and Δ

We give two corollaries of the results obtained in the previous section. They concern the comparison between E and Δ from the point of view of the L^p-dissipativity.

The first one provides a necessary and sufficient condition for the existence of a $k > 0$ such that $E - k\Delta$ is L^p-dissipative. The other one considers the analogue question for $k\Delta - E$.

In all this section we shall assume the smoothness assumptions of Theorem 3.4.

Corollary 3.5. *There exists $k > 0$ such that $E - k\Delta$ is L^p-dissipative if and only if*

$$\left(\frac{1}{2} - \frac{1}{p}\right)^2 < \frac{2(\nu - 1)(2\nu - 1)}{(3 - 4\nu)^2}. \tag{3.21}$$

Proof. Necessity. We remark that if $E - k\Delta$ is L^p-dissipative, then

$$\begin{cases} k \leqslant 1 & \text{if } p = 2 \\ k < 1 & \text{if } p \neq 2. \end{cases} \tag{3.22}$$

In fact, in view of Theorem 3.1, we have the necessary condition

$$-(1 - 2/p)^2[(1 - k)|\xi|^2 + (1 - 2\nu)^{-1}(\xi_j\omega_j)^2](\lambda_j\omega_j)^2 \\ +(1 - k)|\xi|^2|\lambda|^2 + (1 - 2\nu)^{-1}(\xi_j\lambda_j)^2 \geqslant 0 \tag{3.23}$$

for any $\xi, \lambda, \omega \in \mathbb{R}^2$, $|\omega| = 1$. If we take $\xi = (1, 0)$, $\lambda = \omega = (0, 1)$ in (3.23) we find

$$\frac{4}{pp'}(1 - k) \geqslant 0$$

and then $k \leqslant 1$ for any p. If $p \neq 2$ and $k = 1$, taking $\xi = (1, 0)$, $\lambda = (0, 1)$, $\omega = (1/\sqrt{2}, 1/\sqrt{2})$ in (3.23), we find $-(1 - 2/p)^2(1 - 2\nu)^{-1} \geqslant 0$. On the other hand, taking $\xi = \lambda = (1, 0)$, $\omega = (0, 1)$ we find $(1 - 2\nu)^{-1} \geqslant 0$. This is a contradiction and (3.22) is proved.

It is clear that if $E - k\Delta$ is L^p-dissipative, then $E - k'\Delta$ is L^p-dissipative for any $k' < k$. Therefore it is not restrictive to suppose that $E - k\Delta$ is L^p-dissipative for some $0 < k < 1$. Moreover E is also L^p-dissipative.

The L^p-dissipativity of $E - k\Delta$ ($0 < k < 1$) is equivalent to the L^p-dissipativity of the operator

$$E'u = \Delta u + (1 - k)^{-1}(1 - 2\nu)^{-1}\nabla \operatorname{div} u. \tag{3.24}$$

Setting

$$\nu' = \nu(1 - k) + k/2, \tag{3.25}$$

we have $(1 - k)(1 - 2\nu) = 1 - 2\nu'$. Theorem 3.4 shows that

$$\frac{4}{pp'} \geqslant \frac{1}{(3 - 4\nu')^2}. \tag{3.26}$$

Since $3 - 4\nu' = 3 - 4\nu - 2k(1 - 2\nu)$, condition (3.26) means $|3 - 4\nu - 2k(1 - 2\nu)| \geqslant \sqrt{pp'}/2$, i.e.,

$$\left| k - \frac{3 - 4\nu}{2(1 - 2\nu)} \right| \geqslant \frac{\sqrt{pp'}}{4|1 - 2\nu|} \tag{3.27}$$

Note that the L^p-dissipativity of E implies that (3.14) holds. In particular we have $(3 - 4\nu)/(1 - 2\nu) > 0$. Hence (3.27) is satisfied if either

$$k \leqslant \frac{1}{2|1 - 2\nu|} \left(|3 - 4\nu| - \frac{\sqrt{pp'}}{2} \right) \tag{3.28}$$

or

$$k \geqslant \frac{1}{2|1 - 2\nu|} \left(|3 - 4\nu| + \frac{\sqrt{pp'}}{2} \right). \tag{3.29}$$

Since

$$\frac{|3 - 4\nu|}{2|1 - 2\nu|} - 1 = \frac{3 - 4\nu}{2(1 - 2\nu)} - 1 = \frac{1}{2(1 - 2\nu)} \geqslant -\frac{\sqrt{pp'}}{4|1 - 2\nu|}$$

we have

$$\frac{1}{2|1 - 2\nu|} \left(|3 - 4\nu| + \frac{\sqrt{pp'}}{2} \right) \geqslant 1$$

and (3.29) is impossible. Then (3.28) holds. Since $k > 0$, we have the strict inequality in (3.17) and (3.21) is proved.

Sufficiency. Suppose (3.21). Since

$$\frac{4}{pp'} > \frac{1}{(3 - 4\nu)^2},$$

we can take k such that

$$0 < k < \frac{1}{2|1 - 2\nu|} \left(|3 - 4\nu| - \frac{\sqrt{pp'}}{2} \right). \tag{3.30}$$

Note that

$$\frac{|3 - 4\nu|}{2|1 - 2\nu|} - 1 = \frac{3 - 4\nu}{2(1 - 2\nu)} - 1 = \frac{1}{2(1 - 2\nu)} \leqslant \frac{\sqrt{pp'}}{4|1 - 2\nu|}.$$

This means

$$\frac{1}{2|1 - 2\nu|} \left(|3 - 4\nu| - \frac{\sqrt{pp'}}{2} \right) \leqslant 1$$

and then $k < 1$. Let ν' be given by (3.25). The L^p-dissipativity of $E - k\Delta$ is equivalent to the L^p-dissipativity of the operator E' defined by (3.24).

Condition (3.27) (i.e., (3.26)) follows from (3.30) and Theorem 3.4 gives the result. $\square$

Corollary 3.6. *There exists $k < 2$ such that $k\Delta - E$ is L^p-dissipative if and only if*

$$\left(\frac{1}{2} - \frac{1}{p}\right)^2 < \frac{2\nu(2\nu - 1)}{(1 - 4\nu)^2}. \tag{3.31}$$

Proof. We may write $k\Delta - E = \widetilde{E} - \widetilde{k}\Delta$, where $\widetilde{k} = 2 - k$, $\widetilde{E} = \Delta + (1 - 2\widetilde{\nu})^{-1}\nabla\,\mathrm{div}$, $\widetilde{\nu} = 1 - \nu$. Theorem 3.5 shows that $\widetilde{E} - \widetilde{k}\Delta$ is L^p-dissipative if and only if

$$\left(\frac{1}{2} - \frac{1}{p}\right)^2 < \frac{2(\widetilde{\nu} - 1)(2\widetilde{\nu} - 1)}{(3 - 4\widetilde{\nu})^2}. \tag{3.32}$$

Condition (3.32) coincides with (3.31) and the corollary is proved. $\qquad\square$

3.3 L^p-dissipativity of three-dimensional elasticity

As far as the three-dimensional Lamé system is concerned, necessary and sufficient conditions for the L^p-dissipativity are not known. The next theorem shows that condition (3.14) is necessary, even in the case of a non-constant Poisson ratio. Here Ω is a bounded domain in $\mathbb{R}^3$ whose boundary is in the class C^2.

Theorem 3.7. *Suppose $\nu = \nu(x)$ is a continuous function defined in Ω such that*

$$\inf_{x \in \Omega} |2\nu(x) - 1| > 0.$$

If (3.1) is L^p-dissipative in Ω, then

$$\left(\frac{1}{2} - \frac{1}{p}\right)^2 \leqslant \inf_{x \in \Omega} \frac{2(\nu(x) - 1)(2\nu(x) - 1)}{(3 - 4\nu(x))^2}. \tag{3.33}$$

Proof. We have

$$\int_\Omega (\Delta u + (1 - 2\nu(x))^{-1}\nabla\,\mathrm{div}\,u)|u|^{p-2}u\,dx \leqslant 0 \tag{3.34}$$

for any $u \in (W^{2,p}(\Omega) \cap \overset{\circ}{W}{}^{1,p}(\Omega))^3$, in particular for any $u \in (C_0^\infty(\Omega))^3$. Take $v \in (C_0^\infty(\mathbb{R}^2))^2$, $\varphi \in C_0^\infty(\mathbb{R})$, $\varphi \geqslant 0$ and $x^0 \in \Omega$; define $v_\varepsilon(x_1, x_2) = v((x_1 - x_1^0)/\varepsilon, (x_2 - x_2^0)/\varepsilon)$,

$$u(x_1, x_2, x_3) = (v_{\varepsilon,1}(x_1, x_2), v_{\varepsilon,2}(x_1, x_2), 0)\,\varphi(x_3).$$

We suppose that the support of v is contained in the unit ball, $0 < \varepsilon < \mathrm{dist}(x^0, \partial\Omega)$ and the support of φ is contained in $(-\varepsilon, \varepsilon)$. In this way the function u belongs to $(C_0^\infty(\Omega))^3$.

Setting $\gamma(x_1, x_2, x_3) = (1 - 2\,\nu(x_1, x_2, x_3))^{-1}$, we have

$$\Delta u + \gamma\,\nabla\,\mathrm{div}\,u = (\Delta v_\varepsilon + \gamma\,\nabla\,\mathrm{div}\,v_\varepsilon)\,\varphi + v_\varepsilon\varphi''$$

and then

$$(\Delta u + \gamma \nabla \operatorname{div} u)|u|^{p-2}u = (\Delta v_\varepsilon + \gamma \nabla \operatorname{div} v_\varepsilon)|v_\varepsilon|^{p-2}v_\varepsilon \varphi^p + v_\varepsilon^2 \varphi'' \varphi^{p-1}.$$

We can write, in view of (3.34),

$$\int_{\mathbb{R}} \varphi^p dx_3 \iint_{\mathbb{R}^2} (\Delta v_\varepsilon + \gamma \nabla \operatorname{div} v_\varepsilon)|v_\varepsilon|^{p-2}v_\varepsilon \, dx_1 dx_2$$
$$+ \int_{\mathbb{R}} \varphi^{p-1}\varphi'' dx_3 \iint_{\mathbb{R}^2} |v_\varepsilon|^p dx_1 dx_2 \leqslant 0.$$

Noting that

$$\Delta v_\varepsilon + \gamma \nabla \operatorname{div} v_\varepsilon$$
$$= \frac{1}{\varepsilon^2}\left[\Delta v\left(\frac{x_1 - x_1^0}{\varepsilon}, \frac{x_1 - x_1^0}{\varepsilon}\right) + \gamma(x_1, x_2, x_3)\nabla \operatorname{div} v\left(\frac{x_1 - x_1^0}{\varepsilon}, \frac{x_1 - x_1^0}{\varepsilon}\right)\right],$$

a change of variables in the double integral gives

$$\int_{\mathbb{R}} \varphi^p(x_3)\, dx_3 \iint_{\mathbb{R}^2} (\Delta v(t_1, t_2) + \gamma(x_1^0 + \varepsilon\, t_1, x_2^0 + \varepsilon\, t_2, x_3)\nabla \operatorname{div} v(t_1, t_2))$$
$$\times |v(t_1, t_2)|^{p-2}v(t_1, t_2)dt_1 dt_2 + \varepsilon^2 \int_{\mathbb{R}} \varphi^{p-1}\varphi'' dx_3 \iint_{\mathbb{R}^2} |v(t_1, t_2)|^p dt_1 dt_2 \leqslant 0.$$

Letting $\varepsilon \to 0^+$, we get

$$\int_{\mathbb{R}} \varphi^p(x_3)\, dx_3 \iint_{\mathbb{R}^2} (\Delta v(t_1, t_2) + \gamma(x_1^0, x_2^0, x_3)\nabla \operatorname{div} v(t_1, t_2))$$
$$\times |v(t_1, t_2)|^{p-2}v(t_1, t_2)dt_1 dt_2 \leqslant 0.$$

For the arbitrariness of φ, this implies

$$\iint_{\mathbb{R}^2} (\Delta v(t_1, t_2) + \gamma(x_1^0, x_2^0, x_3^0)\nabla \operatorname{div} v(t_1, t_2))|v(t_1, t_2)|^{p-2}v(t_1, t_2)dt_1 dt_2 \leqslant 0$$

for any $v \in (C_0^\infty(B))^2$, B being the unit ball in $\mathbb{R}^2$.

Suppose $p \geqslant 2$. Integrating by parts, we get

$$\mathscr{L}(v, |v|^{p-2}v) \leqslant 0 \tag{3.35}$$

for any $v \in (C_0^\infty(B))^2$.

Given $v \in (C_0^\infty(B))^2$, define $u_\varepsilon = g_\varepsilon^{2/p-1}v$. Since $u_\varepsilon \in (C_0^\infty(B))^2$, in view of (3.35) we write

$$\mathscr{L}(u_\varepsilon, |u_\varepsilon|^{p-2}u_\varepsilon) \leqslant 0.$$

By means of the computations we made in the necessity of Lemma 3.1, letting $\varepsilon \to 0^+$, we find inequality (3.6) for any $v \in (C_0^\infty(B))^2$. This implies that (3.6) holds for any $v \in (C_0^1(B))^2$.

In fact, let $v_m \in (C_0^\infty(B))^2$ such that $v_m \to v$ in C^1-norm. Let us show that

$$\chi_{E_n}|v_m|^{-1}v_m\nabla v_m \to \chi_E|v|^{-1}v\nabla v \quad \text{in } L^2(B), \tag{3.36}$$

where $E_n = \{x \in B \mid v_m(x) \neq 0\}$, $E = \{x \in \Omega \mid v(x) \neq 0\}$. We see that

$$\chi_{E_n}|v_m|^{-1}v_m\nabla v_m \to \chi_E|v|^{-1}v\nabla v \tag{3.37}$$

on the set $E \cup \{x \in B \mid \nabla v(x) = 0\}$. The set $\{x \in B \setminus E \mid \nabla v(x) \neq 0\}$ having zero measure, (3.37) holds almost everywhere. Moreover, since

$$\int_G \chi_{E_n}|v_m|^{-2}|v_m\nabla v_m|^2 dx \leqslant \int_G |\nabla v_m|^2 dx$$

for any measurable set $G \subset \Omega$ and $\{\nabla v_m\}$ is convergent in $L^2(\Omega)$, the sequence $\{|\chi_{E_n}|v_m|^{-1}v_m\nabla v_m - \chi_E|v|^{-1}v\nabla v|^2\}$ has uniformly absolutely continuous integrals. Now we may appeal to Vitali's Theorem to obtain (3.36).

Inequality (3.6) holding for any $v \in (C_0^1(B))^2$, the result follows from Theorem 3.3.

Let now $1 < p < 2$. From the L^p dissipativity of E it follows that the operator $E - \lambda I$ ($\lambda > 0$) is invertible on $L^p(\Omega)$. This means that for any $f \in L^p(\Omega)$ there exists one and only one $u \in W^{2,p}(\Omega) \cap \mathring{W}^{1,p}(\Omega)$ such that $(E - \lambda I)u = f$. Because of well-known regularity results for solutions of elliptic systems (see Agmon, Douglis and Nirenberg [2]), we have also that, if f belongs to $L^{p'}(\Omega)$, the solution u belongs to $W^{2,p'}(\Omega) \cap \mathring{W}^{1,p'}(\Omega)$ and there exists the bounded resolvent $(E^* - \lambda I)^{-1} : L^{p'}(\Omega) \to W^{2,p'}(\Omega) \cap \mathring{W}^{1,p'}(\Omega)$.

Since E is L^p-dissipative and $\|(E^* - \lambda I)^{-1}\| = \|(E - \lambda I)^{-1}\|$, we may write

$$\|(E^* - \lambda I)^{-1}\| \leqslant \frac{1}{\lambda}$$

for any $\lambda > 0$, i.e., we have the $L^{p'}$-dissipativity of E^*, $p' > 2$. We have reduced the proof to the previous case. Therefore (3.33) holds with p replaced by p'. Since

$$\left(\frac{1}{2} - \frac{1}{p}\right)^2 = \left(\frac{1}{2} - \frac{1}{p'}\right)^2$$

the proof is complete. $\qquad\square$

We do not know if condition (3.14) is sufficient for the L^p-dissipativity of the three-dimensional elasticity. The next theorem provides a more strict sufficient condition.

Theorem 3.8. *Let Ω be a domain in $\mathbb{R}^3$. If*

$$(1-2/p)^2 \leqslant \begin{cases} \dfrac{1-2\nu}{2(1-\nu)} & \text{if } \nu < 1/2 \\[2mm] \dfrac{2(1-\nu)}{1-2\nu} & \text{if } \nu > 1. \end{cases} \tag{3.38}$$

the operator (3.1) is L^p-dissipative.

Proof. In view of Lemma 3.1, operator E is L^p-dissipative if and only if inequality (3.6) holds for any $v \in (C_0^1(\Omega))^3$. This can be written as

$$C_p \int_\Omega [|\nabla|v||^2 + \gamma |v|^{-2}|v_h \partial_h|v||^2]\, dx \leqslant \int_\Omega \left[\sum_{j=1}^3 |\nabla v_j|^2 + \gamma |\operatorname{div} v|^2\right] dx. \tag{3.39}$$

Note that the integral on the left-hand side of (3.39) is nonnegative. In fact, setting

$$\xi_{hj} = \partial_h v_j, \quad \omega_j = |v|^{-1} v_j,$$

we have

$$|\nabla|v||^2 + \gamma |v|^{-2}|v_h \partial_h|v||^2 = \omega_i \omega_j (\delta_{hk} + \gamma \omega_h \omega_k)\xi_{hi}\xi_{kj}.$$

Then we can write

$$|\nabla|v||^2 + \gamma |v|^{-2}|v_h \partial_h|v||^2 = |\lambda|^2 + \gamma(\lambda \cdot \omega)^2 \tag{3.40}$$

where λ is the vector whose hth component is $\omega_i \xi_{hi}$. Since ω is a unit vector and $\gamma > -1$ we have

$$|\nabla|v||^2 + \gamma |v|^{-2}|v_h \partial_h|v||^2 \geqslant 0.$$

Also the right-hand side of (3.39) is nonnegative. In fact, denoting by $\widehat{v_j}$ the Fourier transform of v_j,

$$\widehat{v_j}(y) = \int_{\mathbb{R}^3} v_j(x) e^{-iy\cdot x}\, dx,$$

we have

$$\int_\Omega \left[\sum_{j=1}^3 |\nabla v_j|^2 + \gamma |\operatorname{div} v|^2\right] dx = \int_\Omega (\partial_h v_j \partial_h v_j + \gamma \partial_h v_h \partial_j v_j)\, dx \tag{3.41}$$

$$= (2\pi)^{-3} \int_{\mathbb{R}^3} (\widehat{\partial_h v_j}\,\overline{\widehat{\partial_h v_j}} + \gamma \widehat{\partial_h v_h}\,\overline{\widehat{\partial_j v_j}})\, dy = (2\pi)^{-3} \int_{\mathbb{R}^3} (|y|^2|\widehat{v}|^2 + \gamma|y\cdot\widehat{v}|^2)\, dy$$

$$\geqslant \min\{1, 1+\gamma\}(2\pi)^{-3} \int_{\mathbb{R}^3} |y|^2|\widehat{v}|^2 dy = \min\{1, 1+\gamma\} \int_\Omega \sum_{j=1}^3 |\nabla v_j|^2 dx.$$

This implies that (3.39) holds for any v such that the left-hand side vanishes and that E is L^p-dissipative if and only if

$$C_p \leqslant \inf \frac{\displaystyle\int_\Omega \left[\sum_{j=1}^3 |\nabla v_j|^2 + \gamma |\operatorname{div} v|^2 \right] dx}{\displaystyle\int_\Omega [|\nabla|v||^2 + \gamma |v|^{-2} |v_h \partial_h |v||^2] \, dx}, \tag{3.42}$$

where the infimum is taken over all $v \in (C_0^1(\Omega))^3$ such that the denominator is positive.

From (3.40) we get

$$|\nabla|v||^2 + \gamma |v|^{-2} |v_h \partial_h |v||^2 \leqslant \max\{1, 1+\gamma\}|\lambda|^2 \leqslant \max\{1, 1+\gamma\} \sum_{j=1}^3 |\nabla v_j|^2.$$

Keeping in mind also (3.41) we find that

$$\frac{\displaystyle\int_\Omega \left[\sum_{j=1}^3 |\nabla v_j|^2 + \gamma |\operatorname{div} v|^2 \right] dx}{\displaystyle\int_\Omega [|\nabla|v||^2 + \gamma |v|^{-2} |v_h \partial_h |v||^2] \, dx} \geqslant \frac{\min\{1, 1+\gamma\}}{\max\{1, 1+\gamma\}}.$$

Therefore condition (3.42) is satisfied if

$$C_p \leqslant \frac{\min\{1, 1+\gamma\}}{\max\{1, 1+\gamma\}}.$$

This inequality being equivalent to (3.38), the proof is complete. $\qquad\square$

Remark 3.9. The theorems of this section hold in any dimension $n \geqslant 3$ with the same proof.

3.4 Weighted L^p-negativity of elasticity system defined on rotationally symmetric vector functions

Let Φ be a point on the $(n-2)$-dimensional unit sphere S^{n-2} with spherical coordinates $\{\vartheta_j\}_{j=1,\ldots,n-3}$ and φ, where $\vartheta_j \in (0, \pi)$ and $\varphi \in [0, 2\pi)$. A point $x \in \mathbb{R}^n$ is represented as a triple $(\varrho, \vartheta, \Phi)$, where $\varrho > 0$ and $\vartheta \in [0, \pi]$. Correspondingly, a vector u can be written as $u = (u_\varrho, u_\vartheta, u_\Phi)$ with $u_\Phi = (u_{\vartheta_{n-3}}, \ldots, u_{\vartheta_1}, u_\varphi)$. We call $u_\varrho, u_\vartheta, u_\Phi$ the spherical components of the vector u.

Theorem 3.10. *Let the spherical components u_ϑ and u_Φ of the vector u vanish, i.e., $u = (u_\varrho, 0, 0)$, and let u_ϱ depend only on the variable ϱ. Then, if $\alpha \geqslant n - 2$, we have*

$$\int_{\mathbb{R}^n} \left(\Delta u + (1 - 2\nu)^{-1} \nabla \operatorname{div} u \right) |u|^{p-2} u \, \frac{dx}{|x|^\alpha} \leqslant 0 \qquad (3.43)$$

for any $u \in (C_0^\infty(\mathbb{R}^n \setminus \{0\}))^n$ satisfying the aforesaid symmetric conditions, if and only if

$$-(p - 1)(n + p' - 2) \leqslant \alpha \leqslant n + p - 2. \qquad (3.44)$$

If $\alpha < n - 2$ the same result holds replacing $(C_0^\infty(\mathbb{R}^n \setminus \{0\}))^n$ by $(C_0^\infty(\mathbb{R}^n))^n$.

Proof. Setting

$$g_\varepsilon(s) = (s^2 + \varepsilon^2)^{1/2},$$

and denoting by ω_{n-1} the $(n-1)$-dimensional measure of the unit sphere in $\mathbb{R}^n$, we have

$$\int_{\mathbb{R}^n} \Delta u \, g_\varepsilon(|u|)^{p-2} u \, \frac{dx}{|x|^\alpha}$$
$$= \omega_{n-1} \int_0^{+\infty} \left(\frac{1}{\varrho^{n-1}} \partial_\varrho(\varrho^{n-1} \partial_\varrho u_\varrho) - \frac{n-1}{\varrho^2} u_\varrho \right) g_\varepsilon(|u_\varrho|)^{p-2} u_\varrho \varrho^{n-1-\alpha} d\varrho.$$

An integration by parts gives

$$\int_0^{+\infty} \partial_\varrho(\varrho^{n-1} \partial_\varrho u_\varrho) g_\varepsilon(|u_\varrho|)^{p-2} u_\varrho \varrho^{-\alpha} d\varrho$$
$$= -\int_0^{+\infty} \varrho^{n-1} \partial_\varrho u_\varrho \partial_\varrho(g_\varepsilon(|u_\varrho|)^{p-2} u_\varrho \varrho^{-\alpha}) d\varrho$$
$$= -\int_0^{+\infty} \partial_\varrho u_\varrho \partial_\varrho(g_\varepsilon(|u_\varrho|)^{p-2} u_\varrho) \varrho^{n-1-\alpha} d\varrho \qquad (3.45)$$
$$+ \alpha \int_0^{+\infty} g_\varepsilon(|u_\varrho|)^{p-2} u_\varrho \partial_\varrho u_\varrho \varrho^{n-\alpha-2} d\varrho.$$

Since

$$\partial_\varrho(g_\varepsilon(|u_\varrho|)^p) = p \, g_\varepsilon(|u_\varrho|)^{p-2} u_\varrho \partial_\varrho u_\varrho, \qquad (3.46)$$

we have, by means of another integration by parts in the last integral of (3.45),

$$\alpha \int_0^{+\infty} g_\varepsilon(|u_\varrho|)^{p-2} u_\varrho \partial_\varrho u_\varrho \varrho^{n-\alpha-2} d\varrho = \frac{\alpha}{p} \int_0^{+\infty} \partial_\varrho(g_\varepsilon(|u_\varrho|)^p) \varrho^{n-\alpha-2} d\varrho$$
$$= -\frac{\alpha(n-2-\alpha)}{p} \int_K g_\varepsilon(|u_\varrho|)^p \varrho^{n-3-\alpha} d\varrho + \mathcal{O}(\varepsilon^p)$$

where K is the support of u_ϱ.

This proves the identity

$$
\int_{\mathbb{R}^n} \Delta u \, g_\varepsilon(|u|)^{p-2} u \, \frac{dx}{|x|^\alpha} = -\omega_{n-1}\Bigg[(n-1)\int_K g_\varepsilon(|u_\varrho|)^{p-2}u_\varrho^2 \varrho^{n-3-\alpha}d\varrho
$$
$$
+ \frac{\alpha(n-2-\alpha)}{p}\int_K g_\varepsilon(|u_\varrho|)^p \varrho^{n-3-\alpha}d\varrho \tag{3.47}
$$
$$
+ \int_K \partial_\varrho u_\varrho \partial_\varrho(g_\varepsilon(|u_\varrho|)^{p-2}u_\varrho)\varrho^{n-1-\alpha}d\varrho \Bigg] + \mathcal{O}(\varepsilon^p).
$$

We have also

$$
\int_{\mathbb{R}^n} \nabla(\operatorname{div} u)g_\varepsilon(|u|)^{p-2}u \, \frac{dx}{|x|^\alpha} = -\int_{\mathbb{R}^n} \operatorname{div} u \, \operatorname{div}(g_\varepsilon(|u|)^{p-2}u|x|^{-\alpha})dx
$$
$$
= -\omega_{n-1}\int_0^{+\infty} \frac{1}{\varrho^{n-1}}\partial_\varrho(\varrho^{n-1}u_\varrho)\partial_\varrho(\varrho^{n-1-\alpha}g_\varepsilon(|u_\varrho|)^{p-2}u_\varrho)d\varrho. \tag{3.48}
$$

Moreover,

$$
\frac{1}{\varrho^{n-1}}\partial_\varrho(\varrho^{n-1}u_\varrho)\,\partial_\varrho(\varrho^{n-1-\alpha}g_\varepsilon(|u_\varrho|)^{p-2}u_\varrho) \tag{3.49}
$$
$$
= (n-1)(n-1-\alpha)\varrho^{n-3-\alpha}g_\varepsilon(|u_\varrho|)^{p-2}u_\varrho^2 + (n-1)\varrho^{n-2-\alpha}u_\varrho\partial_\varrho(g_\varepsilon(|u_\varrho|)^{p-2}u_\varrho)
$$
$$
+ (n-1-\alpha)\varrho^{n-2-\alpha}g_\varepsilon(|u_\varrho|)^{p-2}u_\varrho\partial_\varrho u_\varrho + \varrho^{n-1-\alpha}\partial_\varrho u_\varrho\partial_\varrho(g_\varepsilon(|u_\varrho|)^{p-2}u_\varrho).
$$

In view of (3.46) we may write

$$
\int_0^{+\infty} \varrho^{n-2-\alpha}g_\varepsilon(|u_\varrho|)^{p-2}u_\varrho\partial_\varrho u_\varrho d\varrho = \frac{1}{p}\int_0^{+\infty} \varrho^{n-2-\alpha}\partial_\varrho(g_\varepsilon(|u_\varrho|)^p)d\varrho \tag{3.50}
$$
$$
= -\frac{n-2-\alpha}{p}\int_K \varrho^{n-3-\alpha}g_\varepsilon(|u_\varrho|)^p d\varrho + \mathcal{O}(\varepsilon^p).
$$

Since

$$
\partial_\varrho(g_\varepsilon(|u_\varrho|)^{p-2}u_\varrho^2) = u_\varrho\partial_\varrho(g_\varepsilon(|u_\varrho|)^{p-2}u_\varrho) + g_\varepsilon(|u_\varrho|)^{p-2}u_\varrho\partial_\varrho u_\varrho
$$

and using again (3.46), we find

$$
\int_0^{+\infty} \varrho^{n-2-\alpha}u_\varrho\partial_\varrho(g_\varepsilon(|u_\varrho|)^{p-2}u_\varrho)d\varrho
$$
$$
= \int_0^{+\infty} \varrho^{n-2-\alpha}\partial_\varrho(g_\varepsilon(|u_\varrho|)^{p-2}u_\varrho^2)d\varrho - \int_0^{+\infty} \varrho^{n-2-\alpha}g_\varepsilon(|u_\varrho|)^{p-2}u_\varrho\partial_\varrho u_\varrho d\varrho
$$
$$
= -(n-2-\alpha)\int_K \varrho^{n-3-\alpha}g_\varepsilon(|u_\varrho|)^{p-2}u_\varrho^2 d\varrho - \frac{1}{p}\int_0^{+\infty} \varrho^{n-2-\alpha}\partial_\varrho(g_\varepsilon(|u_\varrho|)^p)d\varrho
$$
$$
+ \mathcal{O}(\varepsilon^p)
$$

$$= -(n-2-\alpha) \int_K \varrho^{n-3-\alpha} g_\varepsilon(|u_\varrho|)^{p-2} u_\varrho^2 d\varrho$$

$$+ \frac{n-2-\alpha}{p} \int_K \varrho^{n-3-\alpha} g_\varepsilon(|u_\varrho|)^p d\varrho + \mathcal{O}(\varepsilon^p). \tag{3.51}$$

We obtain by (3.48), (3.49), (3.50) and (3.51) that

$$\int_{\mathbb{R}^n} \nabla(\mathrm{div}\, u) g_\varepsilon(|u|)^{p-2} u \frac{dx}{|x|^\alpha}$$

$$= -\omega_{n-1} \left[(n-1) \int_K \varrho^{n-3-\alpha} g_\varepsilon(|u_\varrho|)^{p-2} u_\varrho^2 d\varrho \right.$$

$$+ \frac{\alpha(n-2-\alpha)}{p} \int_K \varrho^{n-3-\alpha} g_\varepsilon(|u_\varrho|)^p d\varrho$$

$$\left. + \int_K \partial_\varrho u_\varrho \partial_\varrho (g_\varepsilon(|u_\varrho|)^{p-2} u_\varrho) \varrho^{n-1-\alpha} d\varrho \right] + \mathcal{O}(\varepsilon^p). \tag{3.52}$$

It follows from (3.47) and (3.52) that

$$\int_{\mathbb{R}^n} \left(\Delta u + \frac{1}{1-2\nu} \nabla \,\mathrm{div}\, u \right) g_\varepsilon(|u|)^{p-2} u \frac{dx}{|x|^\alpha}$$

$$= -\omega_{n-1} \frac{2(1-\nu)}{1-2\nu} \left[(n-1) \int_K \varrho^{n-3-\alpha} g_\varepsilon(|u_\varrho|)^{p-2} u_\varrho^2 d\varrho \right.$$

$$+ \frac{\alpha(n-2-\alpha)}{p} \int_K \varrho^{n-3-\alpha} g_\varepsilon(|u_\varrho|)^p d\varrho$$

$$\left. + \int_K \partial_\varrho u_\varrho \partial_\varrho (g_\varepsilon(|u_\varrho|)^{p-2} u_\varrho) \varrho^{n-1-\alpha} d\varrho \right] + \mathcal{O}(\varepsilon^p).$$

Seeing that, given $a \in \mathbb{R}$, there exists a constant C_α such that $(g_\varepsilon(s))^a \leqslant C_\alpha(s^a + \varepsilon^a)$ $(s \geqslant 0)$, we may apply the dominated convergence theorem and find

$$\int_{\mathbb{R}^n} \left(\Delta u + \frac{1}{1-2\nu} \nabla \,\mathrm{div}\, u \right) |u|^{p-2} u \frac{dx}{|x|^\alpha}$$

$$= -\omega_{n-1} \frac{2(1-\nu)}{1-2\nu} \left\{ \left[n-1 + \frac{\alpha(n-2-\alpha)}{p} \right] \int_K \varrho^{n-3-\alpha} |u_\varrho|^p d\varrho \right.$$

$$\left. + \int_K \partial_\varrho u_\varrho \partial_\varrho (|u_\varrho|^{p-2} u_\varrho) \varrho^{n-1-\alpha} d\varrho \right\}.$$

Keeping in mind that either $\nu > 1$ or $\nu < 1/2$, the last equality shows that

(3.43) holds if and only if

$$\left[n - 1 + \frac{\alpha(n - 2 - \alpha)}{p}\right] \int_K \varrho^{n-3-\alpha}|u_\varrho|^p d\varrho + \int_K \partial_\varrho u_\varrho \partial_\varrho(|u_\varrho|^{p-2})\varrho^{n-1-\alpha} d\varrho \geqslant 0. \tag{3.53}$$

Setting $v_\varrho = |u_\varrho|^{(p-2)/2}u_\varrho$, we see that (3.53) is equivalent to

$$\left[n - 1 + \frac{\alpha(n - 2 - \alpha)}{p}\right] \int_0^{+\infty} |v_\varrho|^2 \varrho^{n-3-\alpha} d\varrho + \frac{4}{pp'} \int_0^{+\infty} (\partial_\varrho v_\varrho)^2 \varrho^{n-1-\alpha} d\varrho \geqslant 0. \tag{3.54}$$

If $\alpha = n - 2$ the inequality (3.54) is obviously satisfied. For $\alpha \neq n - 2$, we recall the Hardy inequality (see, for instance, Maz'ya [73, p. 40])

$$\int_0^{+\infty} \frac{v^2(\varrho)}{\varrho^{\alpha-n+3}} d\varrho \leqslant \frac{4}{(\alpha - n + 2)^2} \int_0^{+\infty} \frac{(\partial_\varrho v(\varrho))^2}{\varrho^{\alpha-n+1}} d\varrho, \tag{3.55}$$

which holds for any $v \in C_0^\infty(\mathbb{R})$ provided $\alpha \neq n - 2$, under the condition $v(0) = 0$ when $\alpha > n - 2$.

Inequality (3.54) can be written as

$$-\frac{pp'}{4}\left[n - 1 + \frac{\alpha(n - 2 - \alpha)}{p}\right] \int_0^{+\infty} |v_\varrho|^2 \varrho^{n-3-\alpha} d\varrho \leqslant \int_0^{+\infty} (\partial_\varrho v_\varrho)^2 \varrho^{n-1-\alpha} d\varrho. \tag{3.56}$$

Now we see that (3.56) holds if, and only if,

$$-\frac{pp'}{4}\left[n - 1 + \frac{\alpha(n - 2 - \alpha)}{p}\right] \leqslant \frac{(\alpha - n + 2)^2}{4}. \tag{3.57}$$

In fact, if (3.57) holds, then (3.56) is true, because of (3.55). Viceversa, if (3.56) holds, thanks to the arbitrariness of v_ϱ and to the sharpness of the constant in (3.55), we get (3.57).

A simple manipulation shows that the latter inequality is equivalent to

$$-\frac{(\alpha - (n + p - 2))(\frac{\alpha}{p-1} + (n + p' - 2))}{pp'} \geqslant 0,$$

which in turn is equivalent to (3.44). The theorem is proved. □

We remark that the inequalities

$$-(p - 1)(n + p' - 2) < 0 < n + p - 2$$

are always satisfied and therefore condition (3.44) is never empty.

3.5   Comments to Chapter 3

The L^p-dissipativity criteria presented in this chapter were obtained in Cialdea and Maz'ya [12, 14].

While the two-dimensional case is completely settled by inequality (3.14), for the n-dimensional elasticity we showed in Section 3.3 that condition (3.14) is necessary for the L^p-dissipativity. For the time being we do not know if it is also sufficient when $n > 2$. In the same section we give also condition (3.38), which is only sufficient.

Sections 3.3 and 3.4 are from Cialdea and Maz'ya [14].

Chapter 4

L^p-dissipativity for Systems of Partial Differential Operators

This chapter is devoted to systems of partial differential operators.

After some auxiliary results in Section 4.1, we give an algebraic necessary condition for the L^p-dissipativity of a general system in the two-dimensional case (Section 4.2). Several results are stated in terms of eigenvalues of the coefficient matrix of the system.

Hinging on these results in Section 4.4 we give an algebraic characterization of the L^p-dissipativity for a certain class of systems of partial differential operators.

The rest of the chapter is devoted to weakly coupled and coupled systems. In particular the relationship between the generation of L^p-contractive semigroups of the corresponding operators and L^2-contractivity of the semigroups generated by certain associated operators are investigated. For operators associated with systems uniformly parabolic in the sense of Petrovskii, algebraic necessary and sufficient conditions for the generation of contraction semigroups on L^p, for all $p \in [1, \infty]$ simultaneously, are presented.

4.1 Technical lemma

The aim of this section is to prove an auxiliary assertion, which can be considered an analogue of Lemma 2.1 for systems.

We shall prove it for operators of the kind

$$Au = \partial_h(\mathscr{A}^{hk}(x)\partial_k u) + \mathscr{B}^h(x)\partial_h u + \partial_h(\mathscr{C}^h u) + \mathscr{A}(x)u \tag{4.1}$$

where $\mathscr{A}^{hk}(x) = \{a_{ij}^{hk}(x)\}$, $\mathscr{B}^h(x) = \{b_{ij}^h(x)\}$, $\mathscr{C}^h(x) = \{c_{ij}^h(x)\}$ and $\mathscr{A}(x) = \{a_{ij}(x)\}$ are $m \times m$ matrices whose elements are complex locally integrable functions defined in an arbitrary domain Ω of $\mathbb{R}^n$ ($1 \leqslant i, j \leqslant m$, $1 \leqslant h, k \leqslant n$).

Let $\mathscr{L}$ be the sesquilinear form related to the operator A,

$$\mathscr{L}(u,v) = -\int_{\Omega} ((\langle \mathscr{A}^{hk}\,\partial_k u, \partial_h v\rangle - \langle \mathscr{B}^h\,\partial_h u, v\rangle + \langle \mathscr{C}^h u, \partial_h v\rangle - \langle \mathscr{A}\,u, v\rangle)) dx.$$

$(\langle\cdot,\cdot\rangle)$ denotes the scalar product in $\mathbb{C}^m$) defined in $(C_0^1(\Omega))^m \times (C_0^1(\Omega))^m$. We consider A as an operator acting from $(C_0^1(\Omega))^m$ to $((C_0^1(\Omega))^*)^m$ through the relation

$$\mathscr{L}(u,v) = \int_{\Omega} \langle Au, v\rangle\,dx$$

for any $u, v \in (C_0^1(\Omega))^m$.

As for scalar operators, the form $\mathscr{L}$ is L^p-dissipative if

$$\mathrm{Re}\,\mathscr{L}(u, |u|^{p-2}u) \leqslant 0 \qquad \text{if } p \geqslant 2, \tag{4.2}$$

$$\mathrm{Re}\,\mathscr{L}(|u|^{p'-2}u, u) \leqslant 0 \qquad \text{if } 1 < p < 2 \tag{4.3}$$

for all $u \in (C_0^1(\Omega))^m$. Unless otherwise stated we assume that the functions are complex vector-valued.

Lemma 4.1. *Let Ω be a domain of $\mathbb{R}^n$. The form $\mathscr{L}$, related to the operator (4.1), is L^p-dissipative if and only if*

$$\int_{\Omega} \Big(\mathrm{Re}\langle \mathscr{A}^{hk}\,\partial_k v, \partial_h v\rangle - (1-2/p)^2|v|^{-4}\,\mathrm{Re}\langle \mathscr{A}^{hk}\,v, v\rangle\,\mathrm{Re}\langle v, \partial_k v\rangle\,\mathrm{Re}\langle v, \partial_h v\rangle$$

$$- (1-2/p)|v|^{-2}\,\mathrm{Re}(\langle \mathscr{A}^{hk}\,v, \partial_h v\rangle\,\mathrm{Re}\langle v, \partial_k v\rangle - \langle \mathscr{A}^{hk}\,\partial_k v, v\rangle\,\mathrm{Re}\langle v, \partial_h v\rangle)$$

$$+ (1-2/p)|v|^{-2}\,\mathrm{Re}\langle \mathscr{B}^h\,v, v\rangle\,\mathrm{Re}\langle v, \partial_h v\rangle - \mathrm{Re}\langle \mathscr{B}^h\,\partial_h v, v\rangle \tag{4.4}$$

$$+ (1-2/p)|v|^{-2}\,\mathrm{Re}\langle \mathscr{C}^h\,v, v\rangle\,\mathrm{Re}\langle v, \partial_h v\rangle + \mathrm{Re}\langle \mathscr{C}^h\,v, \partial_h v\rangle - \mathrm{Re}\langle \mathscr{A}\,v, v\rangle\Big) dx \geqslant 0$$

for any $v \in (C_0^1(\Omega))^m$. Here and in the sequel the integrand is extended by zero on the set where v vanishes.

Proof. Sufficiency. First suppose $p \geqslant 2$. Let $u \in (C_0^1(\Omega))^m$ and set $v = |u|^{(p-2)/2}u$. We have $v \in (C_0^1(\Omega))^m$ and $u = |v|^{(2-p)/p}v$. From the identities

$$\langle \mathscr{A}^{hk}\,\partial_k u, \partial_h(|u|^{p-2}u)\rangle$$

$$= \langle \mathscr{A}^{hk}\,\partial_k v, \partial_h v\rangle - (1-2/p)^2|v|^{-2}\,\mathrm{Re}\langle \mathscr{A}^{hk}\,v, v\rangle\,\partial_k|v|\partial_h|v|$$

$$- (1-2/p)|v|^{-1}\,\mathrm{Re}(\langle \mathscr{A}^{hk}\,v, \partial_h v\rangle\,\partial_k|v| - \langle \mathscr{A}^{hk}\,\partial_k v, v\rangle\partial_h|v|),$$

$$\langle \mathscr{B}^h\,\partial_h u, |u|^{p-2}u\rangle = -(1-2/p)|v|^{-1}\langle \mathscr{B}^h\,v, v\rangle\partial_h|v| + \langle \mathscr{B}^h\,\partial_h v, v\rangle,$$

$$\langle \mathscr{C}^h u, \partial_h(|u|^{p-2}u)\rangle = (1-2/p)|v|^{-1}\langle \mathscr{C}^h\,v, v\rangle\partial_h|v| + \langle \mathscr{C}^h\,v, \partial_h v\rangle,$$

$$\langle \mathscr{A}\,u, |u|^{p-2}u\rangle = \langle \mathscr{A}\,v, v\rangle,$$

$$\partial_k|v| = |v|^{-1}\,\mathrm{Re}\langle v, \partial_k v\rangle,$$

we see that the left-hand side in (4.4) is equal to $\mathscr{L}(u, |u|^{p-2}u)$. Then (4.2) is satisfied for any $u \in (C_0^1(\Omega))^m$.

If $1 < p < 2$, we may write (4.3) as

$$\operatorname{Re} \int_\Omega \langle (\mathscr{A}^{hk})^* \partial_h u, \partial_k(|u|^{p'-2}u)\rangle - \langle (\mathscr{B}^h)^* u, \partial_h(|u|^{p'-2}u)\rangle$$
$$+ \langle (\mathscr{C}^h)^* \partial_h u, |u|^{p'-2}u\rangle - \langle \mathscr{A}^* u, |u|^{p'-2}u\rangle)\, dx \geqslant 0$$

for any $u \in (C_0^1(\Omega))^m$. The first part of the proof shows that this implies

$$\int_\Omega \Big(\operatorname{Re}\langle (\mathscr{A}^{hk})^* \partial_h v, \partial_k v\rangle - (1 - 2/p')^2 |v|^{-4} \operatorname{Re}\langle (\mathscr{A}^{hk})^* v, v\rangle \operatorname{Re}\langle v, \partial_h v\rangle \operatorname{Re}\langle v, \partial_k v\rangle$$
$$- (1 - 2/p')|v|^{-2} \operatorname{Re}(\langle (\mathscr{A}^{hk})^* v, \partial_k v\rangle \operatorname{Re}\langle v, \partial_h v\rangle - \langle (\mathscr{A}^{hk})^* \partial_h v, v\rangle \operatorname{Re}\langle v, \partial_k v\rangle)$$
$$+ (1 - 2/p')|v|^{-2}(- \operatorname{Re}\langle (\mathscr{B}^h)^* v, v\rangle \operatorname{Re}\langle v, \partial_h v\rangle - \operatorname{Re}\langle (\mathscr{B}^h)^* v, \partial_h v\rangle$$
$$- \operatorname{Re}\langle (\mathscr{C}^h)^* v, v\rangle \operatorname{Re}\langle v, \partial_h v\rangle + \operatorname{Re}\langle (\mathscr{C}^h)^* \partial_h v, v\rangle) - \operatorname{Re}\langle \mathscr{A}^* v, v\rangle \Big) dx \geqslant 0 \quad (4.5)$$

for any $v \in (C_0^1(\Omega))^m$. Since $1 - 2/p' = -(1 - 2/p)$, this inequality is exactly (4.4).

Necessity. Let $p \geqslant 2$ and set

$$g_\varepsilon = (|v|^2 + \varepsilon^2)^{1/2}, \quad u_\varepsilon = g_\varepsilon^{2/p-1} v,$$

where $v \in (C_0^1(\Omega))^m$. We have

$$\langle \mathscr{A}^{hk} \partial_k u_\varepsilon, \partial_h(|u_\varepsilon|^{p-2} u_\varepsilon)\rangle$$
$$= |u_\varepsilon|^{p-2} \langle \mathscr{A}^{hk} \partial_k u_\varepsilon, \partial_h u_\varepsilon\rangle + (p-2)|u_\varepsilon|^{p-3} \langle \mathscr{A}^{hk} \partial_k u_\varepsilon, u_\varepsilon\rangle \partial_h |u_\varepsilon|.$$

One checks directly that

$$|u_\varepsilon|^{p-2} \langle \mathscr{A}^{hk} \partial_k u_\varepsilon, \partial_h u_\varepsilon\rangle$$
$$= (1 - 2/p)^2 g_\varepsilon^{-(p+2)} |v|^{p-2} \langle \mathscr{A}^{hk} v, v\rangle \operatorname{Re}\langle v, \partial_k v\rangle \operatorname{Re}\langle v, \partial_h v\rangle$$
$$- (1 - 2/p) g_\varepsilon^{-p} |v|^{p-2} (\langle \mathscr{A}^{hk} v, \partial_h v\rangle \operatorname{Re}\langle v, \partial_k v\rangle + \langle \mathscr{A}^{hk} \partial_k v, v\rangle \operatorname{Re}\langle v, \partial_h v\rangle)$$
$$+ g_\varepsilon^{2-p} |v|^{p-2} \langle \mathscr{A}^{hk} \partial_k v, \partial_h v\rangle,$$

$$|u_\varepsilon|^{p-3} \langle \mathscr{A}^{hk} \partial_k u_\varepsilon, u_\varepsilon\rangle \partial_h |u_\varepsilon|$$
$$= (1 - 2/p)[(1 - 2/p) g_\varepsilon^{-(p+2)} |v|^{p-2}$$
$$- g_\varepsilon^{-p} |v|^{p-4}] \langle \mathscr{A}^{hk} v, v\rangle \operatorname{Re}\langle v, \partial_k v\rangle \operatorname{Re}\langle v, \partial_h v\rangle$$
$$+ [g_\varepsilon^{2-p} |v|^{p-4} - (1 - 2/p) g_\varepsilon^{-p} |v|^{p-2}] \langle \mathscr{A}^{hk} \partial_k v, v\rangle \operatorname{Re}\langle v, \partial_h v\rangle$$

on the set $E = \{x \in \Omega \mid |v(x)| > 0\}$. The inequality $g_\varepsilon^a \leqslant |v|^a$ for $a \leqslant 0$, shows that the right-hand sides are majorized by L^1 functions. Since $g_\varepsilon \to |v|$ pointwise as $\varepsilon \to 0^+$, we find

$$\lim_{\varepsilon \to 0^+} \langle \mathscr{A}^{hk}\, \partial_k u_\varepsilon, \partial_h(|u_\varepsilon|^{p-2}u_\varepsilon)\rangle$$
$$= \langle \mathscr{A}^{hk}\, \partial_k v, \partial_h v\rangle - (1-2/p)^2|v|^{-4}\langle \mathscr{A}^{hk}\, v, v\rangle \operatorname{Re}\langle v, \partial_k v\rangle \operatorname{Re}\langle v, \partial_h v\rangle$$
$$- (1-2/p)|v|^{-2}(\langle \mathscr{A}^{hk}\, v, \partial_h v\rangle \operatorname{Re}\langle v, \partial_k v\rangle - \langle \mathscr{A}^{hk}\, \partial_k v, v\rangle \operatorname{Re}\langle v, \partial_h v\rangle)$$

and dominated convergence gives

$$\lim_{\varepsilon \to 0^+} \int_E \langle \mathscr{A}^{hk}\, \partial_k u_\varepsilon, \partial_h(|u_\varepsilon|^{p-2}u_\varepsilon)\rangle\, dx \qquad (4.6)$$
$$= \int_E [\langle \mathscr{A}^{hk}\, \partial_k v, \partial_h v\rangle - (1-2/p)^2|v|^{-4}\langle \mathscr{A}^{hk}\, v, v\rangle \operatorname{Re}\langle v, \partial_k v\rangle \operatorname{Re}\langle v, \partial_h v\rangle$$
$$- (1-2/p)|v|^{-2}(\langle \mathscr{A}^{hk}\, v, \partial_h v\rangle \operatorname{Re}\langle v, \partial_k v\rangle - \langle \mathscr{A}^{hk}\, \partial_k v, v\rangle \operatorname{Re}\langle v, \partial_h v\rangle)]\, dx.$$

Moreover we have

$$\langle \mathscr{B}^h\, \partial_h u_\varepsilon, |u_\varepsilon|^{p-2}u_\varepsilon\rangle$$
$$= -(1-2/p)g_\varepsilon^{-p}|v|^{p-2}\langle \mathscr{B}^h\, v, v\rangle \operatorname{Re}\langle v, \partial_h v\rangle + g_\varepsilon^{-p+2}|v|^{p-2}\langle \mathscr{B}^h\, \partial_h v, v\rangle,$$

$$\langle \mathscr{C}^h\, u_\varepsilon, \partial_h(|u_\varepsilon|^{p-2}u_\varepsilon)\rangle = g_\varepsilon^{-p}|v|^{p-4}(((1-2/p)(1-p)|v|^2$$
$$+ (p-2)g_\varepsilon^2)\langle \mathscr{C}_h\, v, v\rangle \operatorname{Re}\langle v, \partial_h v\rangle + g_\varepsilon^2|v|^2\langle \mathscr{C}_h\, v, \partial_h v\rangle),$$

$$\langle \mathscr{A}\, u_\varepsilon, |u_\varepsilon|^{p-2}u_\varepsilon\rangle = g_\varepsilon^{-p+2}|v|^{p-2}\langle \mathscr{A}\, v, v\rangle,$$

on the set $E = \{x \in \Omega \mid |v(x)| > 0\}$. As before, the right-hand sides are dominated by L^1 functions and we find

$$\lim_{\varepsilon \to 0^+} \int_E \langle \mathscr{B}^h\, \partial_h u_\varepsilon, |u_\varepsilon|^{p-2}u_\varepsilon\rangle\, dx$$
$$= \int_E (-(1-2/p)|v|^{-2}\langle \mathscr{B}^h\, v, v\rangle \operatorname{Re}\langle v, \partial_h v\rangle + \langle \mathscr{B}^h\, \partial_h v, v\rangle)\, dx,$$

$$\lim_{\varepsilon \to 0^+} \int_E \langle \mathscr{C}^h\, u_\varepsilon, \partial_h(|u_\varepsilon|^{p-2}u_\varepsilon)\rangle\, dx \qquad (4.7)$$
$$= \int_E ((1-2/p)|v|^{-2}\langle \mathscr{C}^h\, v, v\rangle \operatorname{Re}\langle v, \partial_h v\rangle + \langle \mathscr{C}^h\, v, \partial_h v\rangle)\, dx,$$

$$\lim_{\varepsilon \to 0^+} \int_E \langle \mathscr{A}\, u_\varepsilon, |u_\varepsilon|^{p-2}u_\varepsilon\rangle\, dx = \int_E \langle \mathscr{A}\, v, v\rangle\, dx.$$

Formulas (4.6) and (4.7) show that the limit

$$\lim_{\varepsilon \to 0^+} \operatorname{Re} \mathscr{L}(u_\varepsilon, |u_\varepsilon|^{p-2}u_\varepsilon)$$

is equal to the left-hand side in (4.4). The function u_ε being in $(C_0^1(\Omega))^m$, (4.2) implies (4.4).

If $1 < p < 2$, from (4.6) and (4.7) it follows that the limit

$$\lim_{\varepsilon \to 0+} \operatorname{Re} \mathscr{L}(|u_\varepsilon|^{p'-2} u_\varepsilon, u_\varepsilon)$$

coincides with the left-hand side in (4.5). This shows that (4.3) implies (4.5) and the proof is complete.   □

4.2 Necessary condition for the L^p-dissipativity of the system $\partial_h(\mathscr{A}^{hk}(x)\partial_k)$ when $n = 2$

The aim of this section is to give an algebraic necessary condition for the L^p-dissipativity of the form $\mathscr{L}$ related to the operator

$$A = \partial_h(\mathscr{A}^{hk}(x)\partial_k) \tag{4.8}$$

where $\mathscr{A}^{hk}(x) = \{a_{ij}^{hk}(x)\}$ are $m \times m$ matrices whose elements are complex locally integrable functions defined in an arbitrary domain Ω of $\mathbb{R}^2$ ($1 \leqslant i, j \leqslant m$, $1 \leqslant h, k \leqslant 2$).

Theorem 4.2. *Let Ω be a domain of $\mathbb{R}^2$. If the form $\mathscr{L}$, related to operator (4.8), is L^p-dissipative, we have*

$$\begin{aligned}
&\operatorname{Re}\langle(\mathscr{A}^{hk}(x)\xi_h\xi_k)\lambda, \lambda\rangle - (1 - 2/p)^2 \operatorname{Re}\langle(\mathscr{A}^{hk}(x)\xi_h\xi_k)\omega, \omega\rangle(\operatorname{Re}\langle\lambda, \omega\rangle)^2 \\
&- (1 - 2/p) \operatorname{Re}(\langle(\mathscr{A}^{hk}(x)\xi_h\xi_k)\omega, \lambda\rangle - \langle(\mathscr{A}^{hk}(x)\xi_h\xi_k)\lambda, \omega\rangle) \operatorname{Re}\langle\lambda, \omega\rangle \geqslant 0
\end{aligned} \tag{4.9}$$

for almost every $x \in \Omega$ and for any $\xi \in \mathbb{R}^2$, $\lambda, \omega \in \mathbb{C}^m$, $|\omega| = 1$.

Proof. Let us assume that $\mathscr{A}$ is a constant matrix and that $\Omega = \mathbb{R}^2$. Let us fix $\omega \in \mathbb{C}^m$ with $|\omega| = 1$ and take $v(x) = w(x)\,\eta(\log |x|/\log R)$, where

$$w(x) = \mu\omega + \psi(x), \tag{4.10}$$

$\mu, R \in \mathbb{R}^+$, $R > 1$, $\psi \in (C_0^\infty(\mathbb{R}^2))^m$, $\eta \in C^\infty(\mathbb{R})$, $\eta(t) = 1$ if $t \leqslant 1/2$ and $\eta(t) = 0$ if $t \geqslant 1$.

We have

$$\begin{aligned}
\langle\mathscr{A}^{hk}\,\partial_k v, \partial_h v\rangle =\ & \langle\mathscr{A}^{hk}\,\partial_k w, \partial_h w\rangle \eta^2(\log |x|/\log R) \\
&+ (\log R)^{-1}(\langle\mathscr{A}^{hk}\,\partial_k w, w\rangle x_h + \langle\mathscr{A}^{hk}\,w, \partial_h w\rangle x_k) \\
&\times |x|^{-2}\eta(\log |x|/\log R)\,\eta'(\log |x|/\log R) \\
&+ (\log R)^{-2}\langle\mathscr{A}^{hk}\,w, w\rangle x_h x_k |x|^{-4}\,(\eta'(\log |x|/\log R))^2
\end{aligned}$$

and then, choosing δ such that spt $\psi \subset B_\delta(0)$,

$$\int_{\mathbb{R}^2} \langle \mathscr{A}^{hk} \partial_k v, \partial_h v \rangle dx = \int_{B_\delta(0)} \langle \mathscr{A}^{hk} \partial_k w, \partial_h w \rangle dx$$

$$+ \frac{1}{\log^2 R} \int_{B_R(0) \setminus B_{\sqrt{R}}(0)} \langle \mathscr{A}^{hk} w, w \rangle \frac{x_h x_k}{|x|^4} (\eta'(\log |x| / \log R))^2 \, dx$$

provided that $R > \delta^2$. Since

$$\lim_{R \to +\infty} \frac{1}{\log^2 R} \int_{B_R(0) \setminus B_{\sqrt{R}}(0)} \frac{dx}{|x|^2} = 0,$$

we have

$$\lim_{R \to +\infty} \int_{\mathbb{R}^2} \langle \mathscr{A}^{hk} \partial_k v, \partial_h v \rangle dx = \int_{B_\delta(0)} \langle \mathscr{A}^{hk} \partial_k w, \partial_h w \rangle dx.$$

On the set where $v \neq 0$ we have

$$|v|^{-4} \langle \mathscr{A}^{hk} v, v \rangle \operatorname{Re}\langle v, \partial_k v \rangle \operatorname{Re}\langle v, \partial_h v \rangle$$

$$= |w|^{-4} \langle \mathscr{A}^{hk} w, w \rangle \operatorname{Re}\langle w, \partial_k w \rangle \operatorname{Re}\langle w, \partial_h w \rangle \eta^2 (\log |x| / \log R)$$

$$+ (\log R)^{-1} |w|^{-2} \langle \mathscr{A}^{hk} w, w \rangle (\operatorname{Re}\langle w, \partial_h w \rangle x_k + \operatorname{Re}\langle w, \partial_k w \rangle x_h) |x|^{-2}$$

$$\times \eta(\log |x| / \log R) \, \eta'(\log |x| / \log R)$$

$$+ (\log R)^{-2} \langle \mathscr{A}^{hk} w, w \rangle x_h x_k |x|^{-4} (\eta'(\log |x| / \log R))^2$$

and then

$$\lim_{R \to +\infty} \int_{\mathbb{R}^2} |v|^{-4} \langle \mathscr{A}^{hk} v, v \rangle \operatorname{Re}\langle v, \partial_k v \rangle \operatorname{Re}\langle v, \partial_h v \rangle dx$$

$$= \int_{B_\delta(0)} |w|^{-4} \langle \mathscr{A}^{hk} w, w \rangle \operatorname{Re}\langle w, \partial_k w \rangle \operatorname{Re}\langle w, \partial_h w \rangle dx.$$

In the same way we obtain

$$\lim_{R \to +\infty} \int_{\mathbb{R}^2} |v|^{-2} \operatorname{Re}(\langle \mathscr{A}^{hk} v, \partial_h v \rangle \operatorname{Re}\langle v, \partial_k v \rangle - \langle \mathscr{A}^{hk} \partial_k v, v \rangle \operatorname{Re}\langle v, \partial_h v \rangle) dx$$

$$= \int_{B_\delta(0)} |w|^{-2} \operatorname{Re}(\langle \mathscr{A}^{hk} w, \partial_h w \rangle \operatorname{Re}\langle w, \partial_k w \rangle - \langle \mathscr{A}^{hk} \partial_k w, w \rangle \operatorname{Re}\langle w, \partial_h w \rangle) dx.$$

In view of Lemma 4.1, (4.4) holds. Putting v in this formula and letting $R \to +\infty$, we find

$$\int_{B_\delta(0)} \left(\operatorname{Re}\langle \mathscr{A}^{hk} \partial_k w, \partial_h w \rangle \right.$$

$$- (1 - 2/p)^2 |w|^{-4} \operatorname{Re}\langle \mathscr{A}^{hk} w, w \rangle \operatorname{Re}\langle w, \partial_k w \rangle \operatorname{Re}\langle w, \partial_h w \rangle$$

$$- (1 - 2/p) |w|^{-2} \operatorname{Re}(\langle \mathscr{A}^{hk} w, \partial_h w \rangle \operatorname{Re}\langle w, \partial_k w \rangle$$

$$\left. - \langle \mathscr{A}^{hk} \partial_k w, w \rangle \operatorname{Re}\langle w, \partial_h w \rangle) \right) dx \geqslant 0.$$

$$(4.11)$$

On the other hand, keeping in mind (4.10),

$$\mathrm{Re}\langle \mathscr{A}^{hk}\partial_k w, \partial_h w\rangle = \mathrm{Re}\langle \mathscr{A}^{hk}\partial_k \psi, \partial_h \psi\rangle,$$

$$|w|^{-4}\,\mathrm{Re}\langle \mathscr{A}^{hk}w, w\rangle\,\mathrm{Re}\langle w, \partial_k w\rangle\,\mathrm{Re}\langle w, \partial_h w\rangle$$
$$= |\mu\omega + \psi|^{-4}\,\mathrm{Re}\langle \mathscr{A}^{hk}(\mu\omega + \psi), \mu\omega + \psi\rangle\,\mathrm{Re}\langle \mu\omega + \psi, \partial_k \psi\rangle\,\mathrm{Re}\langle \mu\omega + \psi, \partial_h \psi\rangle,$$

$$|w|^{-2}\,\mathrm{Re}(\langle \mathscr{A}^{hk}w, \partial_h w\rangle\,\mathrm{Re}\langle w, \partial_k w\rangle - \langle \mathscr{A}^{hk}\partial_k w, w\rangle\,\mathrm{Re}\langle w, \partial_h w\rangle)$$
$$= |\mu\omega + \psi|^{-2}\,\mathrm{Re}(\langle \mathscr{A}^{hk}(\mu\omega + \psi), \partial_h \psi\rangle\,\mathrm{Re}\langle \mu\omega + \psi, \partial_k \psi\rangle$$
$$- \langle \mathscr{A}^{hk}\partial_k(\mu\omega + \psi), \mu\omega + \psi\rangle\,\mathrm{Re}\langle \mu\omega + \psi, \partial_h \psi\rangle).$$

Letting $\mu \to +\infty$ in (4.11), we obtain

$$\int_{\mathbb{R}^2}\Big(\mathrm{Re}\langle \mathscr{A}^{hk}\partial_k \psi, \partial_h \psi\rangle - (1 - 2/p)^2\,\mathrm{Re}\langle \mathscr{A}^{hk}w, w\rangle\,\mathrm{Re}\langle w, \partial_k \psi\rangle\,\mathrm{Re}\langle w, \partial_h \psi\rangle \quad (4.12)$$
$$- (1 - 2/p)\,\mathrm{Re}(\langle \mathscr{A}^{hk}w, \partial_h \psi\rangle\,\mathrm{Re}\langle w, \partial_k \psi\rangle - \langle \mathscr{A}^{hk}\partial_k \psi, w\rangle\,\mathrm{Re}\langle w, \partial_h \psi\rangle)\Big)\,dx \geqslant 0.$$

Putting in (4.12)

$$\psi(x) = \lambda\,\varphi(x)\,e^{i\mu\langle \xi, x\rangle}$$

where $\lambda \in \mathbb{C}^m$, $\varphi \in C_0^\infty(\mathbb{R}^2)$ and μ is a real parameter, by standard arguments (see, e.g., Fichera [29, pp. 107–108]), we find (4.9).

If the matrix $\mathscr{A}$ is not constant, take $\psi \in (C_0^1(\mathbb{R}^2))^m$ and define

$$v(x) = \psi((x - x_0)/\varepsilon)$$

where x_0 is a fixed point in Ω and $0 < \varepsilon < \mathrm{dist}\,(x_0, \partial\Omega)$.

Putting this particular v in (4.4) and making a change of variables, we obtain

$$\int_{\mathbb{R}^2}\Big(\mathrm{Re}\langle \mathscr{A}^{hk}(x_0 + \varepsilon y)\partial_k \psi, \partial_h \psi\rangle$$
$$- (1 - 2/p)^2|\psi|^{-4}\,\mathrm{Re}\langle \mathscr{A}^{hk}(x_0 + \varepsilon y)\psi, \psi\rangle\,\mathrm{Re}\langle \psi, \partial_k \psi\rangle\,\mathrm{Re}\langle \psi, \partial_h \psi\rangle$$
$$- (1 - 2/p)|\psi|^{-2}\,\mathrm{Re}(\langle \mathscr{A}^{hk}(x_0 + \varepsilon y)\psi, \partial_h \psi\rangle\,\mathrm{Re}\langle \psi, \partial_k \psi\rangle$$
$$- \langle \mathscr{A}^{hk}(x_0 + \varepsilon y)\partial_k \psi, \psi\rangle\,\mathrm{Re}\langle \psi, \partial_h \psi\rangle)\Big)\,dy \geqslant 0.$$

Letting $\varepsilon \to 0^+$ we find

$$\int_{\mathbb{R}^2}\Big(\mathrm{Re}\langle \mathscr{A}^{hk}(x_0)\partial_k \psi, \partial_h \psi\rangle$$
$$- (1 - 2/p)^2|\psi|^{-4}\,\mathrm{Re}\langle \mathscr{A}^{hk}(x_0)\psi, \psi\rangle\,\mathrm{Re}\langle \psi, \partial_k \psi\rangle\,\mathrm{Re}\langle \psi, \partial_h \psi\rangle$$
$$- (1 - 2/p)|\psi|^{-2}\,\mathrm{Re}(\langle \mathscr{A}^{hk}(x_0)\psi, \partial_h \psi\rangle\,\mathrm{Re}\langle \psi, \partial_k \psi\rangle$$
$$- \langle \mathscr{A}^{hk}(x_0)\partial_k \psi, \psi\rangle\,\mathrm{Re}\langle \psi, \partial_h \psi\rangle)\Big)\,dy \geqslant 0$$

for almost every $x_0 \in \Omega$. The arbitrariness of $\psi \in (C_0^1(\mathbb{R}^2))^m$ and what we have proved for constant matrices give the result. $\qquad\square$

4.3 L^p-dissipativity for systems of ordinary differential equations

In order to study L^p-dissipativity for a certain class of systems of partial differential operators, we need some results concerning systems of ordinary differential operators. This is the topic of this section.

Here we consider the operator A defined by

$$Au = (\mathscr{A}(x)u')' \tag{4.13}$$

where $\mathscr{A}(x) = \{a_{ij}(x)\}$ $(i, j = 1, \ldots, m)$ is a matrix with complex locally integrable entries defined in the bounded or unbounded interval (a, b).

In this case the sesquilinear form $\mathscr{L}(u, v)$ is given by

$$\mathscr{L}(u, v) = \int_a^b \langle \mathscr{A} u', v' \rangle \, dx. \tag{4.14}$$

4.3.1   Necessary and sufficient conditions for L^p-dissipativity

Lemma 4.3. *The form $\mathscr{L}$ is L^p-dissipative if and only if*

$$\int_a^b \Big(\operatorname{Re}\langle \mathscr{A} v', v' \rangle - (1 - 2/p)^2 |v|^{-4} \operatorname{Re}\langle \mathscr{A} v, v \rangle (\operatorname{Re}\langle v, v' \rangle)^2 \tag{4.15}$$
$$- (1 - 2/p)|v|^{-2} \operatorname{Re}(\langle \mathscr{A} v, v' \rangle - \langle \mathscr{A} v', v \rangle) \operatorname{Re}\langle v, v' \rangle \Big) \, dx \geqslant 0$$

for any $v \in (C_0^1((a, b)))^m$.

Proof. It is a particular case of Lemma 4.1. $\qquad\square$

Theorem 4.4. *The form $\mathscr{L}$ is L^p-dissipative if and only if*

$$\operatorname{Re}\langle \mathscr{A}(x)\lambda, \lambda \rangle - (1 - 2/p)^2 \operatorname{Re}\langle \mathscr{A}(x)\omega, \omega \rangle (\operatorname{Re}\langle \lambda, \omega \rangle)^2 \tag{4.16}$$
$$- (1 - 2/p) \operatorname{Re}(\langle \mathscr{A}(x)\omega, \lambda \rangle - \langle \mathscr{A}(x)\lambda, \omega \rangle) \operatorname{Re}\langle \lambda, \omega \rangle \geqslant 0$$

for almost every $x \in (a, b)$ and for any $\lambda, \omega \in \mathbb{C}^m$, $|\omega| = 1$.

Proof. Necessity. First we prove the result assuming that the coefficients a_{ij} are constant and that $(a, b) = \mathbb{R}$.

Let us fix λ and ω in $\mathbb{C}^m$, with $|\omega| = 1$, and choose $v(x) = \eta(x/R) w(x)$ where

$$w_j(x) = \begin{cases} \mu\omega_j & \text{if } x < 0 \\ \mu\omega_j + x^2(3 - 2x)\lambda_j & \text{if } 0 \leqslant x \leqslant 1 \\ \mu\omega_j + \lambda_j & \text{if } x > 1, \end{cases}$$

$\mu, R \in \mathbb{R}^+$, $\eta \in C_0^\infty(\mathbb{R})$, $\operatorname{spt} \eta \subset [-1, 1]$ and $\eta(x) = 1$ if $|x| \leqslant 1/2$.

We have

$$\langle \mathscr{A} v', v' \rangle = \langle \mathscr{A} w', w' \rangle (\eta(x/R))^2 + R^{-1}(\langle \mathscr{A} w', w \rangle + \langle \mathscr{A} w, w' \rangle)\eta(x/R)\eta'(x/R)$$
$$+ R^{-2}\langle \mathscr{A} w, w \rangle (\eta'(x/R))^2$$

and then

$$\int_{\mathbb{R}} \langle \mathscr{A} v', v' \rangle \, dx = \int_0^1 \langle \mathscr{A} w', w' \rangle \, dx + \frac{1}{R^2} \int_{-R}^R \langle \mathscr{A} w, w \rangle (\eta'(x/R))^2 dx$$

provided that $R > 2$. Since $\langle \mathscr{A} w, w \rangle$ is bounded, we have

$$\lim_{R \to +\infty} \int_{\mathbb{R}} \langle \mathscr{A} v', v' \rangle \, dx = \int_0^1 \langle \mathscr{A} w', w' \rangle \, dx.$$

On the set where $v \neq 0$ we have

$$|v|^{-4}\langle \mathscr{A} v, v \rangle (\operatorname{Re}\langle v, v' \rangle)^2$$
$$= |w|^{-4}\langle \mathscr{A} w, w \rangle (\operatorname{Re}\langle w, w' \rangle)^2 (\eta(x/R))^2$$
$$+ 2\, R^{-1}|w|^{-2}\langle \mathscr{A} w, w \rangle \operatorname{Re}\langle w, w' \rangle \eta(x/R))\, \eta'(x/R) + R^{-2}\langle \mathscr{A} w, w \rangle (\eta'(x/R))^2$$

from which it follows that

$$\lim_{R \to +\infty} \int_{\mathbb{R}} |v|^{-4}\langle \mathscr{A} v, v \rangle (\operatorname{Re}\langle v, v' \rangle)^2 dx = \int_0^1 |w|^{-4}\langle \mathscr{A} w, w \rangle (\operatorname{Re}\langle w, w' \rangle)^2 dx.$$

In the same way we obtain

$$\lim_{R \to +\infty} \int_{\mathbb{R}} |v|^{-2}(\langle \mathscr{A} v, v' \rangle - \langle \mathscr{A} v', v \rangle) \operatorname{Re}\langle v, v' \rangle \, dx$$
$$= \int_0^1 |w|^{-2}(\langle \mathscr{A} w, w' \rangle - \langle \mathscr{A} w', w \rangle) \operatorname{Re}\langle w, w' \rangle \, dx.$$

Since $v \in (C_0^1(\mathbb{R}))^m$, we can put v in (4.15). Letting $R \to +\infty$, we find

$$\int_0^1 \Big(\operatorname{Re}\langle \mathscr{A} w', w' \rangle - (1 - 2/p)^2 |w|^{-4} \operatorname{Re}\langle \mathscr{A} w, w \rangle (\operatorname{Re}\langle w, w' \rangle)^2$$
$$- (1 - 2/p)|w|^{-2} \operatorname{Re}(\langle \mathscr{A} w, w' \rangle - \langle \mathscr{A} w', w \rangle) \operatorname{Re}\langle w, w' \rangle \Big) \, dx \geqslant 0. \tag{4.17}$$

On the interval $(0, 1)$ we have

$$\langle \mathscr{A} w', w' \rangle = \langle \mathscr{A} \lambda, \lambda \rangle\, 36x^2(1 - x)^2,$$

$$|w|^{-4}\langle \mathscr{A} w, w \rangle (\operatorname{Re}\langle w, w' \rangle)^2 = |\mu\omega + x^2(3 - 2x)\lambda|^{-4}$$
$$\times (\mu^2\langle \mathscr{A} \omega, \omega \rangle + \mu(\langle \mathscr{A} \omega, \lambda \rangle + \langle \mathscr{A} \lambda, \omega \rangle)x^2(3 - 2x) + \langle \mathscr{A} \lambda, \lambda \rangle x^4(3 - 2x)^2)$$
$$\times [\operatorname{Re}(\mu\langle \omega, \lambda \rangle\, 6x(1 - x) + |\lambda|^2 6x^3(3 - 2x)(1 - x))]^2,$$

$$|w|^{-2}(\langle \mathscr{A}\, w, w'\rangle - \langle \mathscr{A}\, w', w\rangle)\operatorname{Re}\langle w, w'\rangle$$
$$= |\mu\omega + x^2(3 - 2x)\lambda|^{-2}\mu(\langle \mathscr{A}\, \omega, \lambda\rangle - \langle \mathscr{A}\, \lambda, \omega\rangle)\, 6x(1 - x)$$
$$\times \operatorname{Re}(\mu\langle \omega, \lambda\rangle\, 6x(1 - x) + |\lambda|^2 6x^3(3 - 2x)(1 - x)).$$

Letting $\mu \to \infty$ in (4.17) we find

$$36\int_0^1 \Big(\operatorname{Re}\langle \mathscr{A}\, \lambda, \lambda\rangle - (1 - 2/p)^2 \operatorname{Re}\langle \mathscr{A}\, \omega, \omega\rangle(\operatorname{Re}\langle \omega, \lambda\rangle)^2$$
$$- (1 - 2/p)\operatorname{Re}(\langle \mathscr{A}\, \omega, \lambda\rangle - \langle \mathscr{A}\, \lambda, \omega\rangle)\operatorname{Re}\langle \omega, \lambda\rangle \Big) x^2(1 - x)^2 dx \geqslant 0$$

and (4.16) is proved.

If a_{hk} are not necessarily constant, consider

$$v(x) = \varepsilon^{-1/2}\psi((x - x_0)/\varepsilon)$$

where x_0 is a fixed point in (a, b), $\psi \in (C_0^1(\mathbb{R}))^m$ and ε is sufficiently small.

In this case (4.15) shows that

$$\int_{\mathbb{R}} \Big(\operatorname{Re}\langle \mathscr{A}(x_0 + \varepsilon y)\psi', \psi'\rangle - (1 - 2/p)^2|\psi|^{-4}\operatorname{Re}\langle \mathscr{A}(x_0 + \varepsilon y)\psi, \psi\rangle(\operatorname{Re}\langle \psi, \psi'\rangle)^2$$
$$- (1 - 2/p)|\psi|^{-2}\operatorname{Re}(\langle \mathscr{A}(x_0 + \varepsilon y)\psi, \psi'\rangle - \langle \mathscr{A}(x_0 + \varepsilon y)\psi', \psi\rangle)\operatorname{Re}\langle \psi, \psi'\rangle \Big) dy \geqslant 0.$$

Letting $\varepsilon \to 0^+$ we find for almost every x_0

$$\int_{\mathbb{R}} \Big(\operatorname{Re}\langle \mathscr{A}(x_0)\psi', \psi'\rangle - (1 - 2/p)^2|\psi|^{-4}\operatorname{Re}\langle \mathscr{A}(x_0)\psi, \psi\rangle(\operatorname{Re}\langle \psi, \psi'\rangle)^2$$
$$- (1 - 2/p)|\psi|^{-2}\operatorname{Re}(\langle \mathscr{A}(x_0)\psi, \psi'\rangle - \langle \mathscr{A}(x_0)\psi', \psi\rangle)\operatorname{Re}\langle \psi, \psi'\rangle \Big) dy \geqslant 0.$$

Because this inequality holds for any $\psi \in C_0^1(\mathbb{R})$, what we have obtained for constant coefficients gives the result.

Sufficiency. It is clear that, if (4.16) holds, then the integrand in (4.15) is nonnegative almost everywhere and Lemma 4.3 gives the result.   □

Corollary 4.5. *If the form $\mathscr{L}$ is L^p-dissipative, then*

$$\operatorname{Re}\langle \mathscr{A}(x)\lambda, \lambda\rangle \geqslant 0$$

for almost every $x \in (a, b)$ and for any $\lambda \in \mathbb{C}^m$.

Proof. Fix $x \in (a, b)$ such that (4.16) holds for any $\lambda, \omega \in \mathbb{C}^m$, $|\omega| = 1$. For any $\lambda \in \mathbb{C}^m$, choose ω such that $\langle \lambda, \omega\rangle = 0$, $|\omega| = 1$. The result follows by putting ω in (4.16).   □

If the operator A has smooth coefficients, we can give necessary and sufficient conditions for the L^p-dissipativity of operator A. We consider A as an operator defined on $W^{2,p}((a,b)) \cap \overset{\circ}{W}^{1,p}((a,b))$.

Theorem 4.6. *Let (a,b) be a bounded interval. Let us suppose $a_{ij} \in C^1([a,b])$ and*

$$\mathrm{Re}\langle \mathscr{A}(x)\lambda, \lambda \rangle > 0$$

for any $x \in [a,b]$ and for any $\lambda \in \mathbb{C}^m \setminus \{0\}$. The operator A is L^p-dissipative if and only if (4.16) holds for any $x \in [a,b]$ and for any $\lambda, \omega \in \mathbb{C}^m$, $|\omega| = 1$.

Proof. The result follows from Theorems 2.20 and 4.4. □

4.3.2 Necessary and sufficient conditions in terms of eigenvalues for real coefficient operators

We start with a lemma about the maximum of a particular quartic form on the unit sphere.

Lemma 4.7. *Let $0 < \mu_1 \leqslant \mu_2 \leqslant \cdots \leqslant \mu_m$. We have*

$$\max_{\substack{\omega \in \mathbb{R}^m \\ |\omega|=1}} [(\mu_h \omega_h^2)(\mu_k^{-1} \omega_k^2)] = \frac{(\mu_1 + \mu_m)^2}{4\,\mu_1 \mu_m}. \tag{4.18}$$

Proof. First we prove by induction on m that

$$\max_{\substack{\omega \in \mathbb{R}^m \\ |\omega|=1}} [(\mu_h \omega_h^2)(\mu_k^{-1} \omega_k^2)] = \max_{1 \leqslant i < j \leqslant m} \frac{(\mu_i + \mu_j)^2}{4\,\mu_i \mu_j}. \tag{4.19}$$

In the case $m = 2$, (4.19) is equivalent to

$$\max_{\varphi \in [0,2\pi]} [\cos^4 \varphi + \sin^4 \varphi + (\mu_1 \mu_2^{-1} + \mu_2 \mu_1^{-1}) \cos^2 \varphi \sin^2 \varphi] = \frac{(\mu_1 + \mu_2)^2}{4\,\mu_1 \mu_2},$$

which can be easily proved.

Let $m > 2$ and suppose $\mu_1 < \mu_2 < \cdots < \mu_m$; the maximum of the left-hand side of (4.19) is the maximum of the function

$$\mu_h \mu_k^{-1} x_h x_k$$

subject to the constraint $x \in K$, where $K = \{x \in \mathbb{R}^m \mid x_1 + \cdots + x_m = 1,\ 0 \leqslant x_j \leqslant 1\ (j = 1, \ldots, m)\}$. To find the constrained maximum, we first examine the system

$$\begin{cases} \gamma_{hk} x_k - \lambda = 0 & h = 1, \ldots, m \\ x_1 + \cdots + x_m = 1 \end{cases} \tag{4.20}$$

with $0 \leqslant x_j \leqslant 1$ $(j = 1, \ldots, m)$, where λ is the Lagrange multiplier and $\gamma_{hk} = \mu_h \mu_k^{-1} + \mu_k \mu_h^{-1}$.

Consider the homogeneous system

$$\gamma_{hk} x_k = 0 \quad (h = 1, \ldots, m). \tag{4.21}$$

One checks directly that the vectors $x^{(k)} = (x_1^{(k)}, \ldots, x_m^{(k)})$,

$$x_1^{(k)} = \frac{\mu_1}{\mu_k} \frac{\mu_k^2 - \mu_2^2}{\mu_2^2 - \mu_1^2}, \quad x_2^{(k)} = \frac{\mu_2}{\mu_k} \frac{\mu_1^2 - \mu_k^2}{\mu_2^2 - \mu_1^2}, \quad x_j^{(k)} = \delta_{jk} \ (j = 3, \ldots, m)$$

for $k = 3, \ldots, m$, are $m - 2$ linearly independent eigensolutions of the system (4.21). On the other hand, the determinant

$$\begin{vmatrix} \gamma_{11} & \gamma_{12} \\ \gamma_{12} & \gamma_{22} \end{vmatrix} = 4 - \gamma_{12}^2 = -\frac{(\mu_1^2 - \mu_2^2)^2}{\mu_1^2 \mu_2^2} < 0$$

and then the rank of the matrix $\{\gamma_{hk}\}$ is 2.

Therefore there exists a solution of the system

$$\gamma_{hk} x_k = \lambda \quad (h = 1, \ldots, m) \tag{4.22}$$

if and only if the vector $(\lambda, \ldots, \lambda)$ is orthogonal to any eigensolution of the adjoint homogeneous system. Since the matrix $\{\gamma_{hk}\}$ is symmetric, there exists a solution of the system (4.22) if and only if

$$\lambda(x_1^{(k)} + \cdots + x_m^{(k)}) = 0 \tag{4.23}$$

for $k = 3, \ldots, m$.

But

$$x_1^{(k)} + \cdots + x_m^{(k)} = -\frac{\mu_1 \mu_2 + \mu_k^2}{\mu_k(\mu_1 + \mu_2)} + 1 = -\frac{(\mu_k - \mu_1)(\mu_k - \mu_2)}{\mu_k(\mu_1 + \mu_2)} < 0$$

and (4.23) are satisfied if and only if $\lambda = 0$. This means that the system (4.22) is solvable only when $\lambda = 0$ and the solutions are given by

$$x = \sum_{k=3}^{m} u_k x^{(k)}$$

for arbitrary $u_k \in \mathbb{R}$. On the other hand we are looking for solutions of (4.20) with $0 \leqslant x_j \leqslant 1$. Since $x_j = u_j$ for $j = 3, \ldots, m$, we have $u_j \geqslant 0$. This implies that

$$x_2 = \sum_{k=3}^{m} \frac{\mu_2}{\mu_k} \frac{\mu_1^2 - \mu_k^2}{\mu_2^2 - \mu_1^2} u_k \leqslant 0$$

and since we require $x_2 \geqslant 0$, we have $u_k = 0$ $(k = 3, \ldots, m)$, i.e., $x = 0$. This solution does not satisfy the last equation in (4.20). This means that there are no extreme points belonging to the interior of K. The maximum is therefore attained on the boundary of K, where at least one of the x_j's is zero. This shows that if (4.19) is true for $m - 1$, then it is true also for m.

We have proved (4.19) assuming $0 < \mu_1 < \cdots < \mu_m$; in case $\mu_i = \mu_j$ for some i, j, the induction hypothesis immediately implies (4.19).

Finally, let us show that

$$\frac{(\mu_i + \mu_j)^2}{4\,\mu_i\mu_j} \leqslant \frac{(\mu_1 + \mu_m)^2}{4\,\mu_1\mu_m} \tag{4.24}$$

for any $1 \leqslant i, j \leqslant m$. Set $\mu_j = \alpha_j\mu_m$ and suppose $i \leqslant j$. We have $0 < \alpha_1 \leqslant \cdots \leqslant \alpha_m = 1$. Inequality (4.24) is equivalent to

$$\alpha_1(\alpha_i + \alpha_j)^2 \leqslant \alpha_i\alpha_j(\alpha_1 + 1)^2,$$

i.e.,

$$\alpha_1\alpha_i(\alpha_i - \alpha_j) + (\alpha_1\alpha_j - \alpha_i)\alpha_j \leqslant 0$$

and this is true, because $\alpha_i \leqslant \alpha_j$ and $\alpha_1\alpha_j \leqslant \alpha_1 \leqslant \alpha_i$. $\qquad\square$

We can now characterize the L^p-dissipative of certain ordinary differential operators in terms of eigenvalues of the corresponding coefficient matrix. We start considering the associated form $\mathscr{L}$.

Theorem 4.8. *Let $\mathscr{A}$ be a real matrix $\{a_{hk}\}$ with $a_{hk} \in L^1_{\text{loc}}((a, b))$, $h, k = 1, \ldots, m$. Let us suppose $\mathscr{A} = \mathscr{A}^t$ and $\mathscr{A} \geqslant 0$ (in the sense $\langle \mathscr{A}(x)\xi, \xi \rangle \geqslant 0$, for almost every $x \in (a, b)$ and for any $\xi \in \mathbb{R}^m$). The form $\mathscr{L}$ is L^p-dissipative if and only if*

$$\left(\frac{1}{2} - \frac{1}{p}\right)^2 (\mu_1(x) + \mu_m(x))^2 \leqslant \mu_1(x)\mu_m(x)$$

almost everywhere, where $\mu_1(x)$ and $\mu_m(x)$ are the smallest and the largest eigenvalues of the matrix $\mathscr{A}(x)$ respectively. In the particular case $m = 2$ this condition is equivalent to

$$\left(\frac{1}{2} - \frac{1}{p}\right)^2 (\operatorname{tr}\mathscr{A}(x))^2 \leqslant \det\mathscr{A}(x)$$

almost everywhere.

Proof. From Theorem 4.4 $\mathscr{L}$ is L^p-dissipative if and only if (4.16) holds for almost every $x \in (a, b)$ and for any $\lambda, \omega \in \mathbb{C}^m$, $|\omega| = 1$. We claim that in the present case this condition is equivalent to

$$\langle \mathscr{A}(x)\xi, \xi \rangle - (1 - 2/p)^2 \langle \mathscr{A}(x)\omega, \omega \rangle (\langle \xi, \omega \rangle)^2 \geqslant 0 \tag{4.25}$$

for almost every $x \in (a, b)$ and for any $\xi, \omega \in \mathbb{R}^m$, $|\omega| = 1$. Indeed, it is obvious that if

$$\langle \mathscr{A}(x)\lambda, \lambda \rangle - (1 - 2/p)^2 \langle \mathscr{A}(x)\omega, \omega \rangle (\mathrm{Re}\langle \lambda, \omega \rangle)^2 \geqslant 0$$

for almost every $x \in (a, b)$ and for any $\lambda, \omega \in \mathbb{C}^m$, $|\omega| = 1$, then (4.25) holds for almost every $x \in (a, b)$ and for any $\xi, \omega \in \mathbb{R}^m$, $|\omega| = 1$. Conversely, fix $x \in (a, b)$ and suppose that (4.25) holds for any $\xi, \omega \in \mathbb{R}^m$, $|\omega| = 1$. Let Q be an orthogonal matrix such that $\mathscr{A}(x) = Q^t D Q$, D being a diagonal matrix. If we denote by μ_j the eigenvalues of $\mathscr{A}(x)$, we have

$$\begin{aligned}
\langle \mathscr{A}(x)\lambda, \lambda \rangle &- (1 - 2/p)^2 \langle \mathscr{A}(x)\omega, \omega \rangle (\mathrm{Re}\langle \lambda, \omega \rangle)^2 \\
&= \langle DQ\lambda, Q\lambda \rangle - (1 - 2/p)^2 \langle DQ\omega, Q\omega \rangle (\mathrm{Re}\langle Q\lambda, Q\omega \rangle)^2 \\
&= \mu_j |(Q\lambda)_j|^2 - (1 - 2/p)^2 (\mu_j |(Q\omega)_j|^2)(\mathrm{Re}\langle Q\lambda, Q\omega \rangle)^2 \\
&\geqslant \mu_j |(Q\lambda)_j|^2 - (1 - 2/p)^2 (\mu_j |(Q\omega)_j|^2)(|(Q\lambda)_k|\,|(Q\omega)_k|)^2.
\end{aligned}$$

The last expression is nonnegative because of (4.25) and the equivalence is proved.

Let us fix $x \in (a, b)$. We may write (4.25) as

$$(1 - 2/p)^2 (\mu_h \omega_h^2)(\xi_k \omega_k)^2 \leqslant \mu_j \xi_j^2 \tag{4.26}$$

for any $\xi, \omega \in \mathbb{R}^m$, $|\omega| = 1$. Let us fix $\omega \in \mathbb{R}^m$, $|\omega| = 1$; inequality (4.26) is true if and only if

$$(1 - 2/p)^2 (\mu_h \omega_h^2) \sup_{\substack{\xi \in \mathbb{R}^n \\ \xi \neq 0}} \frac{(\xi_k \omega_k)^2}{\mu_j \xi_j^2} \leqslant 1.$$

We have

$$\max_{\substack{\xi \in \mathbb{R}^n \\ \xi \neq 0}} \frac{(\xi_k \omega_k)^2}{\mu_j \xi_j^2} = \mu_k^{-1} \omega_k^2;$$

in fact, by Cauchy's inequality, we have $(\xi_k \omega_k)^2 \leqslant (\mu_j \xi_j^2)(\mu_k^{-1} \omega_k^2)$ for any $\xi \in \mathbb{R}^m$ and there is equality if $\xi_j = \mu_j^{-1} \omega_j$.

Therefore (4.26) is satisfied if and only if

$$(1 - 2/p)^2 (\mu_h \omega_h^2)(\mu_k^{-1} \omega_k^2) \leqslant 1$$

for any $\omega \in \mathbb{R}^m$, $|\omega| = 1$, and (4.18) shows that this is true if and only if

$$\left(\frac{1}{2} - \frac{1}{p} \right)^2 \frac{(\mu_1 + \mu_m)^2}{\mu_1 \mu_m} \leqslant 1.$$

The result for $m = 2$ follows from the identities

$$\mu_1(x)\mu_2(x) = \det \mathscr{A}(x), \quad \mu_1(x) + \mu_2(x) = \mathrm{tr}\, \mathscr{A}(x). \tag{4.27}$$

$\square$

Theorem 4.9. *Let (a, b) be a bounded interval and let $\mathscr{A}$ be a real matrix $\{a_{hk}\}$, with $a_{hk} \in C^1([a, b])$, $h, k = 1, \ldots, m$. Let us suppose $\mathscr{A} = \mathscr{A}^t$ and $\mathscr{A} > 0$ (in the sense $\langle \mathscr{A}(x)\xi, \xi \rangle > 0$, for every $x \in [a, b]$ and for any $\xi \in \mathbb{R}^m \setminus \{0\}$). The operator A is L^p-dissipative if and only if*

$$\left(\frac{1}{2} - \frac{1}{p}\right)^2 (\mu_1(x) + \mu_m(x))^2 \leqslant \mu_1(x)\mu_m(x)$$

for any $x \in [a, b]$, where $\mu_1(x)$ and $\mu_m(x)$ are the smallest and the largest eigenvalues of the matrix $\mathscr{A}(x)$ respectively. In the particular case $m = 2$ this condition is equivalent to

$$\left(\frac{1}{2} - \frac{1}{p}\right)^2 (\operatorname{tr} \mathscr{A}(x))^2 \leqslant \det \mathscr{A}(x)$$

for any $x \in [a, b]$.

Proof. This follows from Theorems 2.20 and 4.8. $\qquad\square$

4.3.3 Comparison between A and $I(d^2/dx^2)$

Thanks to characterizations of L^p-dissipativity proved in Subsection 4.3.1, we can now compare the operators A and $I(d^2/dx^2)$ from the point of view of the L^p-dissipativity.

We start considering the question for the relevant forms. Let $\mathscr{L}_0$ be the sesquilinear form associated to $I(d^2/dx^2)$, i.e.,

$$\mathscr{L}_0(u, v) = \int_a^b u' \, \overline{v}' \, dx .$$

As before, $\mathscr{L}$ denotes the sesquilinear form (4.14) related to A.

The next two results are the analogues for systems of ordinary differential equations of the ones obtained in Section 3.2 for Elasticity.

Corollary 4.10. *There exists $k > 0$ such that $\mathscr{L} - k\mathscr{L}_0$ is L^p-dissipative if and only if*

$$\operatorname{ess\,inf}_{\substack{(x, \lambda, \omega) \in (a, b) \times \mathbb{C}^m \times \mathbb{C}^m \\ |\lambda| = |\omega| = 1}} P(x, \lambda, \omega) > 0 \tag{4.28}$$

where

$$P(x, \lambda, \omega) = \operatorname{Re}\langle \mathscr{A}(x)\lambda, \lambda \rangle - (1 - 2/p)^2 \operatorname{Re}\langle \mathscr{A}(x)\omega, \omega \rangle (\operatorname{Re}\langle \lambda, \omega \rangle)^2 \\ - (1 - 2/p) \operatorname{Re}(\langle \mathscr{A}(x)\omega, \lambda \rangle - \langle \mathscr{A}(x)\lambda, \omega \rangle) \operatorname{Re}\langle \lambda, \omega \rangle. \tag{4.29}$$

There exists $k > 0$ such that $k\mathscr{L}_0 - \mathscr{L}$ is L^p-dissipative if and only if

$$\operatorname{ess\,sup}_{\substack{(x, \lambda, \omega) \in (a, b) \times \mathbb{C}^m \times \mathbb{C}^m \\ |\lambda| = |\omega| = 1}} P(x, \lambda, \omega) < \infty. \tag{4.30}$$

Proof. In view of Theorem 4.4, $\mathscr{L} - k\mathscr{L}_0$ is L^p-dissipative if and only if

$$P(x, \lambda, \omega) - k|\lambda|^2 + k(1 - 2/p)^2(\mathrm{Re}\langle\lambda, \omega\rangle)^2 \geqslant 0$$

for almost every $x \in (a, b)$ and for any $\lambda, \omega \in \mathbb{C}^m$, $|\omega| = 1$. Since

$$|\lambda|^2 - (1 - 2/p)^2(\mathrm{Re}\langle\lambda, \omega\rangle)^2 \geqslant \frac{4}{p\,p'}|\lambda|^2, \tag{4.31}$$

we can find a positive k such that this is true if and only if

$$\operatorname{ess\,inf}_{\substack{(x, \lambda, \omega) \in (a, b) \times \mathbb{C}^m \times \mathbb{C}^m \\ |\omega| = 1}} \frac{P(x, \lambda, \omega)}{|\lambda|^2 - (1 - 2/p)^2(\mathrm{Re}\langle\lambda, \omega\rangle)^2} > 0. \tag{4.32}$$

On the other hand, inequality (4.31) shows that

$$\frac{P(x, \lambda, \omega)}{|\lambda|^2} \leqslant \frac{P(x, \lambda, \omega)}{|\lambda|^2 - (1 - 2/p)^2(\mathrm{Re}\langle\lambda, \omega\rangle)^2} \leqslant \frac{p\,p'}{4} \frac{P(x, \lambda, \omega)}{|\lambda|^2} \tag{4.33}$$

and then (4.32) and (4.28) are equivalent. In the same way the operator $k\mathscr{L}_0 - \mathscr{L}$ is L^p-dissipative if and only if

$$-P(x, \lambda, \omega) + k|\lambda|^2 - k(1 - 2/p)^2(\mathrm{Re}\langle\lambda, \omega\rangle)^2 \geqslant 0$$

for almost every $x \in (a, b)$ and for any $\lambda, \omega \in \mathbb{C}^m$, $|\omega| = 1$. We can find a positive k such that this is true if and only if

$$\operatorname{ess\,sup}_{\substack{(x, \lambda, \omega) \in (a, b) \times \mathbb{C}^m \times \mathbb{C}^m \\ |\omega| = 1}} \frac{P(x, \lambda, \omega)}{|\lambda|^2 - (1 - 2/p)^2(\mathrm{Re}\langle\lambda, \omega\rangle)^2} < \infty.$$

This inequality is equivalent to (4.30) because of (4.33). $\qquad\square$

Corollary 4.11. *There exists $k \in \mathbb{R}$ such that $\mathscr{L} - k\mathscr{L}_0$ is L^p-dissipative if and only if*

$$\operatorname{ess\,inf}_{\substack{(x, \lambda, \omega) \in (a, b) \times \mathbb{C}^m \times \mathbb{C}^m \\ |\lambda| = |\omega| = 1}} P(x, \lambda, \omega) > -\infty. \tag{4.34}$$

Proof. As in Corollary 4.10, there exists a real k such that $\mathscr{L} - k\mathscr{L}_0$ is L^p-dissipative if and only if

$$\operatorname{ess\,inf}_{\substack{(x, \lambda, \omega) \in (a, b) \times \mathbb{C}^m \times \mathbb{C}^m \\ |\omega| = 1}} \frac{P(x, \lambda, \omega)}{|\lambda|^2 - (1 - 2/p)^2(\mathrm{Re}\langle\lambda, \omega\rangle)^2} > -\infty. \tag{4.35}$$

Conditions (4.34) and (4.35) are equivalent in view of (4.33). $\qquad\square$

If the coefficients of operator A are real, we can give several comparison results in terms of the eigenvalues of the coefficient matrix of A.

Corollary 4.12. *Let $\mathscr{A}$ be a real and symmetric matrix. Denote by $\mu_1(x)$ and $\mu_m(x)$ the smallest and the largest eigenvalues of $\mathscr{A}(x)$ respectively. There exists $k > 0$ such that $\mathscr{L} - k\,\mathscr{L}_0$ is L^p-dissipative if and only if*

$$\operatorname{ess\,inf}_{x \in (a,b)} \left[(1 + \sqrt{p\,p'}/2)\,\mu_1(x) + (1 - \sqrt{p\,p'}/2)\,\mu_m(x) \right] > 0. \tag{4.36}$$

In the particular case $m = 2$ conditions (4.36) is equivalent to

$$\operatorname{ess\,inf}_{x \in (a,b)} \left[\operatorname{tr}\mathscr{A}(x) - \frac{\sqrt{p\,p'}}{2}\sqrt{(\operatorname{tr}\mathscr{A}(x))^2 - 4\det\mathscr{A}(x)} \right] > 0. \tag{4.37}$$

Proof. Necessity. Corollary 4.5 shows that $\mathscr{A}(x) - kI \geqslant 0$ almost everywhere. In view of Theorem 4.8, we have that $\mathscr{L} - k\,\mathscr{L}_0$ is L^p-dissipative if and only if

$$\left(\frac{1}{p} - \frac{1}{2}\right)^2 (\mu_1(x) + \mu_m(x) - 2k)^2 \leqslant (\mu_1(x) - k)\,(\mu_m(x) - k) \tag{4.38}$$

almost everywhere.

Inequality (4.38) is

$$\frac{4}{p\,p'}(\mu_1(x) + \mu_m(x) - 2k)^2 - (\mu_1(x) - \mu_m(x))^2 \geqslant 0. \tag{4.39}$$

By Corollary 4.10, $\mathscr{L} - k'\,\mathscr{L}_0$ is L^p-dissipative for any $k' \leqslant k$. Therefore inequality (4.39) holds if we replace k by any $k' < k$. This implies that k is less than or equal to the smallest root of the left-hand side of (4.39), i.e.,

$$k \leqslant \frac{1}{2}\left[(1 + \sqrt{p\,p'}/2)\,\mu_1(x) + (1 - \sqrt{p\,p'}/2)\,\mu_m(x)\right] \tag{4.40}$$

and (4.36) is proved.

Sufficiency. Let k be such that

$$0 < k \leqslant \operatorname{ess\,inf}_{x \in (a,b)} \frac{1}{2}\left[(1 + \sqrt{p\,p'}/2)\,\mu_1(x) + (1 - \sqrt{p\,p'}/2)\,\mu_m(x)\right].$$

Since $\mu_1(x) \leqslant \mu_m(x)$ and $\sqrt{p\,p'}/2 \geqslant 1$, we have

$$(1 + \sqrt{p\,p'}/2)\,\mu_1(x) + (1 - \sqrt{p\,p'}/2)\,\mu_m(x) \leqslant 2\,\mu_1(x) \tag{4.41}$$

and then $\mathscr{A}(x) - kI \geqslant 0$ almost everywhere. The constant k satisfies (4.40) and this implies (4.39), i.e., (4.38). Theorem 4.8 gives the result.

The equivalence between (4.36) and (4.37) follows from the identities (4.27).

$\square$

If we require something more about the matrix $\mathscr{A}$ we have also

Corollary 4.13. *Let $\mathscr{A}$ be a real and symmetric matrix. Suppose $\mathscr{A} \geqslant 0$ almost everywhere. Denote by $\mu_1(x)$ and $\mu_m(x)$ the smallest and the largest eigenvalues of $\mathscr{A}(x)$ respectively. If there exists $k > 0$ such that $\mathscr{L} - k\mathscr{L}_0$ is L^p-dissipative, then*

$$\operatorname{ess\,inf}_{x \in (a,b)} \left[\mu_1(x)\mu_m(x) - \left(\frac{1}{2} - \frac{1}{p} \right)^2 (\mu_1(x) + \mu_m(x))^2 \right] > 0. \qquad (4.42)$$

If, in addition, there exists C such that

$$\langle \mathscr{A}(x)\xi, \xi \rangle \leqslant C|\xi|^2 \qquad (4.43)$$

for almost every $x \in (a, b)$ and for any $\xi \in \mathbb{R}^m$, the converse is also true. In the particular case $m = 2$ condition (4.42) is equivalent to

$$\operatorname{ess\,inf}_{x \in (a,b)} \left[\det \mathscr{A}(x) - \left(\frac{1}{2} - \frac{1}{p} \right)^2 (\operatorname{tr} \mathscr{A}(x))^2 \right] > 0.$$

Proof. Necessity. By Corollary 4.12, (4.40) holds. On the other hand we have

$$\left[(1 + \sqrt{p\,p'}/2)\,\mu_1(x) + (1 - \sqrt{p\,p'}/2)\,\mu_m(x) \right]$$
$$\leqslant \left[(1 - \sqrt{p\,p'}/2)\,\mu_1(x) + (1 + \sqrt{p\,p'}/2)\,\mu_m(x) \right]$$

and then

$$4k^2 \leqslant \left[(1 + \sqrt{p\,p'}/2)\,\mu_1(x) + (1 - \sqrt{p\,p'}/2)\,\mu_m(x) \right]$$
$$\times \left[(1 - \sqrt{p\,p'}/2)\,\mu_1(x) + (1 + \sqrt{p\,p'}/2)\,\mu_m(x) \right].$$

This inequality can be written as

$$\frac{4k^2}{p\,p'} \leqslant \mu_1(x)\mu_m(x) - \left(\frac{1}{2} - \frac{1}{p} \right)^2 (\mu_1(x) + \mu_m(x))^2$$

and (4.42) is proved.

Sufficiency. There exists $h > 0$ such that

$$h \leqslant \mu_1(x)\mu_m(x) - \left(\frac{1}{2} - \frac{1}{p} \right)^2 (\mu_1(x) + \mu_m(x))^2$$

almost everywhere, i.e.,

$$p\,p'h \leqslant \left[(1 + \sqrt{p\,p'}/2)\,\mu_1(x) + (1 - \sqrt{p\,p'}/2)\,\mu_m(x) \right]$$
$$\times \left[(1 - \sqrt{p\,p'}/2)\,\mu_1(x) + (1 + \sqrt{p\,p'}/2)\,\mu_m(x) \right]$$

almost everywhere. Since $\mu_1(x) \geqslant 0$, we have also

$$(1 - \sqrt{pp'}/2)\,\mu_1(x) + (1 + \sqrt{pp'}/2)\,\mu_m(x) \leqslant (1 + \sqrt{pp'}/2)\,\mu_m(x) \qquad (4.44)$$

and then

$$(1 + \sqrt{pp'}/2)^{-1}pp'h$$
$$\leqslant \left[(1 + \sqrt{pp'}/2)\,\mu_1(x) + (1 - \sqrt{pp'}/2)\,\mu_m(x)\right] \operatorname{ess\,sup}_{y \in (a,b)} \mu_m(y)$$

almost everywhere. By (4.43) ess sup μ_m is finite and by (4.42) it is greater than zero. Then (4.36) holds and Corollary 4.12 gives the result.   □

Remark 4.14. Generally speaking, assumption (4.43) cannot be omitted, even if $\mathscr{A} \geqslant 0$. Consider, e.g., $(a,b) = (1,\infty)$, $m = 2$, $\mathscr{A}(x) = \{a_{ij}(x)\}$ where $a_{11}(x) = (1 - 2/\sqrt{pp'})x + x^{-1}$, $a_{12}(x) = a_{21}(x) = 0$, $a_{22}(x) = (1 + 2/\sqrt{pp'})x + x^{-1}$. We have

$$\mu_1(x)\mu_2(x) - \left(\frac{1}{2} - \frac{1}{p}\right)^2 (\mu_1(x) + \mu_2(x))^2 = (8 + 4x^{-2})/(pp')$$

and (4.42) holds. But (4.36) is not satisfied, because

$$(1 + \sqrt{pp'}/2)\,\mu_1(x) + (1 - \sqrt{pp'}/2)\,\mu_2(x) = 2x^{-1}.$$

Corollary 4.15. *Let $\mathscr{A}$ be a real and symmetric matrix. Denote by $\mu_1(x)$ and $\mu_m(x)$ the smallest and the largest eigenvalues of $\mathscr{A}(x)$ respectively. There exists $k > 0$ such that $k\mathscr{L}_0 - \mathscr{L}$ is L^p-dissipative if and only if*

$$\operatorname{ess\,sup}_{x \in (a,b)} \left[(1 - \sqrt{pp'}/2)\,\mu_1(x) + (1 + \sqrt{pp'}/2)\,\mu_m(x)\right] < \infty. \qquad (4.45)$$

In the particular case $m = 2$ condition (4.45) is equivalent to

$$\operatorname{ess\,sup}_{x \in (a,b)} \left[\operatorname{tr}\mathscr{A}(x) + \frac{\sqrt{pp'}}{2}\sqrt{(\operatorname{tr}\mathscr{A}(x))^2 - 4\det\mathscr{A}(x)}\right] < \infty.$$

Proof. The proof runs as in Corollary 4.12. We have that $k\mathscr{L}_0 - \mathscr{L}$ is L^p-dissipative if and only if (4.38) holds, provided that

$$kI - \mathscr{A}(x) \geqslant 0$$

almost everywhere. Because of this inequality, we have to replace (4.40) and (4.41) by

$$k \geqslant \frac{1}{2}\left[(1 - \sqrt{pp'}/2)\,\mu_1(x) + (1 + \sqrt{pp'}/2)\,\mu_m(x)\right]$$

and

$$(1 - \sqrt{pp'}/2)\,\mu_1(x) + (1 + \sqrt{pp'}/2)\,\mu_m(x) \geqslant 2\,\mu_m(x) \qquad (4.46)$$

respectively.   □

In the case of a positive matrix $\mathscr{A}$, we have

Corollary 4.16. *Let $\mathscr{A}$ be a real and symmetric matrix. Suppose $\mathscr{A} \geqslant 0$ almost everywhere. Denote by $\mu_1(x)$ and $\mu_m(x)$ the smallest and the largest eigenvalues of $\mathscr{A}(x)$ respectively. There exists $k > 0$ such that $k \mathscr{L}_0 - \mathscr{L}$ is L^p-dissipative if and only if*

$$\operatorname{ess\,sup}_{x \in (a,b)} \mu_m(x) < \infty. \tag{4.47}$$

Proof. The equivalence between (4.45) and (4.47) follows from (4.44) and (4.46).
$\square$

We have also

Corollary 4.17. *Let $\mathscr{A}$ be a real and symmetric matrix. Denote by $\mu_1(x)$ and $\mu_m(x)$ the smallest and the largest eigenvalues of $\mathscr{A}(x)$ respectively. There exists $k \in \mathbb{R}$ such that $\mathscr{L} - k \mathscr{L}_0$ is L^p-dissipative if and only if*

$$\operatorname{ess\,inf}_{x \in (a,b)} \left[(1 + \sqrt{p\,p'}/2)\, \mu_1(x) + (1 - \sqrt{p\,p'}/2)\, \mu_m(x) \right] > -\infty.$$

In the particular case $m = 2$ this condition is equivalent to

$$\operatorname{ess\,inf}_{x \in (a,b)} \left[\operatorname{tr} \mathscr{A}(x) - \frac{\sqrt{p\,p'}}{2} \sqrt{(\operatorname{tr} \mathscr{A}(x))^2 - 4 \det \mathscr{A}(x)} \right] > -\infty.$$

Proof. The proof is similar to that of Corollary 4.12.
$\square$

We now can compare the operators A and $I(d^2/dx^2)$.

Corollary 4.18. *Let (a,b) a bounded interval and let us suppose $a_{ij} \in C^1([a,b])$. There exists $k > 0$ such that $A - kI(d^2/dx^2)$ is L^p-dissipative if and only if*

$$\min_{\substack{(x,\lambda,\omega) \in [a,b] \times \mathbb{C}^m \times \mathbb{C}^m \\ |\lambda| = |\omega| = 1}} P(x, \lambda, \omega) > 0$$

where $P(x, \lambda, \omega)$ is given by (4.29). Moreover there always exists $k > 0$ such that $kI(d^2/dx^2) - A$ is L^p-dissipative.

Proof. It follows from Theorem 2.20 and Corollary 4.10.
$\square$

Other results of this nature can be obtained combining the corollaries of this subsection with Theorem 2.20. We shall not insist on that.

4.4 L^p-dissipativity for a class of systems of partial differential equations

In this section we consider the particular class of operators

$$Au = \partial_h(\mathscr{A}^h(x)\partial_h u) \tag{4.48}$$

where $\mathscr{A}^h(x) = \{a_{ij}^h(x)\}$ $(i,j = 1,\ldots,m)$ are matrices with complex locally integrable entries defined in a domain $\Omega \subset \mathbb{R}^n$ $(h = 1,\ldots,n)$.

Our goal is to prove that A is L^p-dissipative if and only if the algebraic condition

$$\text{Re}\langle \mathscr{A}^h(x)\lambda, \lambda \rangle - (1 - 2/p)^2 \, \text{Re}\langle \mathscr{A}^h(x)\omega, \omega \rangle (\text{Re}\langle \lambda, \omega \rangle)^2$$
$$- (1 - 2/p) \, \text{Re}(\langle \mathscr{A}^h(x)\omega, \lambda \rangle - \langle \mathscr{A}^h(x)\lambda, \omega \rangle) \, \text{Re}\langle \lambda, \omega \rangle \geqslant 0$$

is satisfied for every $x \in \Omega$ and for every $\lambda, \omega \in \mathbb{C}^m$, $|\omega| = 1$, $h = 1, \ldots, n$.

Here y_h denotes the $(n-1)$-dimensional vector $(x_1, \ldots, x_{h-1}, x_{h+1}, \ldots, x_n)$ and we set $\omega(y_h) = \{x_h \in \mathbb{R} \mid x \in \Omega\}$.

Lemma 4.19. *The operator* (4.48) *is L^p-dissipative if and only if the ordinary differential operators*

$$A(y_h)[u(x_h)] = d(\mathscr{A}^h(x)du/dx_h)/dx_h$$

are L^p-dissipative in $\omega(y_h)$ for every $y_h \in \mathbb{R}^{n-1}$ $(h = 1, \ldots, n)$. This condition is void if $\omega(y_h) = \emptyset$.

Proof. Sufficiency. Suppose $p \geqslant 2$. If $u \in (C_0^1(\Omega))^m$ we may write

$$\text{Re} \sum_{h=1}^{n} \int_{\Omega} \langle \mathscr{A}^h(x)\partial_h u, \partial_h(|u|^{p-2}u) \rangle dx$$
$$= \text{Re} \sum_{h=1}^{n} \int_{\mathbb{R}^{n-1}} dy_h \int_{\omega(y_h)} \langle \mathscr{A}^h(x)\partial_h u, \partial_h(|u|^{p-2}u) \rangle dx_h.$$

By assumption

$$\text{Re} \int_{\omega(y_h)} \langle \mathscr{A}^h(x)v'(x_h), (|v(x_h)|^{p-2}v(x_h))' \rangle dx_h \geqslant 0$$

for every $y_h \in \mathbb{R}^{n-1}$ and for any $v \in (C_0^1(\omega(y_h)))^m$, provided $\omega(y_h) \neq \emptyset$ $(h = 1, \ldots, n)$. This implies

$$\text{Re} \sum_{h=1}^{n} \int_{\Omega} \langle \mathscr{A}^h(x)\partial_h u, \partial_h(|u|^{p-2}u) \rangle dx \geqslant 0.$$

The proof for $1 < p < 2$ runs in the same way. We have just to use (4.3) instead of (4.2).

Necessity. Assume first that $\mathscr{A}^h$ are constant matrices and $\Omega = \mathbb{R}^n$. Let $p \geqslant 2$ and fix $1 \leqslant k \leqslant n$.

Take $\alpha \in (C_0^1(\mathbb{R}))^m$ and $\beta \in C_0^1(\mathbb{R}^{n-1})$. Consider

$$u_\varepsilon(x) = \alpha(x_k/\varepsilon)\,\beta(y_k)$$

We have

$$
\sum_{h=1}^{n} \int_{\mathbb{R}^n} \langle \mathscr{A}^h \, \partial_h u_\varepsilon, \partial_h(|u_\varepsilon|^{p-2}u_\varepsilon)\rangle dx
$$

$$
= \varepsilon^{-2} \int_{\mathbb{R}^{n-1}} |\beta(y_k)|^p dy_k \int_{\mathbb{R}} \langle \mathscr{A}^k \, \alpha'(x_k/\varepsilon), \gamma'(x_k/\varepsilon)\rangle \, dx_k
$$

$$
+ \sum_{\substack{h=1 \\ h \neq k}}^{n} \int_{\mathbb{R}^{n-1}} \partial_h \beta(y_k) \, \partial_h(|\beta(y_k)|^{p-2}\beta(y_k)) \, dy_k
$$

$$
\times \int_{\mathbb{R}} \langle \mathscr{A}^h \, \alpha(x_k/\varepsilon), \alpha(x_k/\varepsilon)\rangle \, |\alpha(x_k/\varepsilon)|^{p-2}dx_k
$$

$$
= \varepsilon^{-1} \int_{\mathbb{R}^{n-1}} |\beta(y_k)|^p dy_k \int_{\mathbb{R}} \langle \mathscr{A}^k \, \alpha'(t), (|\alpha(t)|^{p-2}\alpha(t))'\rangle \, dt
$$

$$
+ \varepsilon \sum_{\substack{h=1 \\ h \neq k}}^{n} \int_{\mathbb{R}^{n-1}} \partial_h \beta(y_k) \, \partial_h(|\beta(y_k)|^{p-2}\beta(y_k)) \, dy_k \int_{\mathbb{R}} \langle \mathscr{A}^h \, \alpha(t), \alpha(t)\rangle \, |\alpha(t)|^{p-2}dt
$$

where $\gamma(t) = |\alpha(t)|^{p-2}\alpha(t)$. Keeping in mind (4.2) and letting $\varepsilon \to 0^+$, we find

$$
\mathrm{Re} \int_{\mathbb{R}^{n-1}} |\beta(y_k)|^p dy_k \int_{\mathbb{R}} \langle \mathscr{A}^k \, \alpha'(t), (|\alpha(t)|^{p-2}\alpha(t))'\rangle \, dt \geqslant 0
$$

and then

$$
\mathrm{Re} \int_{\mathbb{R}} \langle \mathscr{A}^k \, \alpha'(t), (|\alpha(t)|^{p-2}\alpha(t))'\rangle \, dt \geqslant 0
$$

for any $\alpha \in C_0^1(\mathbb{R})$. This shows that $A(y_k)$ is L^p-dissipative.

If $\mathscr{A}^h$ are not necessarily constant, consider

$$
v(x) = \varepsilon^{(2-n)/2}\psi((x - x_0)/\varepsilon)
$$

where $x_0 \in \Omega$, $\psi \in (C_0^1(\mathbb{R}^n))^m$ and ε is sufficiently small.

In view of Lemma 4.1 we write

$$
\int_{\Omega} \Big(\mathrm{Re}\langle \mathscr{A}^h \, \partial_h v, \partial_h v\rangle - (1 - 2/p)^2 |v|^{-4} \, \mathrm{Re}\langle \mathscr{A}^h \, v, v\rangle (\mathrm{Re}\langle v, \partial_h v\rangle)^2
$$

$$
- (1 - 2/p)|v|^{-2} \, \mathrm{Re}(\langle \mathscr{A}^h \, v, \partial_h v\rangle - \langle \mathscr{A}^h \, \partial_h v, v\rangle) \, \mathrm{Re}\langle v, \partial_h v\rangle \Big)dx \geqslant 0,
$$

i.e.,

$$
\int_{\mathbb{R}^n} \Big(\mathrm{Re}\langle \mathscr{A}^h(x_0 + \varepsilon z)\partial_h \psi, \partial_h \psi\rangle
$$

$$
- (1 - 2/p)^2 |\psi|^{-4} \, \mathrm{Re}\langle \mathscr{A}^h(x_0 + \varepsilon z)\psi, \psi\rangle (\mathrm{Re}\langle \psi, \partial_h \psi\rangle)^2
$$

$$
- (1 - 2/p)|\psi|^{-2} \, \mathrm{Re}(\langle \mathscr{A}^h(x_0 + \varepsilon z)\psi, \partial_h \psi\rangle
$$

$$
- \langle \mathscr{A}^h(x_0 + \varepsilon z)\partial_h \psi, \psi\rangle) \, \mathrm{Re}\langle \psi, \partial_h \psi\rangle \Big)dz \geqslant 0.
$$

Letting $\varepsilon \to 0^+$, we obtain

$$\int_{\mathbb{R}^n} \Big(\operatorname{Re}\langle \mathscr{A}^h(x_0)\partial_h\psi, \partial_h\psi\rangle - (1-2/p)^2 |\psi|^{-4} \operatorname{Re}\langle \mathscr{A}^h(x_0)\psi, \psi\rangle (\operatorname{Re}\langle \psi, \partial_h\psi\rangle)^2$$
$$- (1-2/p)|\psi|^{-2} \operatorname{Re}(\langle \mathscr{A}^h(x_0)\psi, \partial_h\psi\rangle - \langle \mathscr{A}^h(x_0)\partial_h\psi, \psi\rangle) \operatorname{Re}\langle \psi, \partial_h\psi\rangle \Big) dy \geq 0$$

for every $x_0 \in \Omega$. Because of the arbitrariness of $\psi \in (C_0^1(\mathbb{R}^n))^m$, Lemma 4.1 shows that the constant coefficient operator $\partial_h(\mathscr{A}^h(x_0)\partial_h)$ is L^p-dissipative. From what has already been proved, the ordinary differential operators $(\mathscr{A}^h(x_0)v')'$ are L^p-dissipative $(h = 1, \ldots, n)$. Theorem 4.4 yields

$$\operatorname{Re}\langle \mathscr{A}^h(x_0)\lambda, \lambda\rangle - (1-2/p)^2 \operatorname{Re}\langle \mathscr{A}^h(x_0)\omega, \omega\rangle (\operatorname{Re}\langle \lambda, \omega\rangle)^2$$
$$- (1-2/p) \operatorname{Re}(\langle \mathscr{A}^h(x_0)\omega, \lambda\rangle - \langle \mathscr{A}^h(x_0)\lambda, \omega\rangle) \operatorname{Re}\langle \lambda, \omega\rangle \geq 0 \qquad (4.49)$$

for any $\lambda, \omega \in \mathbb{C}^m$, $|\omega| = 1$, $h = 1, \ldots, n$. Fix h and denote by N the set of $x_0 \in \Omega$ such that (4.49) does not hold for any $\lambda, \omega \in \mathbb{C}^m$, $|\omega| = 1$. Since N has zero measure, for every $y_h \in \mathbb{R}^{n-1}$, the cross-sections $\{x_h \in \mathbb{R} \mid x \in N\}$ are measurable and have zero measure. Hence, for almost every $y_h \in \mathbb{R}^{n-1}$, we have

$$\operatorname{Re}\langle \mathscr{A}^h(x)\lambda, \lambda\rangle - (1-2/p)^2 \operatorname{Re}\langle \mathscr{A}^h(x)\omega, \omega\rangle (\operatorname{Re}\langle \lambda, \omega\rangle)^2$$
$$- (1-2/p) \operatorname{Re}(\langle \mathscr{A}^h(x)\omega, \lambda\rangle - \langle \mathscr{A}^h(x)\lambda, \omega\rangle) \operatorname{Re}\langle \lambda, \omega\rangle \geq 0$$

for almost every $x_h \in \omega(y_h)$ and for any $\lambda, \omega \in \mathbb{C}^m$, $|\omega| = 1$, provided $\omega(y_h) \neq \emptyset$. The conclusion follows from Theorem 4.4.

In the same manner we obtain the result for $1 < p < 2$. $\qquad \square$

Theorem 4.20. *The operator* (4.48) *is L^p-dissipative if and only if* (4.49) *holds for almost every $x_0 \in \Omega$ and for any $\lambda, \omega \in \mathbb{C}^m$, $|\omega| = 1$, $h = 1, \ldots, n$.*

Proof. Necessity. This has been already proved in the necessity part of the proof of Lemma 4.19.

Sufficiency. We have seen that if (4.49) holds for almost every $x_0 \in \Omega$ and for any $\lambda, \omega \in \mathbb{C}^m$, $|\omega| = 1$, the ordinary differential operator $A(y_h)$ is L^p-dissipative for almost every $y_h \in \mathbb{R}^{n-1}$, provided $\omega(y_h) \neq \emptyset$ $(h = 1, \ldots, n)$. By Lemma 4.19, A is L^p-dissipative. $\qquad \square$

Remark 4.21. In the scalar case $(m = 1)$, operator (4.48) falls into the operators considered in Section 2.2. In fact, if $Au = \sum_{h=1}^n \partial_h(a^h \partial_h u)$, a^h being a scalar function, A can be written in the form $\nabla(\mathscr{A}\nabla u)$ with $\mathscr{A} = \{c_{hk}\}$, $c_{hh} = a^h$, $c_{hk} = 0$ if $h \neq k$. The conditions obtained there can be directly compared with (4.49). The results of Section 2.2 show that operator A is L^p-dissipative if and only if (5.2) holds. This means that

$$\frac{4}{pp'} \langle \operatorname{Re} \mathscr{A}\, \xi, \xi\rangle + \langle \operatorname{Re} \mathscr{A}\, \eta, \eta\rangle - 2(1-2/p)\langle \operatorname{Im} \mathscr{A}\, \xi, \eta\rangle \geq 0 \qquad (4.50)$$

almost everywhere and for any $\xi, \eta \in \mathbb{R}^n$ (see Remark 2.8). In this particular case (4.50) is clearly equivalent to the following n conditions

$$\frac{4}{pp'}(\operatorname{Re}a^h)\,\xi^2 + (\operatorname{Re}a^h)\,\eta^2 - 2(1-2/p)(\operatorname{Im}a^h)\,\xi\eta \geqslant 0 \qquad (4.51)$$

almost everywhere and for any $\xi, \eta \in \mathbb{R}$, $h = 1, \ldots, n$. On the other hand, in this case, (4.49) reads as

$$\begin{aligned}(\operatorname{Re}a^h)|\lambda|^2 - (1-2/p)^2(\operatorname{Re}a^h)(\operatorname{Re}(\lambda\overline{\omega})^2 \\ -2(1-2/p)(\operatorname{Im}a^h)\operatorname{Re}(\lambda\overline{\omega})\operatorname{Im}(\lambda\overline{\omega}) \geqslant 0\end{aligned} \qquad (4.52)$$

almost everywhere and for any $\lambda, \omega \in \mathbb{C}$, $|\omega| = 1$, $h = 1, \ldots, n$. Setting $\xi + i\eta = \lambda\overline{\omega}$ and observing that $|\lambda|^2 = |\lambda\overline{\omega}|^2 = (\operatorname{Re}(\lambda\overline{\omega}))^2 + (\operatorname{Im}(\lambda\overline{\omega}))^2$, we see that conditions (4.51) (and then (4.50)) are equivalent to (4.52).

In the case of a real coefficient operator (4.48), we have also

Theorem 4.22. *Let A be the operator (4.48), where $\mathscr{A}^h$ are real matrices $\{a_{ij}^h\}$ with $i, j = 1, \ldots, m$. Let us suppose $\mathscr{A}^h = (\mathscr{A}^h)^t$ and $\mathscr{A}^h \geqslant 0$ ($h = 1, \ldots, n$). The operator A is L^p-dissipative if and only if*

$$\left(\frac{1}{2} - \frac{1}{p}\right)^2 (\mu_1^h(x) + \mu_m^h(x))^2 \leqslant \mu_1^h(x)\,\mu_m^h(x) \qquad (4.53)$$

for almost every $x \in \Omega$, $h = 1, \ldots, n$, where $\mu_1^h(x)$ and $\mu_m^h(x)$ are the smallest and the largest eigenvalues of the matrix $\mathscr{A}^h(x)$ respectively. In the particular case $m = 2$ this condition is equivalent to

$$\left(\frac{1}{2} - \frac{1}{p}\right)^2 (\operatorname{tr}\mathscr{A}^h(x))^2 \leqslant \det\mathscr{A}^h(x)$$

for almost every $x \in \Omega$, $h = 1, \ldots, n$.

Proof. By Theorem 4.20, A is L^p-dissipative if and only if

$$\langle\mathscr{A}^h(x)\lambda, \lambda\rangle - (1-2/p)^2\langle\mathscr{A}^h(x)\omega, \omega\rangle(\operatorname{Re}\langle\lambda, \omega\rangle)^2 \geqslant 0$$

for almost every $x \in \Omega$, for any $\lambda, \omega \in \mathbb{C}^m$, $|\omega| = 1$, $h = 1, \ldots, n$. The proof of Theorem 4.8 shows that these conditions are equivalent to (4.53). $\qquad \square$

4.5   Weakly coupled systems

4.5.1   Preliminary results

In this section we consider an elliptic operator

$$A_p u = \partial_h(a_{hk}\partial_k u) + a_h\partial_h u + \mathscr{A}\,u,$$

where the functions a_{hk}, a_h in $C^1(\overline{\Omega})$ are real-valued and the $n \times n$-matrix $\mathscr{A}$ has $C(\overline{\Omega})$-functions as entries. Without loss of generality, the matrix $\{a_{hk}\}$ is assumed to be pointwise symmetric. We will also use the operator

$$Au = \frac{4}{pp'} \partial_h(a_{hk} \partial_k u) + \frac{1}{2p}(p(\mathscr{A} + \mathscr{A}^*) - 2\partial_h a_h I)u,$$

We consider the operators A_p and A defined in $D(A_p) = (W^{2,p}(\Omega) \cap W_0^{1,p}(\Omega))^N$ and $D(A) = (H^2(\Omega) \cap H_0^1(\Omega))^N$ respectively.

A first result is just a consequence of Lemma 4.1.

Lemma 4.23. *Let Ω be a domain of $\mathbb{R}^n$. The form $\mathscr{L}$, related to the operator A_p, is L^p-dissipative if and only if*

$$\mathrm{Re} \int_{\Omega} \left(a_{ij} \langle \partial_i v, \partial_j v \rangle + \langle (p^{-1}\partial_i a_i I - \mathscr{A})v, v \rangle \right) dx$$

$$- \frac{(p-2)^2}{p^2} \int_{\Omega} a_{ij} \mathrm{Re}\langle \partial_i v, v \rangle \, \mathrm{Re}\langle \partial_j v, v \rangle |v|^{-2} \, dx \geqslant 0 \tag{4.54}$$

for any $v \in (C_0^1(\Omega))^m$.

Proof. Taking $\mathscr{A}^{hk} = \{a_{hk}\delta_{ij}\}$, $\mathscr{B}^h = \{a_h\delta_{ij}\}$ and $\mathscr{C}^h = \{0\}$ in Lemma 4.1, we obtain that $\mathscr{L}$ is L^p-dissipative if and only if

$$\mathrm{Re} \int_{\Omega} (a_{hk}\langle \partial_k v, \partial_h v \rangle - (1 - 2/p)^2 |v|^{-2} a_{hk} \mathrm{Re}\langle v, \partial_k v \rangle \, \mathrm{Re}\langle v, \partial_h v \rangle$$

$$- 2p^{-1} a_h \mathrm{Re}\langle v, \partial_h v \rangle - \langle \mathscr{A} v, v \rangle) \, dx \geqslant 0$$

for any $v \in (C_0^1(\Omega))^m$. We have also

$$-\frac{2}{p} \int_{\Omega} a_h \mathrm{Re}\langle v, \partial_h v \rangle dx = -\frac{1}{p} \int_{\Omega} a_h \partial_h(|v|^2) dx$$

and (4.54) follows by an integration by parts, which completes the proof. $\qquad \square$

Let us suppose that Ω is an open and bounded subset of $\mathbb{R}^n$, having a $C^{2,\alpha}$-boundary for some $\alpha \in (0, 1]$.

By means of the same technique we used in the proof of Theorem 2.20, one can prove that the operator A_p is L^p-dissipative if, and only if, the form $\mathscr{L}$ is L^p-dissipative. In view of Lemma 4.23 we have

Lemma 4.24. *The operator A_p is dissipative in $(L^p(\Omega))^N$ if and only if (4.54) holds for all $v \in (H_0^1(\Omega))^N$.*

In some of the lemmas and theorems below, we will restrict ourselves to operators A_p whose principal part $\mathcal{P}$ is positive, that is, the principal part fulfills $\mathcal{P}(x, \xi) > 0$ for all $x \in \overline{\Omega}$ and all $\xi \in \mathbb{R}^n \setminus \{0\}$. That this is no restriction when dealing with elliptic operators follows from the following lemma.

Lemma 4.25. *If A or A_p is dissipative, then the principal part $\mathcal{P}$ of A_p is positive.*

Proof. Let A_p be dissipative and assume, without loss of generality, that $0 \in \Omega$ and let $u \in C_0^\infty(\mathbb{R}^n)$ be nonzero and have support in $U \subset \Omega$. Write u_i for $\partial_i u$ and define for every $\varepsilon > 0$ the function v_ε by $v_\varepsilon(x) = (u(\varepsilon^{-1}x), 0, \ldots, 0)$. For sufficiently small ε, we have $v_\varepsilon|_\Omega \in (H_0^1(\Omega))^N$. Replacing v by v_ε in (4.54) and multiplying by ε^{2-n}, thanks to Lemma 4.24 we find

$$
0 \leqslant \varepsilon^{2-n} \int_\Omega \left(\varepsilon^{-2} a_{ij}(x)(u_i u_j)(\varepsilon^{-1}x) + b(x)u^2(\varepsilon^{-1}x) \right) dx
$$

$$
- \frac{(p-2)^2}{p^2} \varepsilon^{2-n} \int_\Omega \varepsilon^{-2} a_{ij}(x)(u_i u_j)(\varepsilon^{-1}x)\, dx
$$

$$
= \int_U \left(c a_{ij}(\varepsilon x) u_i(x) u_j(x) + \varepsilon^2 b(\varepsilon x)\, u^2(x) \right) dx
$$

$$
\to c \int_U a_{ij}(0) u_i u_j\, dx = c \int_U \mathcal{P}(0, \nabla u)\, dx, \varepsilon \to 0,
$$

where $b \in C(\overline{\Omega})$ and $c = 4/(pp')$. Since $\mathcal{P}(0, \xi) \neq 0$ for every $\xi \neq 0$ and ∇u is not identically zero, it follows that $\mathcal{P}(0, \xi)$ is positive at some point. But ellipticity and continuity implies then that $\mathcal{P}(0, \xi)$ is positive for every $\xi \neq 0$. A simple translation argument shows that $\mathcal{P}(x, \xi) > 0$ for every $x \in \Omega$ and $0 \neq \xi \in \mathbb{R}^n$. Finally, the continuity of the coefficients and the ellipticity imply positivity for every $x \in \overline{\Omega}$.

The proof that the dissipativity of A implies positivity of the principal part of A_p is completely analogous, but the L^2-dissipativity criterion can be used directly instead of Lemma 4.24. $\square$

In order to investigate the relation between the dissipativity of A_p and the dissipativity of A, we now continue by introducing some functionals and defining two constants associated with A and A_p:

$$
J(v) = \int_\Omega \left(\frac{4}{pp'} a_{ij} \langle \partial_i v, \partial_j v \rangle + \mathrm{Re}\langle (p^{-1}\partial_i a_i I - \mathscr{A})v, v \rangle \right) dx,
$$

$$
J_p(v) = \int_\Omega \left(a_{ij} \langle \partial_i v, \partial_j v \rangle + \mathrm{Re}\langle (p^{-1}\partial_i a_i I - \mathscr{A})v, v \rangle \right) dx
$$

$$
- \frac{(p-2)^2}{p^2} \int_{\{v \neq 0\}} a_{ij} \mathrm{Re}\langle \partial_i v, v \rangle\, \mathrm{Re}\langle \partial_j v, v \rangle |v|^{-2}\, dx,
$$

$$
\mu = \inf\{J(v) : v \in (H_0^1(\Omega))^N, \|v\|_2 = 1\},
$$

$$
\mu_p = \inf\{J_p(v) : v \in (H_0^1(\Omega))^N, \|v\|_2 = 1\}.
$$

Note that the expression for J_p comes from the statement of Lemma 4.24 and that the expression for J comes from the same lemma if it is used on the operator A with $p = 2$ instead of on A_p (since $2\,\mathrm{Re}\langle \mathscr{A} v, v \rangle = \langle (\mathscr{A} + \mathscr{A}^*)v, v \rangle$).

It is clear that J and J_p are real-valued since $\{a_{ij}\}$ is symmetric. Assuming that the principal part of A_p is positive, it follows immediately that J is bounded from below on the set $\{v \in (H_0^1)^N : \|v\|_2 = 1\}$. Hence, μ is finite. This is also the case with μ_p, but the proof is postponed to Lemma 4.27. However, assuming that μ_p is finite, the study of μ and μ_p is justified by the following lemma.

Lemma 4.26. *Let $1 < p < \infty$ and suppose that the principal part of A_p is positive. Then A and A_p generate the semigroups T on $(L^2(\Omega))^N$ and T_p on $(L^p(\Omega))^N$, respectively, fulfilling the inequalities*

$$\|T(t)\| \leqslant e^{-\mu t}, \|T_p(t)\| \leqslant e^{-\mu_p t}, t \geqslant 0. \tag{4.55}$$

The constants μ and μ_p are the best possible.

Proof. By temporarily replacing A_p by $A_p + \mu_p I$, we see that the infimum of J_p is zero. According to Lemma 4.24, this implies that the operator $A_p + \mu_p I$ is dissipative. In Grisvard [34] it is proved that $A_p - \lambda I$ is invertible with a bounded inverse (defined on $(L^p(\Omega))^N$) as soon as $A_p - \lambda I$ is one to one. This is the case when $\lambda > -\mu_p$ since the dissipativity of $A_p + \mu_p I$ implies that

$$-\operatorname{Re} \int_\Omega \langle (A_p - \lambda I)v, v \rangle |v|^{p-2}\, dx$$
$$= -\operatorname{Re} \int_\Omega \langle (A_p + \mu_p I)v, v \rangle |v|^{p-2}\, dx + (\mu_p + \lambda)\|v\|_p^p$$
$$\geqslant (\mu_p + \lambda)\|v\|_p^p, \sigma v \in D(A_p).$$

The theorem by Lumer–Phillips now shows that $A_p + \mu_p I$ generates a contraction semigroup Q_p on $(L^p(\Omega))^N$. Set $T_p(t) = e^{-\mu_p t} Q_p(t)$ for $t \geqslant 0$ and it follows immediately that T_p is a semigroup on $(L^p(\Omega))^N$ with $\|T_p(t)\| \leqslant e^{-\mu_p t}$ for $t \geqslant 0$. Furthermore, from the definition of a generator, it follows that A_p is the generator of T_p. By the definition of J_p and Lemma 4.24, μ_p is the smallest number making $A_p + \mu_p I$ dissipative, hence μ_p is the best constant in (4.55).

The proof of A generating the semigroup T with the indicated properties is completely analogous – just replace A_p, J_p and L^p, by A, J and L^2, respectively. $\square$

4.5.2 The relation between μ and μ_p

Note that it follows from Lemma 4.26 that if $\mu = \mu_p$, then A_p generates a contraction semigroup on $(L^p)^N$ if and only if A generates a contraction semigroup on $(L^2)^N$. We will therefore take an interest in the relation between μ and μ_p.

Lemma 4.27. *Suppose that $1 < p < \infty$ and that the principal part of A_p is positive. Then $\mu \leqslant \mu_p$.*

Proof. Let $v \in (H_0^1(\Omega))^N$ and choose $x \in \Omega$ such that x is a Lebesgue point to all components of v and $\partial_i v$. Let $\lambda_1, \ldots, \lambda_n$ be the eigenvalues of $\{a_{ij}(x)\}$ and let $T : \mathbb{R}^n \to \mathbb{R}^n$ be a linear transformation that takes the corresponding set of orthonormal eigenvectors to the coordinate axes. By introducing the function $u = v \circ T^{-1}$ and denoting the kth component of v and u by v_k and u_k, respectively, it follows that $\nabla v_k = T^{-1} \circ \nabla u_k \circ T$. Set $y = Tx$ and the relation

$$
\begin{aligned}
\big(a_{ij} \langle \partial_i v, \partial_j v \rangle\big)(x) &= (\nabla v_k(x))^t [a_{ij}(x)] \overline{\nabla v_k(x)} \\
&= \sum_k \lambda_i |\partial_i u_k(y)|^2 = \lambda_i |\partial_i u(y)|^2
\end{aligned}
\tag{4.56}
$$

is obtained. Suppose for the moment that the components of v are real-valued, let $\tilde{v}$ be a function that fulfills the same properties as v and set $\tilde{u} = \tilde{v} \circ T^{-1}$. Then

$$
\begin{aligned}
\big(a_{ij} \langle \partial_i v, v \rangle \langle \partial_j \tilde{v}, \tilde{v} \rangle\big)(x) &= \big(v_k \tilde{v}_\ell (\nabla v_k)^t [a_{ij}] \nabla \tilde{v}_\ell\big)(x) \\
&= \lambda_i (u_k \tilde{u}_\ell \partial_i u_k \, \partial_i \tilde{u}_\ell)(y) \\
&= \lambda_i \big(\langle \partial_i u, u \rangle \langle \partial_i \tilde{u}, \tilde{u} \rangle\big)(y).
\end{aligned}
\tag{4.57}
$$

Since in general

$$
\mathrm{Re}\langle \partial_i v, v \rangle = \langle \partial_i \, \mathrm{Re}\, v, \mathrm{Re}\, v \rangle + \langle \partial_i \, \mathrm{Im}\, v, \mathrm{Im}\, v \rangle,
$$

expression (4.57) can be used on each term of $\mathrm{Re}\langle \partial_i v, v \rangle \, \mathrm{Re}\langle \partial_j v, v \rangle$. If $v(x) \neq 0$, this gives

$$
\begin{aligned}
\big(a_{ij} \, \mathrm{Re}\langle \partial_i v, v \rangle \, \mathrm{Re}\langle \partial_j v, v \rangle |v|^{-2}\big)(x) &= \lambda_i \big(\mathrm{Re}\langle \partial_i u(y), u(y) \rangle\big)^2 |u(y)|^{-2} \\
&\leqslant \lambda_i |\partial_i u(y)|^2 = \big(a_{ij} \langle \partial_i v, \partial_j v \rangle\big)(x),
\end{aligned}
\tag{4.58}
$$

the last equality coming from (4.56) and the inequality arising from the Cauchy inequality since the eigenvalues λ_i are positive due to the positivity of the principal part of A_p. The arbitrariness of x now implies that

$$
\int_\Omega a_{ij} \langle \partial_i v, \partial_j v \rangle \, dx - \int_{\{v \neq 0\}} a_{ij} \, \mathrm{Re}\langle \partial_i v, v \rangle \, \mathrm{Re}\langle \partial_j v, v \rangle |v|^{-2} \, dx \geqslant 0.
$$

Defining the functional $v \mapsto K(v)$ by the left-hand side of the inequality above, we immediately get the functional equality

$$
J + \frac{(p-2)^2}{p^2} K = J_p.
\tag{4.59}
$$

Thus $J \leqslant J_p$, implying that $\mu \leqslant \mu_p$. $\square$

Lemma 4.28. *Suppose that $1 < p < \infty$ and that the principal part of A_p is positive. Then $\mu = \mu_p$ if and only if at least one of the nonzero generalized solutions of the equation*

$$
-\frac{4}{pp'} \partial_i (a_{ij} \partial_j v) + \frac{1}{2p} (2 \partial_i a_i I - p(\mathscr{A} + \mathscr{A}^*)) v = \mu v, v \in (H_0^1(\Omega))^N
\tag{4.60}
$$

is of the form $v = fc$ for some real-valued scalar function f and some $c \in \mathbb{C}^N$. Moreover, μ is the least eigenvalue of the left-hand side of the equation.

Proof. We have $2 \operatorname{Re}\langle \mathscr{A} v, v\rangle = \langle (\mathscr{A} + \mathscr{A}^*)v, v\rangle$ for all complex v where $\mathscr{A} + \mathscr{A}^*$ is self-adjoint. It follows from the theory of compact, self-adjoint operators, see e.g. Ladyžhenskaya [52, Ch. II] for the scalar case, that the infimum μ is the least eigenvalue of the operator defined by the right-hand side of (4.60) (whose spectrum only consists of countably many real eigenvalues with $+\infty$ as only possible accumulation point). Furthermore, the infimum is attained by the normalized eigenfunctions corresponding to μ and by no other functions. By general elliptic theory, see Grisvard [34], it follows from the regularity of the coefficients that solutions of (4.60) belong to $(C^1(\overline{\Omega}) \cap H^2(\Omega))^N$.

We go back and adopt the notation used in the proof of Lemma 4.27. Equality in (4.58) holds if and only if $\partial_i u(y) = b_i u(y)$ for every i and some real constants b_i (depending on y), or equivalently, if and only if $\partial_i v(x) = c_i v(x)$, c_i still being real. This shows that $K(v) = 0$ for a $C^1(\Omega)$-function v exactly when $\partial_i v = c_i v$ on $E_v = \{x \in \Omega : v(x) \neq 0\}$ for a collection of real-valued functions $\{c_i\}$. Thus, $K(v) = 0$ and $v \in C^1(\Omega)$ implies that

$$\partial_i \left(|v|^{-1} v \right) = |v|^{-1} \partial_i v - |v|^{-3} (\operatorname{Re}\langle \partial_i v, v\rangle)v$$
$$= c_i |v|^{-1} v - |v|^{-3} c_i |v|^2 v = 0$$

on E_v, from which it follows that $v = |v|c$ for some $c \in \mathbb{C}^N$ with unit length on each component of E_v. Conversely, a straightforward verification shows that for C^1-functions v of this form, $K(v) = 0$. Equation (4.59) gives that $\mu = \mu_p$ if and only if $K(w) = 0$ for one of the minimizers w of J which, by the discussion above, is equivalent to w being of the form $w = |w|c$ on each component of E_w.

Assume that one of the minimizers w is of this form and suppose that E_w consists of at least two components. Denote one of the components by F and let g be the real or imaginary part of any of the N components of w. Then $g \in C^1(\overline{\Omega})$ and g vanishes on the boundary of Ω. Define

$$f_\varepsilon(t) = \min(t + \varepsilon, 0) + \max(t - \varepsilon, 0), t \in \mathbb{R}, \ \varepsilon > 0$$

and set $g_\varepsilon = f_\varepsilon \circ g$. Then $g_\varepsilon|_F \in H_0^1(F)$ since f_ε is Lipschitz and the support of $g_\varepsilon|_F$ is a compact subset of F. It is easy to see that $g_\varepsilon|_F \to g|_F$ in $H^1(F)$ as $\varepsilon \to 0$. Hence, $g|_F \in H_0^1(F)$ and we conclude that $g\chi_F \in H_0^1(\Omega)$. Set $w_1 = \chi_F w$ and $w_2 = w - w_1$ and it follows from the previous discussion that $w_1, w_2 \in (H_0^1(\Omega))^N$. Moreover,

$$\mu = J(w_1) + J(w_2) \geqslant \|w_1\|_2^2 \mu + \|w_2\|_2^2 \mu = \mu,$$

so $J(w_1/\|w_1\|_2) = \mu$. Hence, $w_1/\|w_1\|_2$ is also a minimizer of J and consequently fulfills equation (4.60). Since w_1 is zero on an open subset of Ω, the Aronszajn–Cordes uniqueness theorem, see Hörmander [42], implies that w_1 is identically zero on Ω. This contradiction shows that E_w only consists of one component and we can conclude that $w = fc$ for some nonnegative $f \in C^1$ and $c \in \mathbb{C}^N$. $\square$

From Lemma 4.28, we obtain a necessary condition in algebraic terms for the equality $\mu = \mu_p$ to hold:

Corollary 4.29. *Suppose that the principal part of A_p is positive. If $\mu = \mu_p$, then there is a constant eigenvector to $\mathscr{A} + \mathscr{A}^*$ on $\overline{\Omega}$.*

Proof. By defining the matrix E as

$$E = -\frac{pp'}{4}((p^{-1}\partial_i a_i - \mu)I - 2^{-1}(A + A^*)),$$

equation (4.60) becomes

$$\partial_i(a_{ij}\partial_j v) + Ev = 0. \tag{4.61}$$

Assume that $\mu = \mu_p$ and let $v = fc$ be a solution of equation (4.61) given by Lemma 4.28. Without loss of generality, assume that $|c| = 1$. Consider the problem

$$\inf\{J(gc) : g \in H_0^1(\Omega), \|g\|_2 = 1\}.$$

The infimum is of course attained for $g = f$ and is μ and since

$$J(gc) = \frac{4}{pp'}\int_\Omega \left(a_{ij}\partial_i g\,\partial_j g - \langle Ec, c\rangle g^2\right) dx + \mu\|g\|_2^2,$$

the same theory as used in the proof of Lemma 4.28, or the simple fact that we have $\frac{d}{d\varepsilon}J((f + \varepsilon\varphi)c)|_{\varepsilon=0} = 0$ for all $\varphi \in C_0^\infty(\Omega)$, implies that f satisfies the Euler equation

$$\partial_i(a_{ij}\partial_j f) + \langle Ec, c\rangle f = 0. \tag{4.62}$$

Substituting $v = fc$ into (4.61), we also have the equation

$$\partial_i(a_{ij}\partial_j f)c + fEc = 0. \tag{4.63}$$

By multiplying the scalar equation (4.62) with c and subtracting (4.63), it follows that $fEc = f\langle Ec, c\rangle c$. Suppose that $Ec \neq \langle Ec, c\rangle c$ at some point in $\overline{\Omega}$. Due to the continuity of E, this would imply that f vanishes on some subset of Ω with nonempty interior. As in the proof of Theorem 4.28, the Aronszajn-Cordes uniqueness theorem then would lead to $f = 0$ in Ω. This is not the case since v is nonzero. Hence, $Ec = \langle Ec, c\rangle c$ on $\overline{\Omega}$, which by the definition of E implies that c is a constant eigenvector to $\mathscr{A} + \mathscr{A}^*$ on $\overline{\Omega}$. □

4.5.3   Dissipativity of A and A_p

We are now in the position to state the main results of this section.

Theorem 4.30. *If A generates a contraction semigroup on $(L^2(\Omega))^N$, then A_p generates a contraction semigroup on $(L^p(\Omega))^N$. Conversely, if there is a basis of constant eigenvectors to $\mathscr{A} + \mathscr{A}^*$, then A generates a contraction semigroup on $(L^2(\Omega))^N$ if A_p generates a contraction semigroup on $(L^p(\Omega))^N$. In particular, the converse holds in the scalar case.*

Proof. If A or A_p generate a contraction semigroup, Lemma 4.25 shows that the principal part of A_p is positive.

If A generates a contraction semigroup, Lemma 4.26 implies that $\mu \geqslant 0$ so by Lemma 4.27, it follows that $\mu_p \geqslant 0$. According to Lemma 4.26, A_p generates a contraction semigroup.

The converse is proved by first treating the scalar case and then extending the result to when $\mathscr{A} + \mathscr{A}^*$ has a basis of constant eigenvectors. Suppose that $N = 1$ and that A_p generates a contraction semigroup. Then $\mu_p \geqslant 0$ so $J_p(v) \geqslant 0$ for every $v \in H_0^1(\Omega)$. Consider the functional K defined in the proof of Lemma 4.27. We have for real-valued $v \in H_0^1(\Omega)$

$$K(v) = \int_\Omega a_{ij} \partial_i v \, \partial_j v \, dx - \int_{\{v \neq 0\}} a_{ij} \partial_i v \, \partial_j v \, dx = 0,$$

since $|\nabla v| = 0$ almost everywhere on the set $\{x \in \Omega : v(x) = 0\}$. In view of (4.59) and the nonnegativity of J_p, this gives that $J(v) \geqslant 0$ for all real-valued v. From the equality $J(u + iv) = J(u) + J(v)$, holding for all real-valued functions u and v, it follows that $\mu \geqslant 0$ and we conclude from Lemma 4.26 that A generates a contraction semigroup.

Let N be arbitrary, suppose that A_p generates a contraction semigroup and let T be a linear transformation that diagonalizes $\mathscr{A} + \mathscr{A}^*$ in such a way that T is an isometry. Let $f \in D(A_p)$ and set $g = Tf$. Since T is an isometry,

$$\text{Re} \int_\Omega \langle A_p g, g \rangle |g|^{p-2} \, dx = \text{Re} \int_\Omega \langle T^{-1} A_p T f, f \rangle |f|^{p-2} \, dx,$$

showing that dissipativity of A_p is equivalent to dissipativity of $f \mapsto T^{-1} A_p T f$. This equivalence together with the fact that

$$\text{Re}\langle A_p g, g \rangle = \text{Re}\langle (A_p - \mathscr{A} + 2^{-1}(\mathscr{A} + \mathscr{A}^*)) g, g \rangle$$

for all $g \in D(A_p)$, allows us to assume without loss of generality that $\mathscr{A}$ is a diagonal matrix. Let $u \in W^{2,p} \cap W_0^{1,p}$, fix some $k \in \{1, \ldots, N\}$ and set $w = u e_k$ where e_k is the unit vector in $\mathbb{C}^N$ directed along the kth coordinate axis. Finally, define the scalar operator A_p^k as the "kth row" of A_p, that is,

$$A_p^k \varphi = A_p(\varphi e_k), \varphi \in W^{2,p} \cap W_0^{1,p}.$$

Define A^k analogously. The dissipativity of A_p implies that

$$0 \geqslant \text{Re} \int_\Omega \langle A_p w, w \rangle |w|^{p-2} \, dx = \text{Re} \int_\Omega (A_p^k u) \overline{u} |u|^{p-2} \, dx,$$

so A_p^k is dissipative. From the case $N = 1$ treated above, the dissipativity of A^k follows for every k. Finally, let $v = (v_1, \ldots, v_N) \in D(A)$. Since

$$\int_\Omega \langle Av, v \rangle \, dx = \sum_k \int_\Omega (A^k v_k) \overline{v}_k \, dx,$$

the dissipativity of each A^k implies the dissipativity of A, so $\mu \geqslant 0$ and we conclude that A generates a contraction semigroup. $\square$

Example 4.31. In the scalar case $N = 1$, the equality $\mu = \mu_p$ always holds according to Lemma 4.28. This is however not true in general: consider an operator A_p in $(L^p(\Omega))^2$ with a positive principal part and suppose that the matrix coefficient $\mathscr{A}$ of A_p is given as

$$\mathscr{A}(x) = \begin{pmatrix} 1 & |x| \\ |x| & -1 \end{pmatrix}, x \in \overline{\Omega}.$$

Since $\mathbb{C}^2$ is spanned by the eigenvector

$$\begin{pmatrix} 1 \\ |x|/(1 + \sqrt{1 + |x|^2}) \end{pmatrix}$$

of $\mathscr{A}$ and one eigenvector orthogonal to it, $\mathscr{A}$ has no constant eigenvectors. According to Corollary 4.29, μ is strictly smaller than μ_p. In particular, we can add a suitable multiple of the identity operator to A_p to make A_p generate a contraction semigroup on $(L^p(\Omega))^2$, while A does *not* generate a contraction semigroup on $(L^2(\Omega))^2$.

4.6 L^p-dissipativity of coupled systems for any p's

We will now treat the initial boundary value problem with zero Dirichlet boundary conditions for linear second-order systems which are uniformly parabolic in the sense of Petrovskiĭ. These systems will not necessarily be weakly coupled as in the previous section. Our goal is to give criteria for the L^p-maximum principle to hold for all $p \in [1, \infty]$ simultaneously and then formulate the result in terms of the generation of contraction semigroups.

First we will study a parabolic system and recall some necessary and sufficient conditions for the solution of the parabolic system to not increase in maximum norm as the time increases. Using duality and interpolation, we get criteria for this maximum principle to hold for all $p \in [1, \infty]$ simultaneously. After that, the results will be written in semigroup language.

4.6.1   The maximum modulus principle for a parabolic system

This section is devoted to the L^∞-dissipativity of scalar second-order operators. In this section we recall one of the main results in this topic: for a system which is uniformly parabolic in the sense of Petrovskii and in which the coefficients do not depend on t, the maximum modulus principle holds if and only if the principal part of the system is scalar and the coefficients of the system satisfy a certain algebraic inequality (see Theorem 4.32 below).

Let $\Omega \subset \mathbb{R}^n$ be a bounded domain with a $C^{2,\alpha}$ boundary $(0 < \alpha \leqslant 1)$ and let Q_T be the cylinder $\Omega \times (0, T)$.

Let A be the differential operator

$$Au = \partial_i(\mathscr{A}_{ij}\, \partial_j u) + \mathscr{A}_i\, \partial_i u + \mathscr{A}\, u \tag{4.64}$$

where $\mathscr{A}_{ij}$, $\mathscr{A}_i$ and $\mathscr{A}$ are $N \times N$ matrices whose entries are complex-valued functions. The elements of $\mathscr{A}_{ij}$, $\mathscr{A}_i$ and $\mathscr{A}$ belong to $C^{2,\alpha}(\overline{\Omega})$, $C^{1,\alpha}(\overline{\Omega})$ and $C^{0,\alpha}(\overline{\Omega})$ respectively.

Moreover $\mathscr{A}_{ij} = \mathscr{A}_{ji}$ and there exists $\delta > 0$ such that for every $x \in \overline{\Omega}$ and every $\xi = (\xi_1, \ldots, \xi_N) \in \mathbb{R}^N$, the zeros of the polynomial

$$\lambda \mapsto \det(\xi_i \xi_j\, \mathscr{A}_{ij} + \lambda I)$$

satisfy the inequality $\operatorname{Re} \lambda \leqslant -\delta|\xi|^2$.

Consider the problem

$$\begin{cases} \partial_t u - Au = 0, & \text{on } Q_T, \\ u(\cdot, 0) = \varphi, & \text{on } \Omega, \\ u|_{\partial\Omega \times [0,T]} = 0, \end{cases} \tag{4.65}$$

where $\varphi \in (C^{2,\alpha}(\overline{\Omega}))^N$ and vanishes on $\partial\Omega$.

Theorem 4.32 (Kresin [50]). *Let u be the solution of (4.65). In order that*

$$\|u(\cdot, t)\|_\infty \leqslant \|\varphi\|_\infty, \quad \forall t \in [0, T],$$

for all $\varphi \in (C^{2,\alpha}(\overline{\Omega}))^N$ vanishing on $\partial\Omega$, it is necessary and sufficient that

(a) *there are real-valued scalar functions a_{ij} on $\overline{\Omega}$ such that for every i, j, $\mathscr{A}_{ij} = a_{ij}I$ and the $n \times n$-matrix $\{a_{ij}\}$ is positive definite;*

(b) *for all $\eta_i, \zeta \in \mathbb{C}^N$, $i = 1, \ldots, n$, with $\operatorname{Re}\langle \eta_i, \zeta \rangle = 0$, the inequality*

$$\operatorname{Re}\left\{ a_{ij}\langle \eta_i, \eta_j \rangle - \langle \mathscr{A}_i\, \eta_i, \zeta \rangle - \langle \mathscr{A}\, \zeta, \zeta \rangle \right\} \geqslant 0$$

holds on Ω.

In the scalar case $n = 1$, condition (b) is reduced to the requirement that the inequality

$$-4 \operatorname{Re} \mathscr{A} \geqslant b_{ij}\, \operatorname{Im} \mathscr{A}_i\, \operatorname{Im} \mathscr{A}_j$$

holds on Ω, where $\{b_{ij}\} = \{a_{ij}\}^{-1}$ (cf. (2.41)).

Let us consider now the problem

$$\begin{cases} \partial_t w + A^* w = 0, & \text{on } Q_T, \\ w(\cdot, T) = \psi, & \text{on } \Omega, \\ w|_{\partial\Omega \times [0,T]} = 0, \end{cases} \tag{4.66}$$

where A^* is the formally adjoint operator of A

$$A^* w = \partial_i (\mathscr{A}_{ij}^* \partial_j w) - \mathscr{A}_i^* \partial_i w + (\mathscr{A}^* - \partial_i \mathscr{A}_i^*) w. \tag{4.67}$$

Theorem 4.32 implies

Corollary 4.33. *Let w be the solution of* (4.66). *In order that*

$$\|w(\cdot, t)\|_\infty \leqslant \|\psi\|_\infty, \quad \forall t \in [0, T],$$

for all $\psi \in (C^{2,\alpha}(\overline{\Omega}))^N$ vanishing on $\partial\Omega$, it is necessary and sufficient that

(a) *there are real-valued scalar functions a_{ij} on $\overline{\Omega}$ such that for every i, j, $\mathscr{A}_{ij} = a_{ij} I$ and the $n \times n$-matrix $\{a_{ij}\}$ is positive definite;*

(b) *for all $\eta_i, \zeta \in \mathbb{C}^N$, $i = 1, \dots, n$, with $\mathrm{Re}\langle \eta_i, \zeta \rangle = 0$, the inequality*

$$\mathrm{Re}\left\{ a_{ij}\langle \eta_i, \eta_j \rangle + \langle \mathscr{A}_i \zeta, \eta_i \rangle - \langle (\mathscr{A} - \partial_i \mathscr{A}_i)\zeta, \zeta \rangle \right\} \geqslant 0$$

holds on Ω.

By means of Theorem 4.32 and its Corollary 4.33 and using interpolation, one arrives to the necessary and sufficient conditions for the validity of the L^p maximum principle for all $p \in [1, \infty]$ simultaneously. This will be obtained in the theorem 4.36 below.

4.6.2   The parabolic PDE setting

Let $\alpha \in (0, 1)$, let Ω be a bounded domain in $\mathbb{R}^n$ with $C^{2,\alpha}$-boundary and let for $T > 0$ the cylinder $\Omega \times (0, T)$ be denoted by Q_T. Furthermore, let $C_0^{2,\alpha}$ be the subset of $(C^{2,\alpha}(\overline{\Omega}))^N$ consisting of the functions that vanish on the boundary $\partial\Omega$ of Ω.

If f is a function defined on Q_T, $\partial_i f$ will as before denote the partial derivative of f with respect to the ith space coordinate, whereas $\partial_t f$ will denote the time derivative $\partial_{n+1} f$. Let A be the differential operator (4.64) and A^* its formally adjoint operator (4.67). Let us assume the same smoothness hypothesis made in Section 4.6.1.

We introduce the initial boundary value problem

$$\begin{cases} \partial_t u - Au = 0, & \text{on } Q_T, \\ u(\cdot, 0) = \varphi, & \text{on } \Omega, \\ u|_{\partial\Omega \times [0,T]} = 0, \end{cases} \tag{4.68}$$

where $\varphi \in C_0^{2,\alpha}$. The requirements in the beginning of the section on the coefficients of A imply that the system in (4.68) is uniformly parabolic in the sense of Petrovskiĭ. According to Theorem 10.1 in Ladyžhenskaya, Solonnikov and Ural'ceva [53], there exists a unique classical solution to (4.68) with Hölder continuous second derivatives in the space variables and first derivative in time (in fact, the regularity assumptions on the coefficients and the boundary can be relaxed and were imposed only for simplicity of the presentation).

Hinging on the results of Kresin and Maz'ya described in Section 4.6.1, we have the following theorem

Theorem 4.34. *Let u be the solution of (4.68). In order that*

$$\|u(\,\cdot\,,t)\|_p \leqslant \|\varphi\|_p, \forall t \in [0,T],\tag{4.69}$$

for all $\varphi \in C_0^{2,\alpha}$ and for all $p \in [1,\infty]$, it is necessary and sufficient that

(a) *there are real-valued scalar functions a_{ij} on $\overline{\Omega}$ such that for every i,j, $A_{ij} = a_{ij}I$ and the $n \times n$-matrix $[a_{ij}]$ is positive definite,*

(b) *for all $\eta_i, \zeta \in \mathbb{C}^N$, $i = 1, \ldots, n$, with $\mathrm{Re}\langle \eta_i, \zeta \rangle = 0$, the inequalities*

$$\mathrm{Re}\left\{ a_{ij}\langle \eta_i, \eta_j \rangle - \langle \mathscr{A}_i \eta_i, \zeta \rangle - \langle \mathscr{A} \zeta, \zeta \rangle \right\} \geqslant 0,$$
$$\mathrm{Re}\left\{ a_{ij}\langle \eta_i, \eta_j \rangle + \langle \mathscr{A}_i \zeta, \eta_i \rangle - \langle (\mathscr{A} - \partial_i \mathscr{A}_i)\zeta, \zeta \rangle \right\} \geqslant 0$$

hold on Ω.

Proof. Let u and v be the solutions of (4.68) and (4.66), respectively. Let $\tau \in [0,T)$ and set $\tilde{v}(\,\cdot\,,t) = v(\,\cdot\,,t+\tau)$ for $t \in [0,T-\tau]$. Integrating by parts and applying the equality $\langle Au, \tilde{v} \rangle = \langle u, A^*\tilde{v} \rangle$ gives

$$0 = \int_0^{T-\tau} \int_\Omega \langle \partial_t u - Au, \tilde{v} \rangle \, dx \, dt$$
$$= \int_\Omega \left(\langle u(\,\cdot\,,T-\tau), \tilde{v}(\,\cdot\,,T-\tau) \rangle - \langle u(\,\cdot\,,0), \tilde{v}(\,\cdot\,,0) \rangle \right) dx$$
$$- \int_0^{T-\tau} \int_\Omega \langle u, \partial_t \tilde{v} + A^*\tilde{v} \rangle \, dx \, dt.$$

Since $\tilde{v}$ fulfills the same differential equation as v, the last integral vanishes and we obtain the identity

$$\int_\Omega \langle u(\,\cdot\,,T-\tau), \psi \rangle \, dx = \int_\Omega \langle \varphi, v(\,\cdot\,,\tau) \rangle \, dx.\tag{4.70}$$

We need only prove the necessity of the second inequality in (b), since the necessity of (a) and the necessity of the first inequality in (b) result from Theorem 4.32 along with inequality (4.69) where $p = \infty$. Equality (4.70), together with the Cauchy and Hölder inequalities, imply that

$$\left| \int_\Omega \langle \varphi, v(\,\cdot\,,\tau) \rangle \, dx \right| \leqslant \|\psi\|_\infty \|u(\,\cdot\,,T-\tau)\|_1 \leqslant \|\psi\|_\infty \|\varphi\|_1,$$

where (4.69) with $p = 1$ has been used in the second inequality. The functional on $(L^1)^N$ defined by the integral in the left-hand side of the inequalities above has norm $\|v(\,\cdot\,,\tau)\|_\infty$. By the arbitrariness of φ and the denseness of $C_0^{2,\alpha}$ in $(L^1)^N$,

$$\|v(\,\cdot\,,\tau)\|_\infty \leqslant \|\psi\|_\infty$$

for all ψ. Since τ is arbitrary, this is equivalent to the maximum modulus principle for the problem (4.66). Hence, Corollary 4.33 gives the necessity of the second inequality in condition (b).

We now turn to the sufficiency part. Set $t = T - \tau$. By (4.70) together with the sufficiency part of Corollary 4.33, we have the following inequality:

$$\left| \int_\Omega \langle u(\,\cdot\,,t), \psi \rangle \, dx \right| \leqslant \|\varphi\|_1 \|v(\,\cdot\,,\tau)\|_\infty \leqslant \|\varphi\|_1 \|\psi\|_\infty.$$

As in the L^1-case, the integral on the left in the inequality above defines a functional on $(C_0(\overline{\Omega}))^N$. By using the Cauchy and Hölder inequalities and by applying the functional to $(C_0(\overline{\Omega}))^N$-approximations of the function $|u(\,\cdot\,,t)|^{-1} u(\,\cdot\,,t)$, it is easily shown that the norm of the functional is $\|u(\,\cdot\,,t)\|_1$. Since φ is arbitrary, we conclude that

$$\|u(\,\cdot\,,t)\|_1 \leqslant \|\varphi\|_1. \tag{4.71}$$

Take some $t \in (0, T]$ and define the operator G on $C_0^{2,\alpha}$ by

$$G\varphi = u(\,\cdot\,,t),$$

where u as before is the solution to (4.68). Due to (4.71), G can be extended by continuity to $(L^1)^N$ and

$$\|G\|_{\mathscr{B}((L^1)^N)} \leqslant 1. \tag{4.72}$$

By the sufficiency part of Theorem 4.32, we also have the inequality

$$\|G\varphi\|_\infty \leqslant \|\varphi\|_\infty, \varphi \in C_0^{2,\alpha}.$$

Let $f \in (L^\infty)^N$. By multiplying f with the characteristic function of a suitable precompact set in Ω and then convolving the result with a suitable mollifier, it follows that f can be arbitrarily well approximated in $(L^1)^N$ by a function $h \in (C_0^\infty(\Omega))^N$ in such a way that $\|h\|_\infty \leqslant \|f\|_\infty$. Thus, take a sequence $\{f_k\}_k$ in $(C_0^\infty(\Omega))^N$ with $\|f_k - f\|_1 \to 0$ as $k \to \infty$ and with $\|f_k\|_\infty \leqslant \|f\|_\infty$ for every k. Since

$$\|Gf_k - Gf\|_1 \leqslant \|f_k - f\|_1 \to 0, k \to \infty,$$

there is a subsequence of $\{Gf_k\}_k$ such that the subsequence converges pointwise to Gf almost everywhere on Ω. Rename the subsequence to the original sequence and it follows that

$$|(Gf)(x)| = \lim_{k\to\infty} |(Gf_k)(x)| \leqslant \lim_{k\to\infty} \|Gf_k\|_\infty \leqslant \lim_{k\to\infty} \|f_k\|_\infty \leqslant \|f\|_\infty$$

for almost every x in Ω. Consequently,

$$\|G\|_{\mathscr{B}((L^\infty)^N)} \leqslant 1. \tag{4.73}$$

The norm estimates in (4.72) and (4.73) now enable us to use the Riesz–Thorin interpolation theorem on G and we get

$$\|G\|_{\mathscr{B}((L^p)^N)} \leqslant 1, p \in [1, \infty].$$

The arbitrariness of t finally gives the desired result. $\qquad\square$

4.6.3 The semigroup setting

With Theorem 4.34 in mind, we will now define extensions of the operator A with domain $C_0^{2,\alpha}$ to operators in the various $(L^p)^N$-spaces. Thus, let A_p be the operator A with domain

$$D(A_p) = (W^{2,p}(\Omega) \cap W_0^{1,p}(\Omega))^N, 1 < p < \infty$$

and let A_1 be the operator obtained by taking the closure in $(L^1)^N$ of the operator A with domain $C_0^{2,\alpha}$.

Every A_p is thus a densely defined operator in $(L^p)^N$. In the case $p = \infty$, we have to proceed in another manner since $C_0^{2,\alpha}$ is not dense in $(L^\infty)^N$. Instead, the natural function space in which to define the operator A_∞, is in the closure of $C_0^{2,\alpha}$ in the L^∞-norm; that is, the space $(C_0(\overline{\Omega}))^N$. By Sobolev's embedding theorem, $W^{2,r}(\Omega)$ is a subset of $C(\overline{\Omega})$ for all $r > n/2$. Take some $s > n/2$ and let A_∞ be the operator A with domain

$$D(A_\infty) = \left\{u \in (W^{2,s}(\Omega) \cap W_0^{1,s}(\Omega))^N : Au \in (C_0(\overline{\Omega}))^N\right\}$$

and regard A_∞ as an operator in $(C_0(\overline{\Omega}))^N$.

Before the main result of this section, let us prove a technical lemma. It concerns the approximation of integral expressions arising when checking for dissipativity of partial differential operators.

Lemma 4.35. *Let $p \in (1, \infty)$ and let Ω be a domain in $\mathbb{R}^n$. Furthermore, let α be a multi-index and suppose that $\{f_k\}_1^\infty \subset (W^{|\alpha|,p}(\Omega))^N$ is a sequence with $(W^{|\alpha|,p}(\Omega))^N$-limit f. Then*

$$\lim_{k\to\infty} \int_\Omega \langle \partial^\alpha f_k, f_k \rangle |f_k|^{p-2}\, dx = \int_\Omega \langle \partial^\alpha f, f \rangle |f|^{p-2}\, dx.$$

Proof. By expanding $\langle \partial^\alpha f, f \rangle |f|^{p-2}$ into components, we see that it is enough to show that

$$\lim_{k\to\infty} \int_\Omega f_k g_k h_k^{p-2}\, dx = \int_\Omega f g h^{p-2}\, dx, \tag{4.74}$$

where $\{f_k\}, \{g_k\}, \{h_k\} \subset L^p(\Omega)$ are sequences with L^p-limits f, g and h, respectively, fulfilling $|g_k| \leqslant h_k$ and $|g| \leqslant h$.

After taking subsequences successively, we can assume that $g_k \to g$ and $h_k \to h$ pointwise almost everywhere as $k \to \infty$. Let q be the conjugate exponent to p and note that $\|g_k h_k^{p-2}\|_q \leqslant \|h_k\|_p^{p-1}$, so $\{g_k h_k^{p-2}\}$ is a bounded sequence in L^q with pointwise limit gh^{p-2} almost everywhere. This implies, see Hewitt and Stromberg [36], that $g_k h_k^{p-2} \to gh^{p-2}$ weakly in L^q. An application of Hölder's inequality gives

$$\left| \int_\Omega \left(f_k g_k h_k^{p-2} - fgh^{p-2} \right) ddx \right|$$
$$\leqslant \|f_k - f\|_p \|h_k\|_p^{p-1} + \left| \int_\Omega f(g_k h_k^{p-2} - gh^{p-2})\, dx \right|,$$

which shows that (4.74) holds for a subsequence, since the right-hand side tends to zero due to the convergence of $\{f_k\}$ and the weak convergence of $\{g_k h_k^{p-2}\}$. If there is a subsequence of our original sequence such that the left-hand side of (4.74) does converge to some other value than the right-hand side, we can repeat the proof with this subsequence and get a contradiction. Thus (4.74) holds for our original sequence.   □

Using the results obtained on the parabolic systems in the previous subsection, we can prove the following theorem about the generation of contraction semigroups for all the operators A_p.

Theorem 4.36. *The operators A_p generate contraction semigroups on $(L^p(\Omega))^N$ for all $p \in [1, \infty)$ and on $(C_0(\overline{\Omega}))^N$ for $p = \infty$ simultaneously if and only if*

(a) *there are real-valued scalar functions a_{ij} on $\overline{\Omega}$ such that for every i, j, $\mathscr{A}_{ij} = a_{ij}I$ and the $n \times n$-matrix $[a_{ij}]$ is positive definite,*

(b) *for all $\eta_i, \zeta \in \mathbb{C}^N$, $i = 1, \ldots, n$, with $\mathrm{Re}\langle \eta_i, \zeta \rangle = 0$, the inequalities*

$$\mathrm{Re}\left\{ a_{ij}\langle \eta_i, \eta_j \rangle - \langle \mathscr{A}_i \eta_i, \zeta \rangle - \langle \mathscr{A} \zeta, \zeta \rangle \right\} \geqslant 0,$$
$$\mathrm{Re}\left\{ a_{ij}\langle \eta_i, \eta_j \rangle + \langle \mathscr{A} \zeta, \eta_i \rangle - \langle (\mathscr{A} - \partial_i \mathscr{A}_i)\zeta, \zeta \rangle \right\} \geqslant 0$$

hold on Ω.

Proof. The necessity of (a) and (b) is easy to show. Let $p \in [1, \infty]$. The contraction semigroup generated by A_p gives the solutions to the abstract Cauchy problem associated with the operator A_p. Since $C_0^{2,\alpha}$ is contained in the function space whose elements the semigroup acts on and the solutions to the parabolic system (4.68) are unique, inequality (4.69) holds for the chosen p due to the contractivity of the semigroup. Since p is arbitrary, Theorem 4.34 shows that (a) and (b) follow.

We next show the sufficiency of (a) and (b). This will be achieved by constructing a contraction semigroup for every p and show that its generator is precisely A_p: define the map $t \mapsto Q(t)$, where $t \in [0, T]$ and $Q(t) : C_0^{2,\alpha} \to C_0^{2,\alpha}$, by

$$Q(t)\varphi = u(\,\cdot\,, t)$$

where u is the solution to the parabolic system (4.68) with initial data φ. Defining $Q(t) = Q(T)Q(t - T)$ recursively, Q is extended to $\mathbb{R}^+$. It is easily seen that $Q(s+t) = Q(s)Q(t)$ and that $Q(0)$ is the identity operator on $C_0^{2,\alpha}$. Theorem 4.34 shows that

$$\|Q(t)\varphi\|_p \leqslant \|\varphi\|_p, t \in \mathbb{R}^+,\ p \in [1, \infty],\ \varphi \in C_0^{2,\alpha}.$$

$C_0^{2,\alpha}$ is dense in $(L^p)^N$ for $1 \leqslant p < \infty$ and in $(C_0(\overline{\Omega}))^N$ so extend for each $p \in [1, \infty)$ and $t \in \mathbb{R}^+$ the operator $Q(t)$ by continuity to $Q_p(t) \in \mathscr{B}((L^p(\Omega))^N)$ and to $Q_\infty(t) \in \mathscr{B}((C_0(\overline{\Omega}))^N)$, respectively. Then $\|Q_p(t)\| \leqslant 1$ and the algebraic properties of Q are preserved. Thus Q_p fulfills the semigroup properties every p if the continuity property can be established.

Let $\varphi \in C_0^{2,\alpha}$ and let u be the corresponding solution of (4.68). The solution is classical so, if $N = 1$ and if u is real-valued, the mean value theorem gives

$$\begin{aligned}
\left|\left(t^{-1}(Q(t)\varphi - \varphi) - A\varphi\right)(x)\right| &= |t^{-1}(u(x, t) - u(x, 0)) - \partial_t u(x, 0)| \\
&= |\partial_t u(x, \xi(x, t)) - \partial_t u(x, 0)| \qquad (4.75) \\
&\leqslant C|\xi(x, t)|^\gamma \leqslant Ct^\gamma, x \in \Omega
\end{aligned}$$

for some $C > 0$, $\gamma \in (0, 1)$, coming from the Hölder continuity of $\partial_t u$. If u is complex-valued and N is arbitrary, the inequality between the first and last row in (4.75) still holds, since it can be used on each of the components and real/imaginary parts of u together with the triangle inequality. Since Ω is bounded, (4.75) especially implies that $t \mapsto Q_p(t)\varphi$ is continuous at 0 in the norm of $(L^p(\Omega))^N$ for every $\varphi \in C_0^{2,\alpha}$ and $p \in [1, \infty)$. The same is true for $p = \infty$ since $Q_\infty(t)\varphi$ is zero on the boundary of Ω for every t. Let p be arbitrary, take $f \in (L^p)^N$ or $f \in (C_0(\overline{\Omega}))^N$ if $p = \infty$, let $\varepsilon > 0$ and choose $\varphi \in C_0^{2,\alpha}$ with $\|f - \varphi\|_p < \varepsilon$. Then

$$\begin{aligned}
\|Q_p(t)f - f\|_p &\leqslant \|Q_p(t)\|\|f - \varphi\|_p + \|Q_p(t)\varphi - \varphi\|_p + \|\varphi - f\|_p \\
&\leqslant 2\varepsilon + \|Q_p(t)\varphi - \varphi\|_p,
\end{aligned}$$

showing that the continuity of $t \mapsto Q_p(t)f$ at 0 follows from the continuity on $C_0^{2,\alpha}$. This is actually equivalent to Q_p being a strongly continuous contraction semigroup on $(L^p)^N$ for every $p \in [1, \infty)$ and on $(C_0(\overline{\Omega}))^N$ for $p = \infty$.

Let for each p the operator B_p be the generator of Q_p. Inequality (4.75) shows that the $(L^p)^N$-limit

$$\lim_{t \to 0+} \|t^{-1}(Q_p(t) - I)\varphi - \mathscr{A}\varphi\|_p = 0, \varphi \in C_0^{2,\alpha},\ p \in [1, \infty),$$

so B_p is an extension of the operator $(A, C_0^{2,\alpha})$ for every $p < \infty$. If $p = \infty$, we must in addition require that $A\varphi$ be zero on the boundary of Ω in order for the limit to be zero.

Let $p \in (1, \infty)$. Since B_p is dissipative, so is the operator $(A, C_0^{2,\alpha})$. Since $C_0^{2,\alpha}$ is dense in the domain of A_p, Lemma 4.35 implies that the extension A_p of $(A, C_0^{2,\alpha})$ is dissipative. As in the proof of Lemma 4.26, the injectivity of $A_p - \lambda I$ for $\lambda > 0$ follows from the dissipativity of A_p and the injectivity yields the invertibility of $A_p - \lambda I$ for $\lambda > 0$ (see, e.g., Maz'ya and Shaposhnikova [76, Ch. 14]). By the theorem of Lumer–Phillips, A_p generates a contraction semigroup on $(L^p)^N$.

The next step is to prove that A_1 generates a contraction semigroup in $(L^1)^N$. Since B_1 is a closed extension of $(A, C_0^{2,\alpha})$ and A_1 is the smallest closed extension of the same operator, A_1 is a restriction of B_1. $B_1 - I$ is injective since B_1 is a semigroup generator so if we can show that $A_1 - I$ maps $D(A_1)$ onto $(L^1)^N$, it follows that $A_1 = B_1$ and thus A_1 generates a contraction semigroup.

Denote the space $(W^{2,2} \cap W_0^{1,2})^N$ by F and take $u \in F$. Then there is a sequence $\{u_k\}$ in $C_0^{2,\alpha}$ with limit u in F. We get

$$\|A_1 u_k\|_1 \leqslant C\|u_k\|_{2,1} \leqslant C_1\|u_k\|_{2,2},$$

where the first inequality follows from the smoothness of the coefficients of A and the second follows from the boundedness of Ω. This inequality shows that $\{A_1 u_k\}$ is a Cauchy sequence in $(L^1)^N$ and thus convergent, implying that $u \in D(A_1)$ due to the fact that A_1 is closed. Hence $F \subset D(A_1)$. Since A_2 generates a contraction semigroup, $(A_1 - I)(F) = (A_2 - I)(F) = (L^2)^N$, so the range of $A_1 - I$ includes $(L^2)^N$. Take $f \in (L^1)^N$, let $\{f_k\} \subset (L^2)^N$ have limit f in $(L^1)^N$ and set $u_k = (A_2 - I)^{-1} f_k$. By using that A_r generates a contraction semigroup and that u_k and f_k belong to $(L^r)^N$ for $r \in [1,2]$, the relation

$$\|u_k\|_1 = \lim_{r \to 1+} \|u_k\|_r = \lim_{r \to 1+} \left\|(\mathscr{A}_2 - I)^{-1} f_k\right\|_r$$
$$= \lim_{r \to 1+} \left\|(A_r - I)^{-1} f_k\right\|_r \leqslant \lim_{r \to 1+} \|f_k\|_r = \|f_k\|_1$$

is obtained. It shows that $\{u_k\}$ converges in $(L^1)^N$ to a limit function u. Since $\mathscr{A}_1$ is closed and

$$\lim_{k \to \infty} \|(A_1 - I)u_k - f\|_1 = \lim_{k \to \infty} \|f_k - f\|_1 = 0,$$

we conclude that $u \in D(A_1)$ and $(A_1 - I)u = f$. Thus $A_1 - I$ maps $D(A_1)$ onto $(L^1)^N$.

Finally, we treat the case $p = \infty$. Since A_∞ is a restriction of A_s and A_s generates a contraction semigroup, $A_\infty - \lambda I$ is injective for $\lambda > 0$. From the definition of $D(A_\infty)$, it follows that the range of $A_\infty - \lambda I$ is $(L^s \cap C_0(\overline{\Omega}))^N$ so $A_\infty - \lambda I$ maps $D(A_\infty)$ onto $(C_0(\overline{\Omega}))^N$ due to the boundedness of Ω. Let $\lambda > 0$, take $f \in (C_0(\overline{\Omega}))^N$ and set $u = (A_\infty - \lambda I)^{-1} f$. The operators A_r, $r \in [s, \infty)$, generate contraction semigroups so

$$\|u\|_\infty = \lim_{r \to \infty} \|u\|_r = \lim_{r \to \infty} \left\|(A_s - \lambda I)^{-1} f\right\|_r$$
$$= \lim_{r \to \infty} \left\|(A_r - \lambda I)^{-1} f\right\|_r \leqslant \lim_{r \to \infty} \lambda^{-1} \|f\|_r = \lambda^{-1} \|f\|_\infty,$$

implying that $\|(A_\infty - \lambda I)^{-1}\| \leqslant \lambda^{-1}$. The operator A_∞ is densely defined since $(C_0^\infty(\Omega))^N$ is contained in $D(A_\infty)$ and $C_0^\infty(\Omega)$ is dense in $C_0(\overline{\Omega})$. By the theorem of Hille–Yosida (see, e.g. , Goldstein [31]), the operator A_∞ generates a contraction semigroup on the space $(C_0(\overline{\Omega}))^N$. $\qquad\qquad\qquad\qquad\qquad\qquad\qquad\square$

As an example, consider the Schrödinger operator with magnetic field, see, e.g. , Simon [92] or Cycon et al. [16],

$$-(i\nabla + m)^t(i\nabla + m) - V,$$

i.e. , the scalar operator

$$A = \Delta - 2im \cdot \nabla - i(\nabla \cdot m) - |m|^2 - V,$$

where m is an $\mathbb{R}^n$-valued function on Ω, the function V is complex-valued and the domain Ω and the functions m, V fulfill the smoothness assumptions in the beginning of this section. Using the inequality in the remark following Theorem 4.32 and its dual counterpart, Theorem 4.36 gives necessary and sufficient conditions for A to generate contraction semigroups on all L^p-spaces simultaneously as

$$-4\operatorname{Re}\mathscr{A} \geqslant \sum_{j=1}^{n}(\operatorname{Im}\mathscr{A}_j)^2, \quad -4\operatorname{Re}(\overline{\mathscr{A} - \partial_j \mathscr{A}_j}) \geqslant \sum_{j=1}^{n}(-\operatorname{Im}\overline{\mathscr{A}_j})^2.$$

A simple verification shows that these two equations are equivalent to the condition $\operatorname{Re}V \geqslant 0$ on Ω.

4.7   Comments to Chapter 4

Sections 4.1–4.4 are due to Cialdea and Maz'ya [12].

The topics of Sections 4.5 and 4.6 are weakly coupled and coupled systems respectively. These results, which are contained in [54, 55], are due to Langer and Maz'ya, while the results concerning the L^∞-dissipativity in Section 4.6.1 are due to Kresin and Maz'ya, who widely investigated this topic (see [51]; [49] contains a survey of their results). They considered the case in which the coefficients of the operator depend on t as well, founding necessary and, separately, sufficient conditions for the validity of the maximum modulus principle. They studied also maximum principles in which the norm is understood in a generalized sense, i.e., as the Minkowski functional of a compact convex body in $\mathbb{R}^n$ containing the origin. Also in this general case they gave necessary and (separately if the coefficients of the system depend on t) sufficient conditions for the validity of the maximum norm principle.

We mention that Auschér, Barthélemy, Bénilan and Ouhabaz [4] considered scalar equations under broad hypotheses on the coefficients. They extended Kresin–Maz'ya results for scalar equations with smooth coefficients (see Theorem 4.32) to the case of L^∞ coefficients.

In order to shortly describe Auschér, Barthélemy, Bénilan and Ouhabaz results, we introduce some notation. In the arbitrary domain $\Omega \subset \mathbb{R}^n$, we consider the elliptic differential operator

$$Au = -\partial_j(a_{kj}\partial_k u) + b_k\partial_k u - \partial_k(c_k u) + a_0 u$$

where the complex coefficients a_{kj}, b_k, c_k, a_0 belong to $L^\infty(\Omega)$ and satisfy the ellipticity condition

$$\operatorname{Re} a_{kj}(x)\xi_k\overline{\xi}_j \geqslant c|\xi|^2 \qquad a.e.\ x \in \Omega,\ \forall\,\xi \in \mathbb{C}^n$$

for a constant $c > 0$.

Let $V \subset H^1(\Omega)$ be a closed subspace such that $H_0^1(\Omega) \subset V$.

Let $\mathscr{L}_V$ be the sesquilinear form

$$\mathscr{L}_V(u,v) = \int_\Omega (a_{kj}\partial_k u\,\partial_j\overline{v} + b_j\overline{v}\partial_j u + c_j u\partial_j\overline{v} + a_0 u\overline{v})\,dx$$

whose domain is $D(\mathscr{L}_V) = V$.

Let A_V be the operator defined in the following way. A function u belongs to the domain $D(A_V)$ if and only if there exists $v \in L^2(\Omega)$ such that

$$\mathscr{L}_V(u,\varphi) = \int_\Omega v\overline{\varphi}\,dx, \qquad \forall\,\varphi \in V.$$

We set $A_V u = v$.

It is clear that the Dirichlet condition corresponds to $V = H_0^1(\Omega)$.

The operator $-A_V$ is the generator of a C^0-semigroup on $L^2(\Omega)$ (see Kato [47]). Let us denote it by $(e^{-tA_V})_{t\geqslant0}$.

This semigroup is L^∞-contractive if

$$\|e^{-tA_V}u\|_\infty \leqslant \|u\|_\infty,\ \forall\,u \in L^2(\Omega) \cap L^\infty(\Omega).$$

Given $u \in H^1(\Omega)$, we denote by φ_k the function

$$\varphi_k = \operatorname{Im}(\partial_k u \operatorname{sign}\overline{u})\,|u|^{-1}\chi_{u\neq0}$$

where, as usual,

$$\operatorname{sign} u = |u|^{-1}u\chi_{u\neq0}.$$

Theorem 4.37. *The semigroup* $(e^{-tA_V})_{t\geqslant0}$ *is* L^∞-*contractive if and only if the following conditions are satisfied*

(a) *if* $u \in V$ *then* $(|u| - 1)^+ \operatorname{sign} u \in V$;

(b) *we have*

$$\int_\Omega ((\operatorname{Re} a_{kj})\varphi_k\varphi_j|u| - (\operatorname{Im} a_{kj})\varphi_k\partial_j|u| + \operatorname{Im}(c_j - b_j)\varphi_j|u|$$

$$+(\operatorname{Re} c_j)\partial_j|u| + (\operatorname{Re} a_0)|u|)\,dx \geqslant 0$$

for any $u \in V$ *such that* $\varphi_k\varphi_j|u|$ *and* $\varphi_k\partial_j|u|$ *belong to* $L^1(\Omega)$ ($j,k = 1,\ldots,n$).

In the particular case of Dirichlet conditions, a more explicit result can be obtained:

Theorem 4.38 (Auschér, Barthélemy, Bénilan, Ouhabaz [4]).
The semigroup $(e^{-tA_{H_0^1}})_{t\geqslant 0}$ *is* L^∞*-contractive if and only if the following conditions are satisfied*

(i) $\mathrm{Im}(a_{kj} + a_{jk}) = 0$ *for* $j, k = 1, \ldots, n$;
(ii) $f_0 = \mathrm{Re}\, a_0 - \partial_j(\mathrm{Re}\, c_j)$ *is a positive Radon measure on* Ω;
(iii) $f_k = \partial_j(\mathrm{Im}\, a_{kj}) \in L^1_{\mathrm{loc}}(\Omega)$ *for* $k = 1, \ldots, n$;
(iv) *we have*

$$(\mathrm{Re}\, a_{kj})\xi_k\xi_j + (\mathrm{Im}(c_j - b_j) + f_j)\xi_j + f_{0,r} \geqslant 0$$

a.e. on Ω *and for any* $\xi \in \mathbb{R}^n$, *where* $f_{0,r}$ *is the regular part of the measure* f_0.

Corollary 4.39. *If* $b_j = c_j = a_0 = 0$ $(1 \leqslant j \leqslant n)$, *the following conditions are equivalent*

(i) $(e^{-tA_{H_0^1}})_{t\geqslant 0}$ *is* L^∞*-contractive;*

(ii) $(e^{-tA_{H_0^1}})_{t\geqslant 0}$ *is real;*

(iii) $\mathscr{L}(u, v) = \displaystyle\int_\Omega (\mathrm{Re}\, a_{kj})\partial_k u\, \partial_j\overline{v}\, dx, \quad \forall u, v \in H_0^1(\Omega);$

(iv) $\mathrm{Im}(a_{kj} + a_{jk}) = 0$ *for* $j, k = 1, \ldots, n$ *and* $\partial_j(\mathrm{Im}\, a_{kj}) = 0$ *in the sense of distributions* $(1 \leqslant k \leqslant n)$.

Chapter 5

The Angle of L^p-dissipativity

The analyticity of a contractive semigroup $\{T(t)\}$ is closely connected with the possibility of extending $\{T(t)\}$ to a contractive semigroup $\{T(z)\}$ ($z \in \mathbb{C}$) in an angle, called angle of dissipativity.

The problem then arises: given an L^p-dissipative operator A, to find its angle of L^p-dissipativity, i.e., the set of complex values z such that zA is still L^p-dissipative.

We obtain explicitly such an angle in two cases. The first one, which is discussed in Section 5.1, is about the scalar operator $A = \nabla^t(\mathscr{A}(x)\nabla)$ where $\mathscr{A} = \{a_{ij}(x)\}$ $(i, j = 1, \ldots, n)$ is a matrix with complex entries defined in a domain $\Omega \subset \mathbb{R}^n$.

The second case concerns the system $\partial_h(\mathscr{A}^h(x)\partial_h u)$, where $\mathscr{A}^h(x) = \{a_{ij}^h(x)\}$ $(i, j = 1, \ldots, m)$ are matrices with complex entries defined in a domain $\Omega \subset \mathbb{R}^n$ $(h = 1, \ldots, n)$. This is the object of Section 5.2.

5.1 Angle of dissipativity of the scalar operator $\nabla^t(\mathscr{A}(x)\nabla)$

At first we find the angle of dissipativity of the form $\mathscr{L}$, related to the operator

$$A = \nabla^t(\mathscr{A}(x)\nabla) \tag{5.1}$$

where $\mathscr{A} = \{a_{ij}(x)\}$ $(i, j = 1, \ldots, n)$ is a matrix with complex locally integrable entries defined in a domain $\Omega \subset \mathbb{R}^n$.

This means that we want to find the set of complex values z such that the form $z\mathscr{L}$ is L^p-dissipative, provided $\mathscr{L}$ itself is L^p-dissipative.

We know that, if $\operatorname{Im}\mathscr{A}$ is symmetric, there is the L^p-dissipativity of the form $\mathscr{L}$ if and only if

$$|p - 2|\,|\langle \operatorname{Im}\mathscr{A}(x)\xi, \xi\rangle| \leqslant 2\sqrt{p - 1}\langle \operatorname{Re}\mathscr{A}(x)\xi, \xi\rangle \tag{5.2}$$

for almost every $x \in \Omega$ and for any $\xi \in \mathbb{R}^n$.

We start with the following elementary lemma

Lemma 5.1. *Let P and Q be two real measurable functions defined on a set $\Omega \subset \mathbb{R}^n$. Let us suppose that $P(x) \geqslant 0$ almost everywhere. The inequality*

$$P(x)\cos\vartheta - Q(x)\sin\vartheta \geqslant 0 \qquad (\vartheta \in [-\pi, \pi]) \tag{5.3}$$

holds for almost every $x \in \Omega$ if and only if

$$\operatorname{arccot}\left[\operatorname{ess\,inf}_{x \in \Xi}\left(Q(x)/P(x)\right)\right] - \pi \leqslant \vartheta \leqslant \operatorname{arccot}\left[\operatorname{ess\,sup}_{x \in \Xi}\left(Q(x)/P(x)\right)\right] \tag{5.4}$$

where $\Xi = \{x \in \Omega \mid P^2(x) + Q^2(x) > 0\}$ and we set

$$Q(x)/P(x) = \begin{cases} +\infty & \text{if } P(x) = 0,\ Q(x) > 0 \\ -\infty & \text{if } P(x) = 0,\ Q(x) < 0. \end{cases}$$

Here $0 < \operatorname{arccot} y < \pi$, $\operatorname{arccot}(+\infty) = 0$, $\operatorname{arccot}(-\infty) = \pi$ and

$$\operatorname{ess\,inf}_{x \in \Xi}\left(Q(x)/P(x)\right) = +\infty, \quad \operatorname{ess\,sup}_{x \in \Xi}\left(Q(x)/P(x)\right) = -\infty$$

if Ξ has zero measure.

Proof. If Ξ has positive measure and $P(x) > 0$, inequality (5.3) means

$$\cos\vartheta - (Q(x)/P(x))\sin\vartheta \geqslant 0$$

and this is true if and only if

$$\operatorname{arccot}\left(Q(x)/P(x)\right) - \pi \leqslant \vartheta \leqslant \operatorname{arccot}\left(Q(x)/P(x)\right). \tag{5.5}$$

If $x \in \Xi$ and $P(x) = 0$, (5.3) means

$$-\pi \leqslant \vartheta \leqslant 0,\ \text{if } Q(x) > 0, \quad 0 \leqslant \vartheta \leqslant \pi,\ \text{if } Q(x) < 0.$$

This shows that (5.3) is equivalent to (5.5) provided that $x \in \Xi$. On the other hand, if $x \notin \Xi$, $P(x) = Q(x) = 0$ almost everywhere and (5.3) is always satisfied. Therefore, if Ξ has positive measure, (5.3) and (5.4) are equivalent.

If Ξ has zero measure, the result is trivial. $\square$

The next theorem provides a necessary and sufficient condition for the L^p-dissipativity of $z\,\mathscr{L}$.

Theorem 5.2. *Let the matrix $\mathscr{A}$ be symmetric. Let us suppose that the form $\mathscr{L}$ is L^p-dissipative. Set*

$$\Lambda_1 = \operatorname{ess\,inf}_{(x,\xi)\in\Xi} \frac{\langle \operatorname{Im}\mathscr{A}(x)\xi, \xi\rangle}{\langle \operatorname{Re}\mathscr{A}(x)\xi, \xi\rangle}, \qquad \Lambda_2 = \operatorname{ess\,sup}_{(x,\xi)\in\Xi} \frac{\langle \operatorname{Im}\mathscr{A}(x)\xi, \xi\rangle}{\langle \operatorname{Re}\mathscr{A}(x)\xi, \xi\rangle}$$

where

$$\Xi = \{(x,\xi) \in \Omega \times \mathbb{R}^n \mid \langle \operatorname{Re}\mathscr{A}(x)\xi,\xi \rangle > 0\}. \tag{5.6}$$

The form $z\,\mathscr{L}$ is L^p-dissipative if and only if

$$\vartheta_- \leqslant \arg z \leqslant \vartheta_+ , \tag{5.7}$$

where

$$\vartheta_- = \begin{cases} \operatorname{arccot}\left(\dfrac{2\sqrt{p-1}}{|p-2|} - \dfrac{p^2}{|p-2|}\dfrac{1}{2\sqrt{p-1}+|p-2|\Lambda_1}\right) - \pi & \text{if } p \neq 2 \\ \operatorname{arccot}(\Lambda_1) - \pi & \text{if } p = 2 \end{cases}$$

$$\vartheta_+ = \begin{cases} \operatorname{arccot}\left(-\dfrac{2\sqrt{p-1}}{|p-2|} + \dfrac{p^2}{|p-2|}\dfrac{1}{2\sqrt{p-1}-|p-2|\Lambda_2}\right) & \text{if } p \neq 2 \\ \operatorname{arccot}(\Lambda_2) & \text{if } p = 2. \end{cases}$$

Proof. The matrix $\mathscr{A}$ being symmetric, $\operatorname{Im}(e^{i\vartheta}\mathscr{A})$ is symmetric and in view of (5.2), the form $e^{i\vartheta}\mathscr{L}$ (with $\vartheta \in [-\pi,\pi]$) is L^p-dissipative if and only if

$$|p-2|\,|\langle \operatorname{Re}\mathscr{A}(x)\xi,\xi\rangle \sin\vartheta + \langle \operatorname{Im}\mathscr{A}(x)\xi,\xi\rangle \cos\vartheta|$$
$$\leqslant 2\sqrt{p-1}(\langle \operatorname{Re}\mathscr{A}(x)\xi,\xi\rangle \cos\vartheta - \langle \operatorname{Im}\mathscr{A}(x)\xi,\xi\rangle \sin\vartheta) \tag{5.8}$$

for almost every $x \in \Omega$ and for any $\xi \in \mathbb{R}^n$. Suppose $p \neq 2$. Setting

$$a(x,\xi) = |p-2|\,\langle \operatorname{Re}\mathscr{A}(x)\xi,\xi\rangle, \qquad b(x,\xi) = |p-2|\,\langle \operatorname{Im}\mathscr{A}(x)\xi,\xi\rangle,$$
$$c(x,\xi) = 2\sqrt{p-1}\,\langle \operatorname{Re}\mathscr{A}(x)\xi,\xi\rangle, \quad d(x,\xi) = 2\sqrt{p-1}\,\langle \operatorname{Im}\mathscr{A}(x)\xi,\xi\rangle,$$

the inequality in (5.8) can be written as the system

$$\begin{cases} (c(x,\xi) - b(x,\xi))\cos\vartheta - (a(x,\xi) + d(x,\xi))\sin\vartheta \geqslant 0, \\ (c(x,\xi) + b(x,\xi))\cos\vartheta + (a(x,\xi) - d(x,\xi))\sin\vartheta \geqslant 0. \end{cases} \tag{5.9}$$

Noting that $c(x,\xi) \pm b(x,\xi) \geqslant 0$ because of (5.2), the solutions of the inequalities in (5.9) are given by the ϑ's satisfying both of the following conditions (see Lemma 5.1)

$$\begin{cases} \operatorname{arccot}\left(\operatorname{ess\,inf}_{(x,\xi)\in\Xi_1} \dfrac{a(x,\xi)+d(x,\xi)}{c(x,\xi)-b(x,\xi)}\right) - \pi \\ \qquad \leqslant \vartheta \leqslant \operatorname{arccot}\left(\operatorname{ess\,sup}_{(x,\xi)\in\Xi_1} \dfrac{a(x,\xi)+d(x,\xi)}{c(x,\xi)-b(x,\xi)}\right) \\ \operatorname{arccot}\left(\operatorname{ess\,inf}_{(x,\xi)\in\Xi_2} \dfrac{d(x,\xi)-a(x,\xi)}{c(x,\xi)+b(x,\xi)}\right) - \pi \\ \qquad \leqslant \vartheta \leqslant \operatorname{arccot}\left(\operatorname{ess\,sup}_{(x,\xi)\in\Xi_2} \dfrac{d(x,\xi)-a(x,\xi)}{c(x,\xi)+b(x,\xi)}\right), \end{cases} \tag{5.10}$$

where

$$\Xi_1 = \{(x,\xi) \in \Omega \times \mathbb{R}^n \mid (a(x,\xi)+d(x,\xi))^2 + (c(x,\xi)-b(x,\xi))^2 > 0\},$$
$$\Xi_2 = \{(x,\xi) \in \Omega \times \mathbb{R}^n \mid (a(x,\xi)-d(x,\xi))^2 + (b(x,\xi)+c(x,\xi))^2 > 0\}.$$

We have

$$a(x,\xi)\,d(x,\xi) = b(x,\xi)\,c(x,\xi),$$

$$a^2(x,\xi) + b^2(x,\xi) + c^2(x,\xi) + d^2(x,\xi) = p^2(\langle \operatorname{Re}\mathscr{A}(x)\xi,\xi\rangle^2 + \langle \operatorname{Im}\mathscr{A}(x)\xi,\xi\rangle^2)$$

and then, keeping in mind (5.2), we may write $\Xi_1 = \Xi_2 = \Xi$, where Ξ is given by (5.6).

Moreover

$$\frac{a(x,\xi) + d(x,\xi)}{c(x,\xi) - b(x,\xi)} \geqslant \frac{d(x,\xi) - a(x,\xi)}{c(x,\xi) + b(x,\xi)}$$

and then ϑ satisfies all of the inequalities in (5.10) if and only if

$$\operatorname{arccot}\left(\operatorname{ess\,inf}_{(x,\xi)\in\Xi}\tfrac{d(x,\xi)-a(x,\xi)}{c(x,\xi)+b(x,\xi)}\right) - \pi \leqslant \vartheta \leqslant \operatorname{arccot}\left(\operatorname{ess\,sup}_{(x,\xi)\in\Xi}\tfrac{a(x,\xi)+d(x,\xi)}{c(x,\xi)-b(x,\xi)}\right). \tag{5.11}$$

A direct computation shows that

$$\frac{d(x,\xi) - a(x,\xi)}{c(x,\xi) + b(x,\xi)} = \frac{2\sqrt{p-1}}{|p-2|} - \frac{p^2}{|p-2|}\frac{1}{2\sqrt{p-1} + |p-2|\Lambda(x,\xi)},$$

$$\frac{a(x,\xi) + d(x,\xi)}{c(x,\xi) - b(x,\xi)} = -\frac{2\sqrt{p-1}}{|p-2|} + \frac{p^2}{|p-2|}\frac{1}{2\sqrt{p-1} - |p-2|\Lambda(x,\xi)},$$

where

$$\Lambda(x,\xi) = \frac{\langle \operatorname{Im}\mathscr{A}(x)\xi,\xi\rangle}{\langle \operatorname{Re}\mathscr{A}(x)\xi,\xi\rangle}.$$

Hence condition (5.11) is satisfied if and only if (5.7) holds.

If $p = 2$, (5.8) is simply

$$\langle \operatorname{Re}\mathscr{A}(x)\xi,\xi\rangle \cos\vartheta - \langle \operatorname{Im}\mathscr{A}(x)\xi,\xi\rangle \sin\vartheta \geqslant 0$$

and the result follows directly from Lemma 5.1.   □

In the next corollary Ω is a bounded domain satisfying the same smoothness assumption as in Section 2.4 (see p. 48).

Corollary 5.3. *Let a_{ij} belong to $C^1(\overline{\Omega})$ and let the matrix $\mathscr{A}$ be symmetric. Let us suppose that the operator (5.1) is L^p-dissipative. The operator zA is L^p-dissipative if and only if (5.7) holds.*

Proof. In view of Theorem 2.20, the operator zA is L^p-dissipative if and only if the form $z\mathscr{L}$ is L^p-dissipative. The result follows from Theorem 5.2.   □

Remark 5.4. If $\mathscr{A}$ is a real matrix, then $\Lambda_1 = \Lambda_2 = 0$ and the angle of dissipativity does not depend on the operator. In fact we have

$$\frac{2\sqrt{p-1}}{|p-2|} - \frac{p^2}{2\sqrt{p-1}|p-2|} = -\frac{|p-2|}{2\sqrt{p-1}}$$

and Theorem 5.2 shows that zA is dissipative if and only if

$$\text{arccot}\left(-\frac{|p-2|}{2\sqrt{p-1}}\right) - \pi \leqslant \arg z \leqslant \text{arccot}\left(\frac{|p-2|}{2\sqrt{p-1}}\right),$$

i.e.,

$$|\arg z| \leqslant \arctan\left(\frac{2\sqrt{p-1}}{|p-2|}\right). \tag{5.12}$$

5.2   The angle of dissipativity of the system $\partial_h(\mathscr{A}^h(x)\partial_h u)$

In this section we find the precise angle of dissipativity for the form $\mathscr{L}$ related to operator (4.48) with complex coefficients.

We first consider the ordinary differential operator (4.13) where $\mathscr{A}(x)$ is a matrix whose elements are complex locally integrable functions. Define the functions

$$P(x,\lambda,\omega) = \text{Re}\langle \mathscr{A}\lambda,\lambda\rangle - (1-2/p)^2 \text{Re}\langle \mathscr{A}\omega,\omega\rangle(\text{Re}\langle\lambda,\omega\rangle)^2$$
$$- (1-2/p)\text{Re}(\langle \mathscr{A}\omega,\lambda\rangle - \langle \mathscr{A}\lambda,\omega\rangle)\text{Re}\langle\lambda,\omega\rangle;$$

$$Q(x,\lambda,\omega) = \text{Im}\langle \mathscr{A}\lambda,\lambda\rangle - (1-2/p)^2 \text{Im}\langle \mathscr{A}\omega,\omega\rangle(\text{Re}\langle\lambda,\omega\rangle)^2$$
$$- (1-2/p)\text{Im}(\langle \mathscr{A}\omega,\lambda\rangle - \langle \mathscr{A}\lambda,\omega\rangle)\text{Re}\langle\lambda,\omega\rangle \tag{5.13}$$

and denote by Ξ the set

$$\Xi = \{(x,\lambda,\omega) \in (a,b) \times \mathbb{C}^m \times \mathbb{C}^m \mid |\omega| = 1,\ P^2(x,\lambda,\omega) + Q^2(x,\lambda,\omega) > 0\}.$$

By adopting the conventions introduced in Lemma 5.1, we have

Theorem 5.5. *Let $\mathscr{L}$ be L^p-dissipative. The form $z\mathscr{L}$ is L^p-dissipative if and only if*

$$\vartheta_- \leqslant \arg z \leqslant \vartheta_+$$

where

$$\vartheta_- = \text{arccot}\left(\text{ess inf}_{(x,\lambda,\omega)\in\Xi}(Q(x,\lambda,\omega)/P(x,\lambda,\omega))\right) - \pi,$$

$$\vartheta_+ = \text{arccot}\left(\text{ess sup}_{(x,\lambda,\omega)\in\Xi}(Q(x,\lambda,\omega)/P(x,\lambda,\omega))\right).$$

Proof. In view of Theorem 4.4 the form $e^{i\vartheta}\mathscr{L}$ is L^p-dissipative if and only if

$$\text{Re}\langle e^{i\vartheta}\mathscr{A}\lambda,\lambda\rangle - (1-2/p)^2 \text{Re}\langle e^{i\vartheta}\mathscr{A}\omega,\omega\rangle(\text{Re}\langle\lambda,\omega\rangle)^2$$
$$- (1-2/p)\text{Re}(\langle e^{i\vartheta}\mathscr{A}\omega,\lambda\rangle - \langle e^{i\vartheta}\mathscr{A}\lambda,\omega\rangle)\text{Re}\langle\lambda,\omega\rangle \geqslant 0 \tag{5.14}$$

for almost every $x \in (a,b)$ and for any $\lambda,\omega \in \mathbb{C}^m$, $|\omega| = 1$.

By means of the functions $P(x, \lambda, \omega)$ and $Q(x, \lambda, \omega)$ introduced in (5.13), we can write (5.14) in the form

$$P(x, \lambda, \omega) \cos \vartheta - Q(x, \lambda, \omega) \sin \vartheta \geqslant 0.$$

Lemma 5.1 gives the result.   □

Let now A and $\mathscr{L}$ denote the partial differential operator (4.48) and the related form, respectively. We have

Theorem 5.6. *Let $\mathscr{L}$ be L^p-dissipative. The form $z\mathscr{L}$ is L^p-dissipative if and only if $\vartheta_- \leqslant \arg z \leqslant \vartheta_+$, where*

$$\vartheta_- = \max_{h=1,\dots,n} \operatorname{arccot} \left(\operatorname{ess\,inf}_{(x,\lambda,\omega)\in\Xi_h} (Q_h(x, \lambda, \omega)/P_h(x, \lambda, \omega)) \right) - \pi,$$

$$\vartheta_+ = \min_{h=1,\dots,n} \operatorname{arccot} \left(\operatorname{ess\,sup}_{(x,\lambda,\omega)\in\Xi_h} (Q_h(x, \lambda, \omega)/P_h(x, \lambda, \omega)) \right),$$

and

$$P_h(x, \lambda, \omega) = \operatorname{Re}\langle \mathscr{A}^h(x)\lambda, \lambda \rangle - (1 - 2/p)^2 \operatorname{Re}\langle \mathscr{A}^h(x)\omega, \omega \rangle (\operatorname{Re}\langle \lambda, \omega \rangle)^2$$
$$- (1 - 2/p) \operatorname{Re}(\langle \mathscr{A}^h(x)\omega, \lambda \rangle - \langle \mathscr{A}^h(x)\lambda, \omega \rangle) \operatorname{Re}\langle \lambda, \omega \rangle,$$

$$Q_h(x, \lambda, \omega) = \operatorname{Im}\langle \mathscr{A}^h(x)\lambda, \lambda \rangle - (1 - 2/p)^2 \operatorname{Im}\langle \mathscr{A}^h(x)\omega, \omega \rangle (\operatorname{Re}\langle \lambda, \omega \rangle)^2$$
$$- (1 - 2/p) \operatorname{Im}(\langle \mathscr{A}^h(x)\omega, \lambda \rangle - \langle \mathscr{A}^h(x)\lambda, \omega \rangle) \operatorname{Re}\langle \lambda, \omega \rangle,$$

$$\Xi_h = \{(x, \lambda, \omega) \in \Omega \times \mathbb{C}^m \times \mathbb{C}^m \mid |\omega|$$
$$= 1, \ P_h^2(x, \lambda, \omega) + Q_h^2(x, \lambda, \omega) > 0\}.$$

Proof. By Theorem 4.20, the operator $e^{i\vartheta} A$ is L^p-dissipative if and only if

$$\operatorname{Re}\langle e^{i\vartheta} \mathscr{A}^h(x)\lambda, \lambda \rangle - (1 - 2/p)^2 \operatorname{Re}\langle e^{i\vartheta} \mathscr{A}^h(x)\omega, \omega \rangle (\operatorname{Re}\langle \lambda, \omega \rangle)^2 \tag{5.15}$$
$$- (1 - 2/p) \operatorname{Re}(\langle e^{i\vartheta} \mathscr{A}^h(x)\omega, \lambda \rangle - \langle e^{i\vartheta} \mathscr{A}^h(x)\lambda, \omega \rangle) \operatorname{Re}\langle \lambda, \omega \rangle \geqslant 0$$

for almost every $x \in \Omega$ and for any $\lambda, \omega \in \mathbb{C}^m$, $|\omega| = 1$, $h = 1, \dots, n$.

As in the proof of Theorems 5.5, conditions (5.15) mean $\vartheta_-^{(h)} \leqslant \vartheta \leqslant \vartheta_+^{(h)}$, where

$$\vartheta_-^{(h)} = \operatorname{arccot} \left(\operatorname{ess\,inf}_{(x,\lambda,\omega)\in\Xi_h} (Q_h(x, \lambda, \omega)/P_h(x, \lambda, \omega)) \right) - \pi,$$

$$\vartheta_+^{(h)} = \operatorname{arccot} \left(\operatorname{ess\,sup}_{(x,\lambda,\omega)\in\Xi_h} (Q_h(x, \lambda, \omega)/P_h(x, \lambda, \omega)) \right),$$

and the result follows.   □

5.3 Comments to Chapter 5

The results contained in this chapter are taken from Cialdea and Maz'ya [12].

The fact that, for a real matrix, the angle of dissipativity does not depend on the operator (see Remark 5.4) is well known (see, e.g., Fattorini [26, 27], Okazawa [82]). See also Stein [95], where many years ago a (smaller) angle of dissipativity was used to prove the analyticity of semigroups generated by linear elliptic differential operators.

We remark that generally speaking the angle of analyticity of a semigroup does not coincide with the angle of dissipativity. For example, for the Laplacian, the angle of dissipativity is given by (5.12), while the angle of analyticity is $|\arg z| < \pi/2$ (see, e.g., Ouhabaz [84]).

These two angles may coincide for some particular operators. Chill, Fašangová, Metafune and Pallara [10] proved that this is the case for the Ornstein–Uhlenbeck operator $\Delta + Bx \cdot \nabla$, B being a real nonzero constant matrix.

Chapter 6

Higher-order Differential Operators in L^p

In previous sections we have found conditions for the L^p-dissipativity of second-order scalar equations and systems. One can ask whether these results hold for higher-order operators. In this chapter it is proved that the answer is negative.

In Section 6.1 some counterexamples in dimension one are constructed. They are extended to the multi-dimensional case in Section 6.2.

By means of these counterexamples, it is shown in Section 6.3 that there are no L^p-dissipative operators of higher order for $p \neq 2$.

However there are some positive results when the differential operator is given on the cone of nonnegative functions. Then for a class of operators of fourth order, L^p-dissipativity holds for $3/2 \leqslant p \leqslant 3$ (see Section 6.4).

6.1 Some counterexamples in dimension one

In order to study the dissipativity of higher-order operators, we need to establish at first necessary and sufficient conditions under which the one-dimensional integral

$$\int v^{(k)} |v|^{p-1} \operatorname{sgn} v \, dx$$

preserves sign as v ranges over real-valued elements of $C_0^\infty(\mathbf{R})$. We will also consider the same integral for functions ranging over the more narrow class $(C_0^\infty(\mathbf{R}))^+$. The proof of these necessary and sufficient conditions hinges on some counterexamples which will be given in the next lemmas.

Lemma 6.1. *Let $k \geqslant 2$, let I be a nonempty open interval and suppose that the function $v : \mathbf{R} \to \mathbf{R}$ is infinitely differentiable on I, $v|_I > 0$, $v^{(k)}|_I = 0$ and $(v^{p-1})^{(k)}$ is nonzero at some point in I. Then there exist functions v_1 and v_2,* *infinitely differentiable and nonnegative on* I, *such that* $\operatorname{supp}(v - v_1) \cup \operatorname{supp}(v - v_2) \subset I$ *and*

$$\int_I v_1^{(k)} v_1^{p-1} \, dx < 0, \qquad \int_I v_2^{(k)} v_2^{p-1} \, dx > 0.$$

Proof. Since $(v^{p-1})^{(k)}$ is continuous and not identically zero on I, there is a nonzero function $\varphi \in (C_0^\infty(\mathbf{R}))^+$ with $\operatorname{supp} \varphi \subset I$ such that $(v^{p-1})^{(k)}$ is either positive or negative on the support of φ. Define the function f by

$$f(\varepsilon) = \int_I (v + \varepsilon \varphi)^{(k)} (v + \varepsilon \varphi)^{p-1} \, dx.$$

Then f is well defined on a small neighborhood of 0, is infinitely differentiable there and fulfills $f(0) = 0$.

Since $v^{(k)}|_I = 0$, it follows that

$$f'(\varepsilon) = \int_I \varphi^{(k)} (v + \varepsilon \varphi)^{p-1} \, dx + (p-1)\varepsilon \int_I \varphi^{(k)} \varphi (v + \varepsilon \varphi)^{p-2} \, dx,$$

$$f'(0) = \int_I \varphi^{(k)} v^{p-1} \, dx = (-1)^k \int_I \varphi (v^{p-1})^{(k)} \, dx.$$

Hence, by our choice of φ, $f'(0) \neq 0$ so for some small ε_1 and ε_2 with different signs, the desired functions can be given by $v_i = v + \varepsilon_i \varphi$, $i = 1, 2$. $\qquad \square$

Lemma 6.2. *Let* $p > 1$ *and* $k \geqslant 2$, *and suppose that* $v \in C^\infty(\mathbf{R})$, $v > 0$, $(v^{p-1})^{(k)} = 0$ *and* $v^{(k)}$ *is nonzero at some point. If* $v^{(k)} v^{p-1} \in L^1(\mathbf{R})$ *and*

$$\int v^{(k)} v^{p-1} \, dx = 0,$$

there exist functions v_1 *and* v_2, *infinitely differentiable and nonnegative, such that* $\operatorname{supp}(v - v_i)$ *is compact* $(i = 1, 2)$ *and*

$$\int v_1^{(k)} v_1^{p-1} \, dx < 0, \qquad \int v_2^{(k)} v_2^{p-1} \, dx > 0.$$

Proof. Choose a nonzero function $\varphi \in (C_0^\infty(\mathbf{R}))^+$ such that $v^{(k)}$ is either positive or negative on the support of φ and define as above, for small ε, the differentiable function f by

$$f(\varepsilon) = \int (v + \varepsilon \varphi)^{(k)} (v + \varepsilon \varphi)^{p-1} \, dx.$$

By hypothesis, $f(0) = 0$, and also

$$f'(\varepsilon) = \int \varphi^{(k)} (v + \varepsilon \varphi)^{p-1} \, dx + (p-1) \int (v + \varepsilon \varphi)^{(k)} \varphi (v + \varepsilon \varphi)^{p-2} \, dx,$$

$$f'(0) = (-1)^k \int \varphi (v^{p-1})^{(k)} \, dx + (p-1) \int \varphi v^{(k)} v^{p-2} \, dx.$$

Since the first integral in the expression for $f'(0)$ vanishes and the second is nonzero by our choice of φ, $f'(0)$ is nonzero. As in the proof of Lemma 6.1, the existence of the desired functions v_1 and v_2 now follows. $\qquad\square$

Lemma 6.3. *If $p > 1$, $p \neq 2$ and $k \geqslant 3$ is odd, then the integral*

$$\int v^{(k)} v^{p-1} \, dx \tag{6.1}$$

does not preserve sign as v ranges over $(C_0^\infty(\mathbf{R}))^+$.

Proof. That the integral (6.1) can not preserve sign over $(C_0^\infty(\mathbf{R}))^+$ unless it is identically zero, follows directly by observing that the substitution $u(x) = v(-x)$ changes the sign of (6.1) since k is odd. To see that (6.1) is not identically zero, take a $v \in (C_0^\infty(\mathbf{R}))^+$ whose restriction to the interval $[1, 2]$ is $x \mapsto x$ if $p < 2$ and $x \mapsto x^{k-1}$ otherwise. If $p \neq 2$, the assumptions of Lemma 6.1 are clearly satisfied with $I = (1, 2)$, so let v_1 be as in the conclusion of the lemma. We get

$$\int v^{(k)} v^{p-1} \, dx = \int v_1^{(k)} v_1^{p-1} \, dx - \int_1^2 v_1^{(k)} v_1^{p-1} \, dx$$

and since the last term is nonzero, (6.1) is not identically zero. $\qquad\square$

Lemma 6.4. *Let $k \geqslant 6$ be even and suppose that $p > 1$, $p \neq 2$. Then the integral*

$$\int v^{(k)} v^{p-1} \, dx$$

assumes both negative and positive values as v ranges over $(C_0^\infty(\mathbf{R}))^+$.

Proof. We treat the case $p > 2$ first. Let $u \in C^\infty(\mathbb{R})$ be defined by

$$u(x) = (1 + x^2)^{1/(p-1)}, \quad x \in \mathbb{R}.$$

An induction argument applied to the terms of $u^{(j)}$ gives the estimate

$$|u^{(j)}(x)| \leqslant C|x|^{2/(p-1)-j}, \quad |x| > 1, \ j = 0, \ldots, k, \tag{6.2}$$

so $u^{(k)} u^{p-1}$ is an L^1-function. Repeated integration by parts implies that

$$\int_{-\omega}^{\omega} u^{(k)} u^{p-1} \, dx = \left[u^{(k-1)} u^{p-1} - u^{(k-2)}(u^{p-1})' + u^{(k-3)}(u^{p-1})'' \right]_{-\omega}^{\omega},$$

where, by (6.2), the right-hand side tends to zero as ω tends to infinity, so we can apply Lemma 6.2 to the function u. Denote the resulting two functions by u_1 and u_2. Choose an even function $\phi \in C_0^\infty(\mathbf{R})$ that satisfies $0 \leqslant \phi \leqslant 1$, $\operatorname{supp} \phi \subset (-2, 2)$ and is identically 1 on a neighborhood of $[-1, 1]$. Choose $i \in \{0, 1\}$ and define for $\omega \geqslant 2$ the functions $v_\omega \in (C_0^\infty(\mathbf{R}))^+$ by

$$v_\omega(x) = u_i(x)\phi(x/\omega), \quad x \in \mathbb{R}.$$

With C being a generic constant not depending on ω, it follows by (6.2) that $v_\omega^{(k)}$ is estimated as

$$|v_\omega^{(k)}(x)| = \left| \sum_{j=0}^{k} a_j u^{(j)}(x)\omega^{j-k}\phi^{(k-j)}(x/\omega) \right| \tag{6.3}$$

$$\leqslant C\omega^{2/(p-1)-k}, \quad \omega \leqslant |x| \leqslant 2\omega$$

for ω large enough. We have for all large ω,

$$\int v_\omega^{(k)} v_\omega^{p-1}\, dx = \int_{-\omega}^{\omega} u_i^{(k)} u_i^{p-1}\, dx + 2\int_{\omega}^{2\omega} v_\omega^{(k)} v_\omega^{p-1}\, dx,$$

where, using (6.3), the modulus of the second integral on the right-hand side is majorized by

$$\int_{\omega}^{2\omega} C\omega^{2/(p-1)-k}\omega^2\, dx = C\omega^{2/(p-1)+3-k}.$$

This tends to zero as ω tends to infinity since the hypothesis implies that the exponent in the right-hand side is less than -1. Hence, we conclude that

$$\lim_{\omega \to \infty} \int v_\omega^{(k)} v_\omega^{p-1}\, dx = \int u_i^{(k)} u_i^{p-1}\, dx$$

and since $v_\omega \in (C_0^\infty(\mathbf{R}))^+$ for each ω and the sign of the right-hand side can be chosen arbitrarily, we are done with the case $p > 2$.

Suppose that $1 < q < 2$ and let p be the conjugate exponent to q. Then $p > 2$ so let u, u_i and ϕ be as above and define, for some $i \in \{1,2\}$, the functions $v_\omega \in (C_0^\infty(\mathbf{R}))^+$ for $\omega \geqslant 2$ by

$$v_\omega(x) = u_i^{p-1}(x)\phi(x/\omega), \quad x \in \mathbb{R}.$$

It is easily verified that

$$v_\omega(x) \leqslant Cx^2, \quad |v_\omega^{(k)}(x)| \leqslant C\omega^{2-k}, \quad \omega \leqslant |x| \leqslant 2\omega,$$

and we have, observing that $(p-1)(q-1) = 1$,

$$\int v_\omega^{(k)} v_\omega^{q-1}\, dx = \int_{-\omega}^{\omega} (u_i^{p-1})^{(k)} u_i\, dx + 2\int_{\omega}^{2\omega} v_\omega^{(k)} v_\omega^{q-1}\, dx. \tag{6.4}$$

Using the estimates of v_ω and $v_\omega^{(k)}$ above, we see that the second integral on the right in (6.4) is majorized by $C\omega^{2q+1-k}$ so it tends to zero as $\omega \to \infty$, since $k \geqslant 6$. Integrating the first term on the right-hand side of (6.4) by parts, the equality

$$\int_{-\omega}^{\omega} (u_i^{p-1})^{(k)} u_i\, dx$$

$$= \int_{-\omega}^{\omega} u_i^{(k)} u_i^{p-1}\, dx + \left[-u^{(k-3)}(u^{p-1})'' + u^{(k-2)}(u^{p-1})' - u^{(k-1)}u^{p-1} \right]_{-\omega}^{\omega}$$

is obtained. The last term tends to zero as $\omega \to \infty$ due to (6.2) so, by letting $\omega \to \infty$ in (6.4), it follows that

$$\lim_{\omega \to \infty} \int v_\omega^{(k)} v_\omega^{q-1} \, dx = \int u_i^{(k)} u_i^{p-1} \, dx,$$

showing that the integral given in the statement of the lemma can be both negative and positive as v ranges over $(C_0^\infty(\mathbf{R}))^+$. $\qquad\square$

Lemma 6.5. *Suppose that* $1 < p < \frac{3}{2}$ *or* $p > 3$. *Then the integral*

$$\int v^{(4)} v^{p-1} \, dx$$

assumes both negative and positive values as v *ranges over* $(C_0^\infty(\mathbf{R}))^+$.

Proof. Suppose that $p > 3$ and define the function u on $\mathbf{R}$ by

$$u(x) = x(1 - x)\chi_{[0,1]}(x),$$

where $\chi_{[0,1]}$ is the characteristic function of the set $[0, 1]$. It is straightforward to verify that u fulfills the requirements of Lemma 6.1 with $I = (0, 1)$ and $k = 4$ so let u_1 and u_2 be the two functions corresponding to u. Let ψ be a mollifier that is even and define, for $\varepsilon > 0$,

$$\varphi_\varepsilon(x) = \varepsilon^{-1} \psi(\varepsilon^{-1} x), \ x \in \mathbf{R}.$$

From now on, let C be a generic constant not depending on ε. We have the following estimate:

$$|\varphi_\varepsilon^{(j)}(x)| \leqslant C\varepsilon^{-j-1}, \ x \in \mathbf{R}, \ j = 0, \ldots, 4. \tag{6.5}$$

Choose $i \in \{1, 2\}$ and set $v_\varepsilon = u_i * \varphi_\varepsilon$. Then v_ε is a regularization of u_i and it is well known, see for instance Hörmander [41], that this implies that for each fixed $j \in \mathbf{N}$, $v_\varepsilon^{(j)}$ tends to $u_i^{(j)}$ uniformly on compact subsets of $(0, 1)$ as $\varepsilon \to 0+$ and that each v_ε belongs to $(C_0^\infty(\mathbf{R}))^+$. We now proceed by choosing $\eta > 0$ such that $\mathrm{supp}(u - u_i) \subset (\eta, 1 - \eta)$. Since $u(x) = u(1 - x)$ and φ_ε is even, it follows that $v_\varepsilon(x) = v_\varepsilon(1 - x)$ and we get, with ε small enough,

$$\int v_\varepsilon^{(4)} v_\varepsilon^{p-1} \, dx = 2 \int_{-\eta}^{\eta} v_\varepsilon^{(4)} v_\varepsilon^{p-1} \, dx + \int_{\eta}^{1-\eta} v_\varepsilon^{(4)} v_\varepsilon^{p-1} \, dx, \tag{6.6}$$

where the last term, by our remark on uniform convergence above, tends to

$$\int_{\eta}^{1-\eta} u_i^{(4)} u_i^{p-1} \, dx = \int_0^1 u_i^{(4)} u_i^{p-1} \, dx$$

as ε tends to zero. The final step is to show that the first term in the right-hand side of (6.6) tends to zero as $\varepsilon \to 0+$. Note that $u = u_i$ on a neighborhood of $[-\eta, \eta]$. Hence, using (6.5) and letting ε be small enough, it follows that

$$
\begin{aligned}
|v_\varepsilon^{(j)}(x)| &= \left| \int_{-\varepsilon}^{\varepsilon} u(x-y)\varphi_\varepsilon^{(j)}(y)\, dy \right| \\
&\leqslant \int_{-\varepsilon}^{\varepsilon} C\varepsilon\varepsilon^{-j-1}\, dy = 2C\varepsilon^{1-j}, \quad |x| < \varepsilon,\ j = 0,\ldots,4
\end{aligned}
\tag{6.7}
$$

and that $v_\varepsilon^{(4)}(x) = 0$ for $\varepsilon \leqslant |x| \leqslant \eta$, so we arrive at

$$
\left| \int_{-\eta}^{\eta} v_\varepsilon^{(4)} v_\varepsilon^{p-1}\, dx \right| \leqslant \int_{-\varepsilon}^{\varepsilon} C\varepsilon^{-3}\varepsilon^{p-1}\, dx = 2C\varepsilon^{p-3},
$$

which tends to zero as ε tends to zero. Thus, taking limits in the expression (6.6), we finally obtain

$$
\lim_{\varepsilon \to 0+} \int v_\varepsilon^{(4)} v_\varepsilon^{p-1}\, dx = \int_0^1 u_i^{(4)} u_i^{p-1}\, dx
$$

and since the right-hand side is negative for $i = 1$ and positive for $i = 2$, the result follows for $p > 3$.

Suppose that $1 < q < \frac{3}{2}$ and let, as in the proof of the previous lemma, p be the conjugate exponent to q. It follows that $p > 3$, so let u_i and η be as defined above. Choose $i \in \{1,2\}$ and define for all small $\varepsilon > 0$ the functions $w_\varepsilon \in (C_0^\infty(\mathbf{R}))^+$ by $w_\varepsilon = u_i^{p-1} * \varphi_\varepsilon$. For $j = 0,\ldots,4$, we immediately get

$$
|w_\varepsilon^{(j)}(x)| \leqslant
\begin{cases}
\displaystyle\int_{-\varepsilon}^{\varepsilon} |x-y|^{p-1}|\varphi_\varepsilon^{(j)}(y)|\, dy \leqslant C\varepsilon^{p-1-j}, & |x| \leqslant 2\varepsilon, \\[2mm]
\displaystyle\int_{-\varepsilon}^{\varepsilon} C(x-y)^{p-1-j}\varphi_\varepsilon(y)\, dy \leqslant C_1 x^{p-1-j}, & 2\varepsilon < x < \eta,
\end{cases}
$$

implying that the functions $w_\varepsilon^{(4)} w_\varepsilon^{q-1}$ are majorized by the $L^1(\mathbf{R})$-function

$$
x \mapsto C\mathcal{X}_{(-1,2)}(x)(1 + |x|^{p-4} + |x-1|^{p-4}), \quad x \notin \{0,1\}.
$$

Furthermore, $w_\varepsilon^{(4)} w_\varepsilon^{q-1}$ tends to $(u_i^{p-1})^{(4)} u_i$ almost everywhere, the exceptional set being $\{0,1\}$, so a direct application of Lebesgue's dominated convergence theorem gives

$$
\lim_{\varepsilon \to 0+} \int w_\varepsilon^{(4)} w_\varepsilon^{q-1}\, dx = \int_0^1 (u_i^{p-1})^{(4)} u_i\, dx = \int_0^1 u_i^{(4)} u_i^{p-1}\, dx,
$$

where the second equality follows by integrating by parts, having in mind that u_i is equal to u on two neighborhoods of 0 and 1. The sign of the right-hand side can, by construction of u_i, be chosen arbitrarily by choosing i, and we are done. $\square$

Lemma 6.6. *If $\frac{3}{2} \leqslant p \leqslant 3$, $p \neq 2$, the integral*

$$\int v^{(4)} |v|^{p-1} \operatorname{sgn} v \, dx$$

changes sign as v ranges over real-valued elements of $C_0^\infty(\mathbf{R})$.

Proof. Suppose that $p > 2$ and let the function u be defined as in the proof of Lemma 6.5:

$$u(x) = x(1-x)\chi_{[0,1]}(x), \quad x \in \mathbf{R}.$$

As before, Lemma 6.1 guarantees the existence of two functions u_1 and u_2 with properties stated in the same lemma. Choose the constant $\eta \in (0, 1/4)$ so that $\operatorname{supp}(u - u_i) \subset (\eta, 1 - \eta)$ and let h be a nonnegative function which is infinitely differentiable on $(0, \infty)$, coincides with u on $[0, 2\eta]$ and fulfills $\operatorname{supp} h \subset [0, 1 - 2\eta]$. Let $i \in \{1, 2\}$ and define the constants

$$A = \int_0^1 u_i^{(4)} u_i^{p-1} \, dx, \, \sigma B = \int_0^1 h^{(4)} h^{p-1} \, dx,$$

choose $\ell \in \mathbb{N}$ so that $\ell |A| > 2|B|$ and let the function v be given by

$$v(x) = -h(-x) + \sum_{j=0}^{\ell-1} (-1)^j u_i(x - j) + (-1)^\ell h(x - \ell), \quad x \in \mathbf{R}.$$

Let φ_ε be defined exactly as in the proof of Lemma 6.5 and set $v_\varepsilon = v * \varphi_\varepsilon$ and $h_\varepsilon = h * \varphi_\varepsilon$. For ε small enough, we have

$$\int v_\varepsilon^{(4)} |v_\varepsilon|^{p-1} \operatorname{sgn} v_\varepsilon \, dx \tag{6.8}$$

$$= (\ell + 1) \int_{-\eta}^{\eta} v_\varepsilon^{(4)} |v_\varepsilon|^{p-1} \operatorname{sgn} v_\varepsilon \, dx + \ell \int_{\eta}^{1-\eta} v_\varepsilon^{(4)} v_\varepsilon^{p-1} \, dx + 2 \int_{\eta}^{1} h_\varepsilon^{(4)} h_\varepsilon^{p-1} \, dx,$$

where as before, the second integral on the right tends to

$$\int_{\eta}^{1-\eta} v^{(4)} v^{p-1} \, dx = \int_0^1 u_i^{(4)} u_i^{p-1} \, dx$$

as ε tends to zero due to the uniform convergence of v_ε and its derivatives on $[\eta, 1 - \eta]$. The same can be said about h_ε and its derivatives, so the third integral on the right-hand side of (6.8) tends to

$$\int_0^1 h^{(4)} h^{p-1} \, dx$$

as $\varepsilon \to 0+$. Now consider the first term on the right-hand side of (6.8). Similarly as in (6.7), it follows easily that $|v_\varepsilon(x)| \leqslant 2\varepsilon$ if $|x| < \varepsilon$. Let δ denote the Dirac distribution $\varphi \mapsto \varphi(0)$. If ε is small enough, a straightforward calculation together with (6.5) gives

$$v^{(4)}\big|_{(-\eta,\eta)} = -4\delta',$$

$$|v_\varepsilon^{(4)}(x)| = 4|\varphi_\varepsilon'(x)| \leqslant C\varepsilon^{-2}\chi_{(-\varepsilon,\varepsilon)}(x), \ |x| < \eta.$$

Hence

$$\left|\int_{-\eta}^{\eta} v_\varepsilon^{(4)} |v_\varepsilon|^{p-1} \operatorname{sgn} v_\varepsilon \, dx\right| \leqslant \int_{-\varepsilon}^{\varepsilon} C\varepsilon^{-2}\varepsilon^{p-1} \, dx = 2C\varepsilon^{p-2},$$

which tends to zero as $\varepsilon \to 0+$, so we can finally deduce that

$$\lim_{\varepsilon \to 0+} \int v_\varepsilon^{(4)} |v_\varepsilon|^{p-1} \operatorname{sgn} v_\varepsilon \, dx = \ell A + 2B.$$

But the right-hand side is by construction negative if $i = 1$ and positive if $i = 2$, so since $v_\varepsilon \in C_0^\infty(\mathbf{R})$ for every ε, the claim of the lemma is proven for $p > 2$.

Now suppose that $\frac{3}{2} \leqslant q < 2$ and let p be the conjugate exponent to q. Then $p \in (2, 3]$ so let u, u_i and η be as above. Integration by parts gives

$$\int_\mu^{1-\mu} u_i''(u_i^{p-1})'' \, dx = \left[u''(u^{p-1})' - u^{(3)}u^{p-1}\right]_\mu^{1-\mu} + \int_\mu^{1-\mu} u_i^{(4)}u_i^{p-1} \, dx$$

$$= O(\mu^{p-2}) + \int_0^1 u_i^{(4)}u_i^{p-1} \, dx, \ 0 < \mu < \eta. \tag{6.9}$$

Choose $\mu \in (0, \eta)$ such that, for $i = 1, 2$, the integral on the left-hand side has the same sign as the integral on the right. Let p be a polynomial of degree one which joins u^{p-1} in a C^1-manner at μ, let $a \in \mathbf{R}$ fulfill $p(a) = 0$ and define for $i = 1, 2$ the functions g_i on $\mathbf{R}$ by

$$g_i(x) = \begin{cases} u_i^{p-1}(x), & \mu \leqslant x \leqslant 1-\mu, \\ p(x), & a < x < \mu, \\ p(1-x), & 1-\mu < x < 1-a, \\ 0, & \text{otherwise.} \end{cases}$$

To simplify notation, let f_i be g_i composed with an affine transformation that maps a to 0 and $1 - a$ to 1. Redefine h as to satisfy the properties $\operatorname{supp} h = [0, 1]$, $h|_{(0,2)} \in C^\infty(0, 2)$, $h \geqslant 0$ and $h|_{[0,\lambda]} = f_i|_{[0,\lambda]}$, where $\lambda > 0$ is so small that f_i is linear on $[0, \lambda]$. Let $i \in \{0, 1\}$ and, as with v above, set w to

$$w(x) = -h(-x) + \sum_{j=0}^{\ell-1} (-1)^j f_i(x-j) + (-1)^\ell h(x-\ell), \ x \in \mathbf{R},$$

where ℓ will be chosen later, and define $w_\varepsilon = f_i * \varphi_\varepsilon$ for $\varepsilon > 0$. If ε is small enough it follows that

$$\int w_\varepsilon^{(4)} |w_\varepsilon|^{q-1} \operatorname{sgn} w_\varepsilon \, dx = \ell \int_0^1 w_\varepsilon''(w_\varepsilon^{q-1})'' \, dx + 2 \int_{\lambda/2}^2 h_\varepsilon^{(4)} h_\varepsilon^{q-1} \, dx, \tag{6.10}$$

since there are two neighborhoods of 0 and 1 where w_ε'' is identically zero for all small ε. By the same reason, and since f_i'' only has finitely many jump discontinuities, $w_\varepsilon''(w_\varepsilon^{q-1})''$ is uniformly bounded with respect to ε. But $w_\varepsilon''(w_\varepsilon^{q-1})''$ converges almost everywhere to $f_i''(f_i^{q-1})''$ on $[0,1]$ and $h_\varepsilon^{(4)}h_\varepsilon^{q-1}$ converges uniformly to $h^{(4)}h^{q-1}$ on $[\lambda/2, 2]$ as $\varepsilon \to 0+$ so the dominated convergence theorem applied to (6.10) implies that

$$
\lim_{\varepsilon \to 0+} \int w_\varepsilon^{(4)} |w_\varepsilon|^{q-1} \operatorname{sgn} w_\varepsilon \, dx = \int_0^1 f_i''(f_i^{q-1})'' \, dx + 2 \int_0^1 h^{(4)} h^{q-1} \, dx
$$
$$
= J \int_\mu^{1-\mu} (u_i^{p-1})'' u_i'' \, dx + 2 \int h^{(4)} h^{q-1} \, dx, \tag{6.11}
$$

where J is a positive constant resulting from the affine transformation. By the choice of μ above, ℓ can now be chosen sufficiently large that the sign of the right-hand side of (6.11) coincides with the sign of the left-hand side of (6.9) and the proof is complete.   □

Lemma 6.7. *If $k \geqslant 3$, the integral*

$$
\int v^{(k)} \operatorname{sgn} v \, dx
$$

changes sign as v ranges over the real-valued elements of $C_0^\infty(\mathbf{R})$.

Proof. Let $f \in (C_0^\infty(\mathbf{R}))^+$ and set $v(x) = xf(x)$ on $\mathbf{R}$. Then

$$
\int v^{(k)} \operatorname{sgn} v \, dx = \left[\frac{d^{k-1}}{dx^{k-1}}(xf(x)) \right]_0^\infty - \left[\frac{d^{k-1}}{dx^{k-1}}(xf(x)) \right]_{-\infty}^0
$$
$$
= -2(k-1)f^{(k-2)}(0),
$$

which can be made to attain arbitrary sign by a suitable choice of f.   □

We are now in a position to give the aforementioned necessary and sufficient conditions.

Theorem 6.8. *Let $k \in \mathbf{N}$ and $p \in [1, \infty)$. The integral*

$$
\int v^{(k)} |v|^{p-1} \operatorname{sgn} v \, dx \tag{6.12}
$$

preserves sign as v ranges over real-valued elements of $C_0^\infty(\mathbf{R})$ if and only if $p = 2$ or $k \in \{0, 1, 2\}$.

Proof. The necessity of the stated conditions follows immediately from the counterexamples in Lemmas 6.3–6.7 so we have only to prove sufficiency. In the rest of the proof, let v be an arbitrary real-valued $C_0^\infty(\mathbf{R})$-function. If $p = 2$ we obtain, by integrating by parts,

$$\int v^{(k)} v \, dx = \begin{cases} -\int vv^{(k)} \, dx, & k \text{ odd}, \\ (-1)^{k/2} \int (v^{(k/2)})^2 \, dx, & k \text{ even}, \end{cases}$$

which shows that (6.12) preserves sign since the integral vanishes for odd k and the right-hand side has a nonnegative integrand for even k. The case $k = 0$ is trivial and the case $k = 1$ follows by observing that the function $p^{-1}|v|^p$ is absolutely continuous with derivative $v'|v|^{p-1} \operatorname{sgn} v$ almost everywhere. Thus, the integral (6.12) is zero. The only remaining case to investigate is $k = 2$. Let $\varepsilon > 0$ and consider the equalities

$$\int v''(v^2 + \varepsilon)^{p/2-1} v \, dx$$
$$= -\int (v')^2 (v^2 + \varepsilon)^{p/2-1} \, dx - (p-2) \int (vv')^2 (v^2 + \varepsilon)^{p/2-2} \, dx$$
$$= -\int ((p-1)v^2 + \varepsilon)(v')^2 (v^2 + \varepsilon)^{p/2-2} \, dx,$$

obtained by integration by parts and a simple rearrangement. The integrand in the left-hand side is uniformly bounded with respect to ε, so an application of Lebesgue's dominated convergence theorem implies that the first integral converges to (6.12) as $\varepsilon \to 0+$. Since the right-hand side is nonpositive for every $\varepsilon > 0$, the theorem is proved. $\qquad \square$

Theorem 6.9. *Let $k \in \mathbb{N}$ and $p \in (1, \infty)$. The integral*

$$\int v^{(k)} v^{p-1} \, dx \qquad (6.13)$$

preserves sign as v ranges over $(C_0^\infty(\mathbf{R}))^+$ if and only if $p = 2$ or $k \in \{0, 1, 2\}$ or $k = 4$ and $\frac{3}{2} \leqslant p \leqslant 3$.

Proof. That the stated conditions are necessary follows from Lemmas 6.3, 6.4 and 6.5. For sufficiency, Theorem 6.8 covers the cases $p = 2$ or $k \in \{0, 1, 2\}$, so the remaining case is when $k = 4$ and $\frac{3}{2} \leqslant p \leqslant 3$. Let $v \in (C_0^\infty(\mathbf{R}))^+$ be arbitrary and define for every $\varepsilon > 0$ the function $v_\varepsilon = v + \varepsilon$. We have

$$\int v^{(4)} v_\varepsilon^{p-1} \, dx = (p-1) \int (v'')^2 v_\varepsilon^{p-2} \, dx + (p-1)(p-2) \int (v')^2 v'' v_\varepsilon^{p-3} \, dx, \quad (6.14)$$

obtained by integrating by parts twice. By dominated convergence, the left-hand side tends to the integral (6.13) as ε tends to zero, so it is enough to show that the right-hand side of (6.14) is nonnegative for each ε in order to show that (6.13) is nonnegative. The first term on the right is clearly nonnegative, so let us examine the second term:

$$\int (v')^2 v'' v_\varepsilon^{p-3}\, dx = (3-p) \int (v')^4 v_\varepsilon^{p-4}\, dx - 2 \int (v')^2 v'' v_\varepsilon^{p-3}\, dx$$
$$= \frac{1}{3}(3-p) \int (v')^4 v_\varepsilon^{p-4}\, dx. \tag{6.15}$$

Substituting (6.15) into (6.14), the nonnegativity immediately follows if $2 \leqslant p \leqslant 3$. If $\frac{3}{2} \leqslant p < 2$, we estimate the integrals in (6.15) by using the Cauchy–Schwarz inequality:

$$\int (v')^4 v_\varepsilon^{p-4}\, dx = \frac{3}{3-p} \int (v'' v_\varepsilon^{p/2-1})((v')^2 v_\varepsilon^{p/2-2})\, dx$$
$$\leqslant \frac{3}{3-p} \left(\int (v'')^2 v_\varepsilon^{p-2}\, dx \right)^{1/2} \left(\int (v')^4 v_\varepsilon^{p-4}\, dx \right)^{1/2}.$$

This implies that

$$\int (v')^4 v_\varepsilon^{p-4}\, dx \leqslant \frac{9}{(3-p)^2} \int (v'')^2 v_\varepsilon^{p-2}\, dx,$$

which, together with (6.15) and (6.14), finally gives

$$\int v^{(4)} v_\varepsilon^{p-1}\, dx \geqslant \frac{(p-1)(2p-3)}{3-p} \int (v'')^2 v_\varepsilon^{p-2}\, dx.$$

This clearly shows that the right-hand side is nonnegative if $\frac{3}{2} \leqslant p < 2$ and completes the proof.   □

6.2   The multi-dimensional case

We will now turn our attention to the multi-dimensional extension of the results above and we will therefore study expressions of the form

$$\mathrm{Re} \int_\Omega (Pu)|u|^{p-2}\overline{u}\, dx_n, \tag{6.16}$$

where $\Omega \subset \mathbb{R}^n$ is open, $1 \leqslant p < \infty$ and $u \in C_0^\infty(\Omega)$. In the case $p = 1$, the expression $|u|^{-1}\overline{u}$ is interpreted as being zero where u is zero. The operator P is a linear partial differential operator; writing

$$P = \sum_{|\alpha|\leqslant k} a_\alpha \partial^\alpha, \tag{6.17}$$

we will require that all of the coefficient functions a_α are elements of $L_{\mathrm{loc}}^1(\Omega)$ in order for expression (6.16) to make sense. With coefficients belonging to this function class, P is of order k if (6.17) holds and at least one of the functions in $\{a_\alpha\}_{|\alpha|=k}$ is nonzero on a set of positive measure.

The two main results are Theorems 6.12 and 6.13, where necessary conditions on the order of P are given in order for (6.16) to preserve sign as u ranges over $C_0^\infty(\Omega)$ and $(C_0^\infty(\Omega))^+$, respectively. To that end, some preliminary lemmas are needed.

Lemma 6.10. *Any linear partial differential operator P of order k defined on an open nonempty set $\Omega \subset \mathbb{R}^n$ can by a linear transformation T be transformed into*

$$Q(y, \partial) = b(y)\partial_1^k + \sum_{\substack{|\alpha| \leqslant k \\ \alpha_1 < k}} b_\alpha(y)\partial^\alpha$$

where $b, b_\alpha \in L_{\mathrm{loc}}^1(T(\Omega))$ for $|\alpha| \leqslant k$ and b is nonzero on a set of positive measure.

Proof. Let P be given by

$$P(x, \partial_x) = \sum_{|\alpha| \leqslant k} a_\alpha(x)\partial_x^\alpha. \tag{6.18}$$

Since almost every point of Ω is a Lebesgue point of all functions in $\{a_\alpha\}_{|\alpha|=k}$ and at least one of the functions is nonzero, one of the points, say $x_0 \in \Omega$, can be chosen such that $a_\beta(x_0) \neq 0$ for some β of order k. Now consider the multivariate polynomial

$$\xi \mapsto \sum_{|\alpha|=k} a_\alpha(x_0)\xi^\alpha, \ \xi \in \mathbb{R}^n, \tag{6.19}$$

where now at least one of the coefficients is nonzero. Therefore, we can find a $\xi \in \mathbb{R}^n$ such that the polynomial (6.19) is nonzero at ξ. Since the mapping (6.19) is continuous, the choice of $\xi = (\xi_1, \ldots, \xi_n)$ can be made so that all the coordinates $\{\xi_j\}$ are nonzero. Now define a linear transformation $T : \mathbb{R}^n \to \mathbb{R}^n$ by

$$y_j = \sum_{\ell=j}^n \xi_\ell x_\ell, \ j = 1, \ldots, n,$$

$$Tx = y.$$

The determinant of the transformation equals $\xi_1 \cdots \xi_n$ and hence, by our choice of ξ, T is an admissible change of coordinates. We get

$$\partial_{x_j} = \xi_j(\partial_{y_1} + \cdots + \partial_{y_j}), \ j = 1, \ldots, n,$$

$$\partial_x^\alpha = \xi^\alpha \partial_{y_1}^{|\alpha|} + \sum_{\substack{|\beta| \leqslant |\alpha| \\ \beta_1 < |\alpha|}} c_{\alpha,\beta}\partial_y^\beta \tag{6.20}$$

for any multi-index α, where $\{c_{\alpha,\beta}\}$ are suitably chosen constants. Substituting (6.20) into (6.18) we obtain, with $f \in C^\infty(T(\Omega))$,

$$P(x, \partial_x)(f \circ T)(x) = \left(\sum_{|\alpha|=k} a_\alpha(x)\xi^\alpha \right) (\partial_{y_1}^k f)(Tx) + \sum_{\substack{|\alpha| \leqslant k \\ \alpha_1 < k}} c_\alpha(x)(\partial_y^\alpha f)(Tx) \quad (6.21)$$

for some $L^1_{\mathrm{loc}}(\Omega)$-functions $\{c_\alpha\}$. Define the functions b and b_α on $T(\Omega)$ by

$$b = \sum_{|\alpha|=k} (a_\alpha \circ T^{-1})\xi^\alpha, \sigma b_\alpha = c_\alpha \circ T^{-1}.$$

Then (6.21) becomes

$$P(x, \partial)(f \circ T)(x) = (Q(Tx, \partial)f)(Tx), \quad x \in \Omega,$$

with Q as in the statement of the lemma. By our choice of ξ and x_0, $b \circ T$ is nonzero on a set of positive measure and since T is linear and nondegenerate, it follows that b is nonzero. $\qquad\square$

Lemma 6.11. *If $k \geqslant 3$ and $r \geqslant -\frac{1}{2}$, $r \neq 0$, there are real-valued functions v_1 and v_2, both belonging to $C_0^\infty(\mathbf{R})$, such that*

$$\int_{\{v_1^2+v_2^2\neq 0\}} (v_1^{(k)} v_2 - v_1 v_2^{(k)})(v_1^2 + v_2^2)^r \, dx_1 \quad (6.22)$$

is nonzero.

Proof. Given v_1 and then choosing $v_2 \in C_0^\infty(\mathbf{R})$ such that v_2 is constant, say $v_2 = \omega > 0$, on the support of v_1, the integral (6.22) reduces to

$$\omega \int v_1^{(k)} (v_1^2 + \omega^2)^r \, dx_1.$$

Setting $\varepsilon = \omega^{-1/2}$, it is therefore enough to show that for some $v \in C_0^\infty(\mathbf{R})$, the function f defined by

$$f(\varepsilon) = \int v^{(k)}(1 + \varepsilon v^2)^r \, dx_1$$

is nonzero for some $\varepsilon > 0$. This follows by observing that f is well defined in a small neighborhood of the origin and differentiable there with

$$f'(0) = r \int v^{(k)} v^2 \, dx_1,$$

which can be made nonzero by Theorem 6.9 for some $v \in (C_0^\infty(\mathbf{R}))^+$ if $k \neq 4$. If $k = 4$, it follows by integrating by parts that

$$f'(0) = \int 2rv''((v')^2 + vv'') \, dx_1 = 2r \int v(v'')^2 \, dx_1,$$

which of course is nonzero if v is nonzero. Thus, for some small positive ε, $f(\varepsilon) \neq 0$ and we are done. $\qquad\square$

Theorem 6.12. *Suppose that $p \in [1, \infty)$, $p \neq 2$ and that P is a linear partial differential operator defined on an open nonempty set $\Omega \subset \mathbb{R}^n$. If*

$$\mathrm{Re} \int_\Omega (Pu)|u|^{p-2}\overline{u}\,dx_n \tag{6.23}$$

does not change sign as u ranges over $C_0^\infty(\Omega)$, then P is of order 0, 1 or 2.

Proof. The idea of the proof is to scale the coordinates around a Lebesgue point in order to reduce the problem to the one-dimensional case with constant coefficients.

Let Q be the operator given by Lemma 6.10 and let T be the linear transformation that takes P into Q. Since, with $\det T'$ denoting the Jacobian of T,

$$\int_{T(\Omega)} (Qu)|u|^{p-2}\overline{u}\,dx_n = |\det T'| \int_\Omega P(u \circ T)|u \circ T|^{p-2}\overline{u \circ T}\,dx_n,$$

it follows that the assumptions of the theorem hold if and only if the same assumptions hold with P and Ω replaced by Q and $T(\Omega)$, respectively. We can therefore, without loss of generality, assume that P is of the form

$$P(x, \partial) = b(x)\partial_1^k + \sum_{\alpha \in J} b_\alpha(x)\partial^\alpha,$$

$$J = \{\alpha \in \mathbb{N}^n : |\alpha| \leqslant k, \alpha_1 < k\}.$$

We will now assume that $k \geqslant 3$ and show that the integral in the hypothesis does not preserve sign. Since b is nonzero, there is a function $\varphi \in (C_0^\infty(\Omega))^+$ such that

$$\int_\Omega b\varphi^p\,dx_n \neq 0.$$

This is a well-known fact if $p = 1$, see, e.g. , Hörmander [41, Th. 1.2.5], but the proof holds with obvious modifications for any $p \geqslant 1$. Extend b and φ by zero outside of Ω and define the function a on $\mathbb{R}$ by

$$a(x_1) = \int b(x_1, \cdot)\varphi^p(x_1, \cdot)\,dx_{n-1}. \tag{6.24}$$

Since the integrand above belongs to $L^1(\mathbb{R}^n)$, the Fubini Theorem implies that $a \in L^1(\mathbb{R})$ and that

$$\int a\,dx_1 = \int_\Omega b\varphi^p\,dx_n,$$

so a is nonzero by our choice of φ. Hence, there exists a Lebesgue point $y \in \mathbb{R}$ of a with $a_0 = a(y) \neq 0$. Write a_0 as $a_0 = a_1 + ia_2$ where a_1 and a_2 are real numbers. We will divide the rest of the proof into two parts – one where we assume that $a_1 = 0$ and one where a_1 is assumed to be nonzero.

Suppose that $a_1 = 0$ and consequently that $a_2 \neq 0$. Let v_1 and v_2 be the two functions given by Lemma 6.11 with $r = p/2 - 1$ and define for $\omega \geqslant 1$ the functions $w_\omega \in C_0^\infty(\mathbf{R})$, $u_\omega \in C_0^\infty(\Omega)$ and the sets K_ω by

$$w_\omega(t) = (v_1 + iv_2)(y + \omega t), \ t \in \mathbf{R}, \tag{6.25}$$

$$u_\omega(x) = w_\omega(x_1)\varphi(x), \ x = (x_1, \ldots, x_n) \in \Omega, \tag{6.26}$$

$$K_\omega = \operatorname{supp} \varphi \cap ([-B/\omega, B/\omega] \times \mathbf{R}^{n-1}), \tag{6.27}$$

where B has been chosen so that $\operatorname{supp} w_1 \subset (-B, B)$. With these definitions,

$$|\partial^\alpha u_\omega(x)| \leqslant C\omega^{\alpha_1}\chi_{K_\omega}(x), \ |\alpha| \leqslant k, \ x \in \Omega, \tag{6.28}$$

where χ_{K_ω} is the characteristic function of the set K_ω and C is a constant not depending on ω. Now consider the equalities

$$
\begin{aligned}
&\operatorname{Re} \int_\Omega (Pu_\omega)|u_\omega|^{p-2}\overline{u}_\omega \, dx_n \\
&= \operatorname{Re} \int_\Omega b(x)\varphi^p(x)w_\omega^{(k)}(x_1)|w_\omega(x_1)|^{p-2}\overline{w}_\omega(x_1) \, dx + \operatorname{Re} H_1(\omega) \\
&= \operatorname{Re} \int a w_\omega^{(k)}|w_\omega|^{p-2}\overline{w}_\omega \, dx_1 + \operatorname{Re} H_1(\omega) \\
&= a_2\omega^{k-1} \int_{\{v_1^2 + v_2^2 \neq 0\}} (v_1^{(k)}v_2 - v_1 v_2^{(k)})(v_1^2 + v_2^2)^{p/2-1} \, dx_1 \\
&\quad + \operatorname{Re} H_1(\omega) + \operatorname{Re} H_2(\omega).
\end{aligned}
\tag{6.29}
$$

Here we have made use of (6.24) together with an application of the Fubini Theorem in the second equality. The functions H_1 and H_2 are defined by

$$
\begin{aligned}
H_1(\omega) &= \int_\Omega \sum_{\alpha \in J} b_\alpha \partial^\alpha u_\omega \, |u_\omega|^{p-2}\overline{u}_\omega \, dx_n \\
&\quad + \int_\Omega \sum_{j=0}^{k-1} b(x)c_j u_\omega^{(j)}(x_1)\big((\partial_1^{k-j}\varphi)|u_\omega|^{p-2}\overline{u}_\omega\big)(x) \, dx, \\
H_2(\omega) &= \int (a - a_0)w_\omega^{(k)}|w_\omega|^{p-2}\overline{w}_\omega \, dx_1.
\end{aligned}
$$

These functions can be estimated using (6.28) to obtain:

$$\omega^{1-k}|H_1(\omega)| \leqslant C\Big(\sum_{\alpha \in J} \|b_\alpha \chi_{K_\omega}\|_{L^1(\Omega)} + \|b\chi_{K_\omega}\|_{L^1(\Omega)}\Big), \tag{6.30}$$

$$\omega^{1-k}|H_2(\omega)| \leqslant C\omega \int_{-B/\omega}^{B/\omega} |a(y+t) - a(y)| \, dt. \tag{6.31}$$

The collection $\{b\chi_{K_\omega}\}_{\omega \geqslant 1}$ is majorized by the $L^1(\Omega)$-function $|b\chi_{K_1}|$ and $b\chi_{K_\omega}$ tends to zero almost everywhere as $\omega \to \infty$. For each fixed multi-index α, the same can be said about the collection $\{b_\alpha \chi_{K_\omega}\}_{\omega \geqslant 1}$ and $|b_\alpha \chi_{K_1}|$, so by Lebesgue's theorem of dominated convergence, the right-hand side of (6.30) tends to zero as $\omega \to \infty$. Since y is a Lebesgue point of a, the right-hand side of (6.31) also tends to zero as $\omega \to \infty$ and by using these limits together with (6.29), we obtain

$$\lim_{\omega \to \infty} \omega^{1-k} \operatorname{Re} \int_\Omega (Pu_\omega)|u_\omega|^{p-2}\overline{u}_\omega \, dx_n$$
$$= a_2 \int_{\{v_1^2 + v_2^2 \neq 0\}} (v_1^{(k)} v_2 - v_1 v_2^{(k)})(v_1^2 + v_2^2)^{p/2-1} \, dx_1. \tag{6.32}$$

By construction, the right-hand side is nonzero. But the integrand on the right is antisymmetric with respect to v_1 and v_2 so by changing places of v_1 and v_2 in the definition (6.25), the right-hand side of (6.32) changes sign. This proves that (6.23) can not preserve sign over $C_0^\infty(\Omega)$ and we are done with the first part.

Assume instead that a_1 is nonzero. Let $v \in C_0^\infty(\mathbf{R})$ be real valued and define the functions w_ω for $\omega \geqslant 1$ by

$$w_\omega(t) = v(y + \omega t), \ t \in \mathbb{R}. \tag{6.33}$$

Letting u_ω and K_ω be defined by (6.26) and (6.27), respectively, we immediately see that (6.28) still holds. We get as above

$$\operatorname{Re} \int_\Omega (Pu_\omega)|u_\omega|^{p-2}\overline{u}_\omega \, dx_n$$
$$= \omega^{k-1} a_1 \int v^{(k)}|v|^{p-1} \operatorname{sgn} v \, dx_1 + \operatorname{Re} H_1(\omega) + \operatorname{Re} H_2(\omega),$$

where H_1 and H_2 are as previously defined, still fulfilling the estimates (6.30) and (6.31), respectively. Passing to the limit, it therefore follows that

$$\lim_{\omega \to \infty} \omega^{1-k} \operatorname{Re} \int_\Omega (Pu_\omega)|u_\omega|^{p-2}\overline{u}_\omega \, dx_n = a_1 \int v^{(k)}|v|^{p-1} \operatorname{sgn} v \, dx_1. \tag{6.34}$$

But from Theorem 6.8 it follows that the right-hand side can assume arbitrary sign by choosing v properly since $k \geqslant 3$. Hence, (6.23) can not preserve sign over $C_0^\infty(\Omega)$. $\qquad\square$

Theorem 6.13. *Suppose that $p \in (1, \infty)$, $p \neq 2$ and that P is a linear partial differential operator with real-valued coefficient functions, defined on an open nonempty set $\Omega \subset \mathbb{R}^n$. Assume that*

$$\int_\Omega (Pu)u^{p-1} \, dx_n$$

does not change sign as u ranges over $(C_0^\infty(\Omega))^+$. Then either P is of order 0, 1 or 2, or P is of order 4 and $\frac{3}{2} \leqslant p \leqslant 3$.

Proof. This is proved in exactly the same way as in the second part of the proof of Theorem 6.12. Just note that the definitions (6.33) and (6.26) imply that $u_\omega \in (C_0^\infty(\Omega))^+$ for every ω if $v \in (C_0^\infty(\mathbf{R}))^+$ and that (6.34) now reduces to

$$\lim_{\omega \to \infty} \omega^{1-k} \int_\Omega (Pu_\omega)u_\omega^{p-1}\, dx_n = a_1 \int v^{(k)} v^{p-1}\, dx_1.$$

The result now follows from Theorem 6.9. □

6.3 Absence of L^p-dissipativity for higher-order operators

Dissipativity is a necessary condition for an operator to generate a contraction semigroup. Hence, the dissipativity criterion together with Theorem 6.12 will lead us to one of the main objectives of this paper. We formulate the result for partial differential operators acting on vector-valued functions.

Let, as in previous subsection, the differential operator P be given by

$$P = \sum_{|\alpha| \leq k} a_\alpha \partial^\alpha, \tag{6.35}$$

where the coefficients a_α now are allowed to be $N \times N$-matrices with entries belonging to $L_{\text{loc}}^1(\Omega)$ for some positive integer N. The order of P is k if at least one of the matrices in $\{a_\alpha\}_{|\alpha|=k}$ has an entry which is nonzero on a set of positive measure.

For differential operators of this form, we now state the following theorem.

Theorem 6.14. *If $\Omega \subset \mathbf{R}^n$ is open and $1 \leq p < \infty$, $p \neq 2$, no linear partial differential operator of order higher than two which contains $(C_0^\infty(\Omega))^N$ in its domain of definition can generate a contraction semigroup on $(L^p(\Omega))^N$.*

Proof. Suppose that the operator P generates a contraction semigroup on $(L^p(\Omega))^N$ and that it is written as in (6.35). Let a_β be a matrix in $\{a_\alpha\}_{|\alpha|=k}$ which has a nonvanishing element. Then there is some $c \in \mathbf{C}^N$ with $|c| = 1$ such that $\langle a_\beta c, c \rangle$ is nonzero on a set of positive measure.

Define the partial differential operator Q on $C_0^\infty(\Omega)$ by

$$Qu = \langle P(uc), c \rangle = \sum_{|\alpha| \leq k} \langle a_\alpha c, c \rangle \partial^\alpha u, \ u \in C_0^\infty(\Omega).$$

By our choice of c, this scalar operator is still of order k. From the dissipativity criterion it follows that

$$0 \geq \text{Re} \int_\Omega \langle P(uc), uc \rangle |uc|^{p-2}\, dx_n = \text{Re} \int_\Omega (Qu)|u|^{p-2}\bar{u}\, dx_n$$

for all $u \in C_0^\infty(\Omega)$. By Theorem 6.12, the order of Q is 0, 1, or 2, so since the order of P and Q are equal, the theorem is proved. □

6.4   Contractivity on the cone of nonnegative functions

In applications, solutions to the Cauchy problem

$$\begin{cases} s'(t) = As(t), & t \in \mathbb{R}^+, \\ s(0) = x, & x \in D(A), \end{cases} \tag{6.36}$$

are sometimes known to be nonnegative functions on some interval. That is, for each t belonging to some interval, the function $s(t) \in L^p(\Omega)$ in (6.36) is nonnegative. It is therefore natural to ask if there is an analogue to the contractivity property of dissipative operators in this case.

In the rest of this section, let Ω be an open subset of $\mathbb{R}^n$ and write L^r instead of $L^r(\Omega)$. All L^r-spaces, $r \in [1, \infty)$, will be real. Indices will appear extensively in this section; the letters i, j, k and ℓ are used for indices ranging over $\{1, \ldots, n\}$.

Lemma 6.15. *Suppose that the Cauchy problem* (6.36) *is well posed. If* $1 < p < \infty$, *then*

$$\frac{d}{dt}\|s(t)\|_p\bigg|_{t=0+} \leqslant 0$$

for every $s(0) \in (D(A))^+$ *if and only if*

$$\int_\Omega (Au)u^{p-1}\, dx_n \leqslant 0 \tag{6.37}$$

for every $u \in (D(A))^+$. *In the case* $p = 1$, (6.37) *holds for every* $u \in (D(A))^+$ *if*

$$\liminf_{t \to 0+} t^{-1}(\|s(t)\|_1 - \|s(0)\|_1) \leqslant 0 \tag{6.38}$$

for every $s(0) \in (D(A))^+$.

Proof. Let $1 < p < \infty$. Since s and the map $L^p \to L^1$ given by $u \mapsto |u|^p$ are differentiable, the composition $t \mapsto |s(t)|^p$, taking $\mathbb{R}^+$ into L^1, is differentiable and the right-hand derivative at $t = 0$ is

$$p(s(0))^{p-1}s'(0).$$

Thus we have the relations

$$\lim_{t \to 0+} \frac{\|s(t)\|_p^p - \|s(0)\|_p^p}{t} = \lim_{t \to 0+} \int_\Omega \frac{|s(t)|^p - (s(0))^p}{t}\, dx_n$$

$$= p \int_\Omega s'(0)(s(0))^{p-1}\, dx_n = p \int_\Omega (As(0))(s(0))^{p-1}\, dx_n,$$

implying that the one-sided derivative of $\|s(t)\|_p$ at $t = 0$ exists and is nonpositive for all nonnegative initial data if and only if (6.37) holds on $(D(A))^+$.

Let $p = 1$. Consider the relations

$$\frac{\|s(t)\|_1 - \|s(0)\|_1}{t} = \int_\Omega \frac{|s(t)| - s(0)}{t}\, dx_n \geqslant \int_\Omega \frac{s(t) - s(0)}{t}\, dx_n$$

$$\to \int_\Omega s'(0)\, dx_n = \int_\Omega As(0)\, dx_n, \ t \to 0+.$$

By assumption, the limes inferior of the left-hand side is nonpositive for every $s(0) \in (D(A))^+$, showing that (6.37) is nonpositive for all $u \in (D(A))^+$. □

Theorem 6.16. *Let* $1 < p < \infty$, $p \neq 2$ *and suppose that* $C_0^\infty(\Omega)$ *is a subset of the domain* $D(A)$ *of the linear partial differential operator* A. *Assume furthermore that* A *has* $L_{loc}^1(\Omega)$-*coefficients and that the Cauchy problem* (6.36) *is well posed for all nonnegative initial data in* $D(A)$. *If*

$$\frac{d}{dt}\|s(t)\|_p\bigg|_{t=0+} \leqslant 0$$

for every $s(0) \in (D(A))^+$, *then either* A *is of order 0, 1 or 2, or* A *is of order 4 and* $\frac{3}{2} \leqslant p \leqslant 3$.

Proof. In view of Lemma 6.15, this theorem is an immediate corollary of Theorem 6.13. □

Remark 6.17. The case $p = 1$ is not covered by Theorem 6.16, but we can make the observation that if the operator

$$A = \sum_{|\alpha| \leqslant k} a_\alpha \partial^\alpha,$$

having $L_{loc}^1(\Omega)$-coefficients and satisfying the hypotheses of Theorem 6.16, fulfills inequality (6.38) for every $s(0) \in (C_0^\infty(\Omega))^+$, then

$$-\sum_{|\alpha| \leqslant k} (-1)^{|\alpha|} \partial^\alpha a_\alpha$$

is a positive measure in the sense of distributions. This follows by noting that Lemma 6.15 implies that the functional

$$u \mapsto -\int_\Omega a_\alpha \partial^\alpha u\, dx_n, \ u \in C_0^\infty(\Omega)$$

defines a positive distribution and hence, see Hörmander [41], can be represented by a positive measure μ on Ω through

$$u \mapsto \int_\Omega u\, d\mu, \ u \in C_0^\infty(\Omega).$$

Comparing the two different expressions for the same functional the statement immediately follows.

Theorem 6.18. *Suppose that $1 < p < \infty$, that $\Omega \subset \mathbb{R}^n$ is open, bounded and has C^∞-boundary and that the real constant coefficients $\{a_{ijk\ell}\}$ fulfill*

$$a_{ijk\ell} = a_{jki\ell} = a_{j\ell ik}$$

for all i, j, k and ℓ, and also fulfill the relation

$$\sum_{1 \leqslant i,j,k,\ell \leqslant n} a_{ijk\ell} \xi_{ij} \xi_{k\ell} \geqslant 0$$

for all real symmetric $n \times n$-matrices $\xi = [\xi_{ij}]$. Then

$$\int_\Omega (a_{ijk\ell} \partial_i \partial_j \partial_k \partial_\ell u) u^{p-1} \, dx_n \geqslant 0$$

for all nonnegative functions $u \in W^{4,p}(\Omega) \cap W_0^{2,p}(\Omega)$ if and only if $\frac{3}{2} \leqslant p \leqslant 3$.

Proof. The necessity of $\frac{3}{2} \leqslant p \leqslant 3$ follows immediately from Theorem 6.13.

Let X consist of all functions in $C^4(\overline{\Omega})$ that, together with their gradients, vanish on the boundary of Ω, and suppose that $u \in X^+$. Denote $\partial_i u$ by u_i, $\partial_i \partial_j u$ by $u_{ij}, \ldots$ and define the functions $v_\varepsilon = u + \varepsilon$ for $\varepsilon > 0$. We have, for fixed indices i, j, k and ℓ,

$$\int_\Omega u_{ijk\ell} v_\varepsilon^{p-1} \, dx_n = R(\varepsilon) - (p-1) \int_\Omega u_{ijk} u_\ell v_\varepsilon^{p-2} \, dx_n \tag{6.39}$$

$$= R(\varepsilon) + (p-1) \int_\Omega u_{ij} u_{k\ell} v_\varepsilon^{p-2} \, dx_n + (p-1)(p-2) \int_\Omega u_{ij} u_k u_\ell v_\varepsilon^{p-3} \, dx_n,$$

where the second equality follows by integration by parts and the first equality follows from Gauss' theorem, producing the boundary term

$$R(\varepsilon) = \int_{\partial\Omega} u_{ijk} v_\varepsilon^{p-1} \hat{e}_\ell \cdot \hat{n} \, dS,$$

S being the $(n-1)$-dimensional Lebesgue measure on $\partial\Omega$, $\hat{e}_\ell$ being the ℓth unit vector and $\hat{n}$ denoting the outward unit normal on $\partial\Omega$. But $v_\varepsilon = \varepsilon$ on the boundary and since Ω is bounded and the boundary is of class C^∞, the surface area of $\partial\Omega$ is finite, implying that $R(\varepsilon) = O(\varepsilon^{p-1})$. Furthermore,

$$\int_\Omega u_{ij} u_k u_\ell v_\varepsilon^{p-3} \, dx_n \tag{6.40}$$

$$= (3-p) \int_\Omega u_i u_j u_k u_\ell v_\varepsilon^{p-4} \, dx_n - \int_\Omega u_{jk} u_i u_\ell v_\varepsilon^{p-3} \, dx_n - \int_\Omega u_{j\ell} u_i u_k v_\varepsilon^{p-3} \, dx_n,$$

and combining this with (6.39) and permuting indices, we arrive at

$$
\int_\Omega u_{ijk\ell} v_\varepsilon^{p-1} \, dx_n + O(\varepsilon^{p-1})
$$

$$
= \frac{p-1}{3} \int_\Omega (u_{ij}u_{k\ell} + u_{jk}u_{i\ell} + u_{j\ell}u_{ik}) v_\varepsilon^{p-2} \, dx_n
$$

$$
+ \frac{(p-1)(p-2)}{3} \int_\Omega (u_{ij}u_k u_\ell + u_{jk}u_i u_\ell + u_{j\ell}u_i u_k) v_\varepsilon^{p-3} \, dx_n
$$

$$
= \frac{p-1}{3} \int_\Omega (u_{ij}u_{k\ell} + u_{jk}u_{i\ell} + u_{j\ell}u_{ik}) v_\varepsilon^{p-2} \, dx_n
$$

$$
+ \frac{(p-1)(p-2)(3-p)}{3} \int_\Omega u_i u_j u_k u_\ell v_\varepsilon^{p-4} \, dx_n. \tag{6.41}
$$

Summing and letting ε tend to 0, using dominated convergence, (6.41) becomes

$$
\int_\Omega a_{ijk\ell} u_{ijk\ell} u^{p-1} \, dx_n = (p-1) \lim_{\varepsilon \to 0+} \left(\int_\Omega a_{ijk\ell} u_{ij} u_{k\ell} v_\varepsilon^{p-2} \, dx_n \right. \tag{6.42}
$$

$$
\left. + \frac{(p-2)(3-p)}{3} \int_\Omega a_{ijk\ell} u_i u_j u_k u_\ell v_\varepsilon^{p-4} \, dx_n \right),
$$

where the first hypothesis on the coefficients $\{a_{ijk\ell}\}$ has been used in the first integral on the right-hand side. By setting $\xi_{ij} = u_{ij}$, ξ becomes pointwise symmetric and hence, the second hypothesis on $\{a_{ijk\ell}\}$ implies that the first integrand on the right-hand side of (6.42) is nonnegative for every ε. The same holds for the second integrand, since defining $\xi_{ij} = u_i u_j$ makes ξ pointwise symmetric. Therefore, if $2 \leqslant p \leqslant 3$, the whole right-hand side is nonnegative and the theorem is proven for $2 \leqslant p \leqslant 3$ and $u \in X^+$.

Assume that $\frac{3}{2} \leqslant p < 2$, let Z be the space of $n \times n$-matrices equipped with the standard scalar product, let Y be the subspace of symmetric matrices and define the linear operator $A : Y \to Z$ by

$$
A[\xi_{ij}] = \left[\sum_{k\ell} a_{ijk\ell} \xi_{k\ell} \right], \quad \xi \in Y.
$$

We see that the hypothesis on the constants $\{a_{ijk\ell}\}$ implies that A is a positive operator on Y, that is, $(A\xi, \xi)_Z \geqslant 0$ for every $\xi \in Y$. This means that the map $(\xi, \lambda) \mapsto (A\xi, \lambda)_Z$ defines a (possibly degenerated) scalar product on Y, enabling us to use the Cauchy–Schwarz inequality to get

$$
(A\xi, \lambda)_Z \leqslant \sqrt{(A\xi, \xi)_Z} \sqrt{(A\lambda, \lambda)_Z}, \quad \xi, \lambda \in Y.
$$

This makes it possible to estimate the integrals on the right-hand side of equation (6.42) in terms of each other. Going back to (6.40), summing and using the inequality just stated with $\xi_{ij} = u_{ij}$ and $\lambda_{k\ell} = u_k u_\ell$, we have

$$\frac{3-p}{3} \int_\Omega a_{ijk\ell} u_i u_j u_k u_\ell v_\varepsilon^{p-4} \, dx_n = \int_\Omega a_{ijk\ell} u_{ij} u_k u_\ell v_\varepsilon^{p-3} \, dx_n$$

$$\leqslant \int_\Omega \left(\sum_{i,j,k,\ell} a_{ijk\ell} u_{ij} u_{k\ell} \right)^{1/2} \left(\sum_{i,j,k,\ell} a_{ijk\ell} u_i u_j u_k u_\ell \right)^{1/2} v_\varepsilon^{p/2-1} v_\varepsilon^{p/2-2} \, dx_n$$

$$\leqslant \left(\int_\Omega a_{ijk\ell} u_{ij} u_{k\ell} v_\varepsilon^{p-2} \, dx_n \right)^{1/2} \left(\int_\Omega a_{ijk\ell} u_i u_j u_k u_\ell v_\varepsilon^{p-4} \, dx_n \right)^{1/2},$$

so

$$\int_\Omega a_{ijk\ell} u_i u_j u_k u_\ell v_\varepsilon^{p-4} \, dx_n \leqslant \frac{9}{(3-p)^2} \int_\Omega a_{ijk\ell} u_{ij} u_{k\ell} v_\varepsilon^{p-2} \, dx_n.$$

Substituting this estimate into (6.42), it follows that

$$\int_\Omega a_{ijk\ell} u_{ijk\ell} u^{p-1} \, dx_n \geqslant \frac{(p-1)(2p-3)}{3-p} \liminf_{\varepsilon \to 0+} \int_\Omega a_{ijk\ell} u_{ij} u_{k\ell} v_\varepsilon^{p-2} \, dx_n,$$

showing that the left-hand side is nonnegative, since the integrand on the right, as before, is nonnegative for each ε.

We have now shown that the conclusion of the theorem holds for all $u \in X^+$. X^+ being dense in $(W^{4,p}(\Omega) \cap W_0^{2,p}(\Omega))^+$, the theorem holds for the given function class. $\qquad\square$

Corollary 6.19. *Suppose that* $\frac{3}{2} \leqslant p \leqslant 3$ *and that* Ω *and the coefficients of the operator*

$$A = -a_{ijk\ell} \partial_{ijk\ell},$$

with domain $W^{4,p}(\Omega) \cap W_0^{2,p}(\Omega)$, *fulfill the hypotheses of Theorem 6.18. Then any differentiable solution* s *of the Cauchy problem* (6.36) *with nonnegative initial value* $s(0) \in D(A)$ *fulfills*

$$\frac{d}{dt} \|s(t)\|_p \Big|_{t=0+} \leqslant 0.$$

Proof. This follows directly by combining the sufficiency part of Lemma 6.15 with Theorem 6.18. $\qquad\square$

Example 6.20. The biharmonic operator $\Delta^2 = \sum_{i,j} \partial_{ii} \partial_{jj}$ meets the condition on the coefficients in Theorem 6.18: write Δ^2 as

$$a'_{ijk\ell} = \partial_{ij} \partial_{k\ell}, \quad a_{ijk\ell} = 3^{-1}(a'_{ijk\ell} + a'_{jki\ell} + a'_{j\ell ik}), \quad \Delta^2 = a_{ijk\ell} \partial_{ijk\ell},$$

where ∂_{ij} is one if $i = j$ and zero otherwise. Then

$$\sum_{i,j,k,\ell} a_{ijk\ell} \xi_{ij} \xi_{k\ell} = \frac{1}{3} \left(\sum_{i,k} \xi_{ii} \xi_{kk} + \sum_{j,k} \xi_{jk} \xi_{jk} + \sum_{j,\ell} \xi_{j\ell} \xi_{j\ell} \right)$$

$$= \frac{1}{3} \left(\sum_i \xi_{ii} \right)^2 + \frac{2}{3} \sum_{i,j} \xi_{ij}^2 \geqslant 0$$

for all real $n \times n$-matrices ξ. Hence, Corollary 6.19 holds with $A = -\Delta^2$ and we have the inequality

$$\|s(t)\| \leqslant \|s(0)\|, \quad t \in \mathbb{R}^+$$

$s(t)$ being the nonnegative solution of Cauchy problem

$$\begin{cases} s'(t) = -\Delta^2 s(t) & t \in \mathbb{R}^+ \\ s(0) = s_0, \end{cases} \tag{6.43}$$

with a nonnegative initial value s_0 in $W^{4,p}(\Omega) \cap W_0^{2,p}(\Omega)$ $(\frac{3}{2} \leqslant p \leqslant 3)$.

We remark that the class of positive solutions to this problem is not empty (see Comments below).

6.5 Comments to Chapter 6

The topic of the present chapter for $p = \infty$ was considered by Kresin and Maz'ya [50], who proved that arbitrary higher-order differential operators fail to generate contraction semigroups on $(L^\infty(\Omega))^N$, where $\Omega \subset \mathbb{R}^n$ and the norm is given by

$$\left\| \left(\sum_{i=1}^N |u_i|^2 \right)^{1/2} \right\|_{L^\infty(\Omega)}.$$

The result was extended to any $p \in [1, \infty]$ by Langer and Maz'ya [54, 55]. Our exposition follows these papers.

Concerning the question of existence of positive solutions to (6.43), Gazzola [30, Corollary 4, p. 3589] gives an example of initial data s_0 for which the global positivity takes place. He proved that, if $n = 1$, there exists β_0 such that if $\beta \in (0, \beta_0)$ and $s_0(x) = |x|^{-\beta}$, then the solution s to (6.43) is positive a.e. in $\mathbb{R}^+ \times \mathbb{R}$.

On the other hand Gazzola shows that if s_0 is positive a.e. on $\mathbb{R}^n$ and $(1 + |x|^4)s_0$ belongs to $L^1(\mathbb{R}^n)$, then the solution s to (6.43) changes sign.

We mention also that Berchio [7] considers the problem

$$\begin{cases} s'(t) = -\Delta^2 s(t) + g(x, t) & t \in \mathbb{R}^+ \\ s(0) = s_0, \end{cases}$$

and gave several sufficient conditions for the eventual local positivity of the solution s. This means that, for any compact set $K \subset \mathbb{R}^n$, there exists $T_K > 0$ such that $s(x, t) > 0$, for all $x \in K$ and $t \geqslant T_K$. Moreover Berchio considers the problem of the global positivity of the solution and shows that if the source g has compact support with respect to one or both of its variables, the negativity may always occur, and therefore, in general, no global positivity can be expected.

Chapter 7

Weighted Positivity and Other Related Results

Most of the results in the present chapter concern the L^2-weighted positivity of different operators. In the case of functions taking scalar values, by this positivity we mean the inequality

$$\text{Re} \int_\Omega \langle Lu, u \rangle \, \Psi dx \geqslant 0, \quad \forall \, u \in C_0^\infty(\Omega), \tag{7.1}$$

where Ψ is a weight.

First, in Section 7.1, examples of linear ordinary differential operators with variable coefficients, either satisfying or not satisfying (7.1), are presented.

Various questions related to the positivity of second-order elliptic systems, 3D-Lamé system and polyharmonic operator are considered in Sections 7.2, 7.3 and 7.4 respectively.

More precisely the main result of Section 7.2 describes the only possible matrix weight homogeneous of degree $2 - n$, which provides the L^2-positivity of a rather general elliptic system, next in Section 7.3 one can find either sufficient or necessary conditions for the weighted positivity of the Lamé system and finally in Section 7.4 we show that the polyharmonic operator is L^2-positive only for dimensions in a certain interval.

Section 7.5 is devoted to necessary and sufficient conditions for the L^2-positivity of real positive powers of the Laplacian. These results extend the ones obtained in Subsection 7.4.1 with a direct simpler argument.

The topic of Section 7.6 is the L^p-positivity of the fractional powers $(-\Delta)^\alpha$ $(0 < \alpha < 1)$ for any $p \in (1, \infty)$.

It is shown in Section 7.7 how the best constant in the Hardy inequality improves when the vectors considered are divergence-free and axisymmetric. This gives a new sharp lower bound for the quadratic form of the Stokes operator.

In the last Section 7.8 the semi-boundedness below of a pseudo-differential operator is obtained by proving a refinement of the sharp Gårding inequality. Here we treat the situation in which the symbol of the operator under consideration is not smooth.

7.1   Weighted positivity of ordinary differential operators

We will be concerned with operators $p(D)$ in $\mathbb{R}^1$, where p is a positive polynomial and $Du(x) = -idu/dx$.

In Section 7.1.1 we prove that there exist operators of arbitrary even order satisfying (7.1) in the one-dimensional case. In fact, we prove that if the sequence (a_j) grows sufficiently fast then (7.1) holds for

$$p(D) = (a_1 + D^2)(a_2 + D^2)\cdots(a_m + D^2)$$

and $\Psi = \overline{\Gamma}$, Γ being the fundamental solution. We also give explicit examples of such operators. In Section 7.1.2 we find some necessary conditions for operators to satisfy (7.1), and deduce examples of operators not having this property, for instance $1 + D^4$. Finally, in Section 7.1.3, we study the operators $(1 + D^2)^m$. We prove that they satisfy (7.1) if and only if $m = 0, 1, 2, 3$. The case $m = 3$ is more complicated than the others. For this case, an important step in the proof is the identity (7.22). In the cited papers it was essential to have a certain minorant (instead of 0) on the right of (7.1). We will also see in Section 7.1.3 that the operator $(1+D^2)^3$ has a different behavior, with respect to this, than the operators $1 + D^2$ and $(1 + D^2)^2$.

By Parseval's formula, these results can also be interpreted as results for certain integral operators. For instance, it follows from Proposition 7.12 that if $m = 1, 2, 3$ then

$$\iint_{\mathbb{R}^2} \frac{(1 + x^2)^m}{(1 + (x - y)^2)^m} f(x) f(y)\, dx\, dy \geq 0, \quad f \text{ real in } C_0^\infty(\mathbb{R}^1),$$

with equality only for $f = 0$, while for $m \geq 4$, the double integral can take on negative values.

Some notation: Φ denotes the Fourier transform,

$$(\Phi u)(\xi) = \hat{u}(\xi) = \int e^{-i\xi x} u(x)\, dx.$$

We write $\int$ instead of $\int_\infty^\infty$. Let $\mathscr{S}$ denote the Schwartz space of rapidly decreasing C^∞-functions on $\mathbb{R}^1$. We also write C_0^∞ instead of $C_0^\infty(\mathbb{R}^1)$. The letter c denotes positive constants. The notation $a \sim b$ means that there exists c such that $c^{-1}a \leq b \leq ca$.

7.1.1 Positivity

For a positive polynomial p we let Γ be defined by $\widehat{\Gamma} = 1/p$. Thus Γ is a fundamental solution of the operator $p(D)$. By Parseval's formula we have

$$\int p(D)u \cdot \overline{u\Gamma}\, dx = (2\pi)^{-2} \iint \frac{p(x)}{p(x-y)} \hat{u}(x)\overline{\hat{u}(y)}\, dx\, dy, \qquad (7.2)$$

for $u \in \Sigma$. We define $\mathcal{P}$ to be the class of those positive polynomials p for which the real part of (7.2) is nonnegative for all $u \in \Sigma$.

Lemma 7.1. *For any polynomial p of degree $2n$ or $2n+1$ there are polynomials q_j such that*

$$p(x) + p(-y) = \sum_{j=0}^{n} (xy)^j q_j(x-y). \qquad (7.3)$$

In $q_j(t)$ and $p(x)$ the coefficients for t^m and x^{m+2j} are proportional and have the same sign.

Proof. With the new variables $s = (x+y)/2$, $t = (x-y)/2$ and $u = xy$ we can write $x^m + (-y)^m$ as

$$(t+s)^m + (t-s)^m = 2 \sum_{\substack{k=0 \\ k \text{ even}}}^{m} \binom{m}{k} s^k t^{m-k} = 2 \sum_{k=0}^{\lfloor m/2 \rfloor} \binom{m}{2k} (t^2+u)^k t^{m-2k}$$

$$= 2 \sum_{k=0}^{\lfloor m/2 \rfloor} \sum_{j=0}^{k} \binom{m}{2k} \binom{k}{j} u^j t^{m-2j}.$$

The statement follows. $\qquad\qquad\square$

Remark 7.2. It can be shown that

$$q_j(x) = \frac{1}{j!} \left(e^j (p(x) + p(-y)) \right)\big|_{y=0} = \sum_{m=0}^{N-2j} c_{m,j} b_{m+2j} x^m, \qquad (7.4)$$

where the operator e is given by

$$e = (x+y)^{-1}(\partial/\partial x + \partial/\partial y),$$

N is the degree of p, b_m is the coefficient for x^m in $p(x)$, and the coefficients $c_{m,j}$ are given by

$$c_{m,j} = 2^{1-m} \sum_{k=0}^{\lfloor m/2 \rfloor} \binom{m+2j}{2(k+j)} \binom{k+j}{j}.$$

Corollary 7.3. *If $p > 0$ is an even polynomial of degree $2n$ and $F(q_j/p) \geq 0$ for $j = 0, \ldots, n$, then $p \in \mathcal{P}$ and for all $u \in \mathcal{S}$ it holds that*

$$2 \operatorname{Re} \int P(D)u \cdot \overline{u}\Gamma \, dx \geq |u(0)|^2 + \int \left(b_0|u|^2 + 2b_{2n}|u^{(n)}|^2\right)\Gamma \, dx, \qquad (7.5)$$

where $b_0 = p(0)$ and b_{2n} is the leading coefficient of p. (The hypotheses imply that $\Gamma > 0$.)

Proof. Using (7.2) and expanding $p(x) + p(y)$ according to the lemma we see that the left-hand side equals

$$(2\pi)^{-2} \iint \frac{p(x) + p(y)}{p(x - y)} \hat{u}(x)\overline{\hat{u}(y)} \, dx \, dy = (2\pi)^{-1} \sum_{j=0}^{n} \int \Phi(q_j/p)|u^{(j)}|^2 \, d\xi.$$

If we put $y = 0$ in (7.3) we get $q_0(x) = p(x) + p(0)$, so

$$F(q_0/p) = 2\pi\delta + p(0)\Phi(1/p).$$

Similarly, if we let $x = y \to \infty$, we get $q_n = 2b_{2n}$. Since for an even p we have $F(1/p) = 2\pi\Gamma$, this proves the assertion. $\square$

Proposition 7.4. *For each integer $n \geq 1$ there is an $\varepsilon > 0$ such that if the positive constants $a_1, \ldots, a_n$ satisfy $a_j/a_{j+1} \leq \varepsilon$, then the polynomial*

$$p(x) = (a_1 + x^2)(a_2 + x^2) \ldots (a_n + x^2)$$

belongs to $\mathcal{P}$ and satisfies the inequality (7.5).

Proof. Define the polynomials p_j and the constants b_k^j by

$$p_j(x) = (a_1 + x^2)(a_2 + x^2) \ldots (a_j + x^2) = x^{2j} + b_1^j x^{2(j-1)} + \cdots + b_j^j.$$

Thus $p = p_n$ and we have by Lemma 7.1 that the corresponding q_j's are of the form

$$q_{n-j}(x) = c_0^j x^{2j} + c_1^j b_1^n x^{2(j-1)} + \cdots + c_j^j b_j^n, \qquad (7.6)$$

where c_k^j are positive constants.

Now, writing

$$q_{n-j}(x) = c_0^j p_j(x) + b_1^n(c_1^j + d_1^j)p_{j-1} + \cdots + b_j^n(c_j^j + d_j^j), \qquad (7.7)$$

we claim that $d_l^j = \mathcal{O}(\varepsilon)$, as $\varepsilon \to 0$. If we assume that this has been proved for $l = 1, 2 \ldots, k-1$, and identify the coefficients for $x^{2(j-k)}$ in (7.7) and (7.6), we get

$$c_0^j b_k^j + b_1^n(c_1^j + \mathcal{O}(\varepsilon))b_{k-1}^{j-1} + \cdots + b_{k-1}^n(c_{k-1}^j + \mathcal{O}(\varepsilon))b_1^{j-(k-1)} + b_k^n d_k^j = 0. \qquad (7.8)$$

Observe that if $\varepsilon \leq 1$ (so that (a_j) is increasing) there is a number M such that we have the following estimation for b_k^j:

$$a_{j-k+1}a_{j-k+2}\ldots a_j \leq b_k^j \leq M a_{j-k+1}a_{j-k+2}\ldots a_j.$$

Therefore, if $0 \leq l \leq k-1$,

$$\frac{b_l^n b_{k-l}^{j-l}}{b_k^n} \leq M^2 \frac{a_{j-k+1}\ldots a_{j-l}}{a_{n-k+1}\ldots a_{n-l}} = \mathcal{O}(\varepsilon).$$

Hence (7.8) shows that $d_k^j = \mathcal{O}(\varepsilon)$. Since for $k = 1$ no assumptions were used, the claim is proved.

Since p_j/p has positive Fourier transform and the coefficients in (7.7) are positive for small ε, the proof is completed by Corollary 7.3. □

Example 7.5. For any polynomial p of degree $2n$, we can easily compute (for instance by using (7.4)) the following:

$$q_0(x) = p(x) + p(0),$$
$$q_1(x) = (p'(x) - p'(0))/x,$$
$$q_{n-1}(x) = n^2 b_{2n} x^2 + (2n-1)b_{2n-1}x + 2b_{2n-2},$$
$$q_n(x) = 2b_{2n},$$

where b_m is the coefficient for x^m in $p(x)$. Now let p be as in the proposition. It follows immediately that q_0/p and q_n/p have positive Fourier transforms. The same is true for q_1/p since

$$\frac{p'(x)}{xp(x)} = 2\sum_{j=1}^{n} \frac{1}{a_j + x^2}.$$

As for q_{n-1} we now have

$$q_{n-1}(x) = n^2(a_1 + x^2) + 2\sum_{k=1}^{n} a_j - n^2 a_1,$$

so the condition $2\sum_1^n a_j \geq n^2 a_1$ is sufficient for the Fourier transform of q_{n-1}/p to be positive.

Taking $n = 1, 2$ and 3, we have proved that the polynomials

$$a + x^2, \qquad a > 0,$$
$$(a + x^2)(b + x^2), \qquad a, b > 0,$$
$$(a + x^2)(b + x^2)(c + x^2), \qquad a, b, c > 0, \quad 7a \leq 2(b + c)$$

are in $\mathcal{P}$ and satisfy the inequality (7.5).

7.1.2 Non-positivity

It is quite immediate that a necessary and sufficient condition for a positive polynomial p to belong to $\mathcal{P}$ is the condition

$$\sum_{j,k=1}^{n} c_j c_k \frac{p(t_j)}{p(t_j - t_k)} \geq 0, \quad \text{for all } c_j, t_j \in \mathbb{R} \text{ and } n = 1, 2, \ldots. \quad (7.9)$$

We can consider (7.9) as the limit of the right-hand side of (7.2) as the function $\hat{u}$ tends to the distribution $2\pi \sum_1^n c_j \delta_{t_j}$, where δ_{t_j} is the Dirac measure at t_j.

If we instead let $\hat{u}$ tend to the distribution $2\pi L(-iD)\delta_t$, where L is a polynomial with real coefficients, we obtain the necessary condition

$$L(\partial/\partial x)L(\partial/\partial y) \left.\frac{p(x)}{p(x-y)}\right|_{x=y=t} \geq 0, \quad t \in \mathbb{R},$$

in which only the two points 0 and t occur. The following proposition provides an equivalent form of this condition.

Proposition 7.6. *Let L be any polynomial with real coefficients. The condition*

$$\big(L(-iD)(p(t+iD)L)(iD)\big)(1/p)(0) \geq 0, \quad t \in \mathbb{R} \quad (7.10)$$

is necessary for $p \in \mathcal{P}$.

Proof. Let $\phi \in C_0^\infty$ have $\phi(0) = 1$. Taking $u(x) = e^{itx}L(ix)\phi(\varepsilon x)$ and letting $\varepsilon \to 0$, the real part of (7.2) tends to

$$\int \overline{p(t+D)(L(ix))}L(ix)\Gamma(x)\,dx = \int L(ix)(p(t+iD)L)(-ix)\Gamma(x)\,dx.$$

(When passing to the limit, we notice that $\Gamma(x)$ decreases exponentially as $|x| \to \infty$.) Since $\widehat{\Gamma} = 1/p$, the last integral equals the left-hand side of (7.10). $\square$

Corollary 7.7. *The condition*

$$4p(t)^2\big(p(0)p''(0) - p'(0)^2\big) \geq \big(2p(t)p'(0) - p(0)p'(t)\big)^2, \quad t \in \mathbb{R} \quad (7.11)$$

is necessary for $p \in \mathcal{P}$.

Proof. If we take $L(x) = a + x$, the operator that acts on $1/p$ in (7.10) becomes

$$a^2 p(t) + a p'(t) - \big(p(t)(iD)^2 + p'(t)iD\big).$$

Thus the left-hand side of (7.10) becomes a quadratic form in a. This form being nonnegative for all real a is equivalent to

$$-4\big(p(t)/p(0)\big)\big(p(t)(iD)^2 + p'(t)iD\big)(1/p)(0) \geq \big(p'(t)/p(0)\big)^2.$$

The last inequality, multiplied by $p(0)^4$, can be written as (7.11). $\square$

Example 7.8. Condition (7.11) implies that if $p''(0) \le 0$ then p is either constant or does not belong to $\mathcal{P}$.

Example 7.9. Let $p(x) = (1 + x^2)^m$. For this polynomial (7.11) reads

$$2(1 + t^2)^2 \ge mt^2,$$

which is equivalent to $m \le 8$.

   If we take $L(x) = ax + x^2$ and $t = 1$, Proposition 7.6 leads to

$$12m(m + 1) - 2m(m^2 - ma - a^2) \ge 0.$$

This is equivalent to $24(m + 1) \ge 5m^2$, that is, $m \le 5$.

   It is also possible to prove non-positivity for $(1 + x^2)^5$ and $(1 + x^2)^4$ with the aid of Proposition 7.6, but then one has to use a polynomial L of degree 3 in the former and of degree 4 in the latter case. In Section 7.1.3 we give another proof of the non-positivity when $m \ge 4$.

Proposition 7.10. *If $p \in P$ then the real part of $F(1/p)$ is nonnegative.*

Proof. Assume that $\operatorname{Re} \Phi(1/p) < 0$ at the point ξ_0 and hence also at the point $-\xi_0$. Let ϕ be a real, even function with $\phi(0) = 1$ and $\hat{\phi} \in C_0^\infty$. Put $f(x) = \cos(\xi_0 x)\phi(\varepsilon x)$, so that $\operatorname{supp} \hat{f} \to \{-\xi_0, \xi_0\}$ as $\varepsilon \to 0$.

   Let q_j be as in Lemma 7.1. Thus q_n is a positive constant so, by the continuity of $F(1/p)$, there is an $a > 0$ such that $\operatorname{Re} F(q_n/p) \le -a$ in $\operatorname{supp} \hat{f}$, if ε is small enough. Also, there is a number A such that $\operatorname{Re} F(q_j/p)(\xi) \le A$, for $j = 0, \ldots, n-1$ and $\xi \ne 0$.

   Now, using the inequality

$$\int g(K * g)\, dx \le \sup_{\operatorname{supp} \hat{g}} \left(\operatorname{Re} \hat{K}\right) \int g^2\, dx, \quad g, K \text{ real},$$

we get, since f is even, that (7.2) with f in place of $\hat{u}$ can be estimated by a constant times

$$\iint \frac{p(x) + p(-y)}{p(x - y)} f(x)f(y)\, dx\, dy = \int \sum_{j=0}^{n} x^j f(x)\big((q_j/p) * y^j f(y)\big)(x)\, dx$$

$$\le \int \left(A \sum_{j=0}^{n-1} x^{2j} - ax^{2n}\right) f(x)^2\, dx.$$

The last expression is clearly negative for small ε. $\qquad \square$

Example 7.11. Let $p > 0$ be a non-constant even polynomial with $F(1/p) \ge 0$ (for instance, p can be any non-constant even polynomial in Ψ). Since

$$\operatorname{Re} F(1/p(x - \varepsilon))(\xi) = F(1/p)(\xi)\cos(\varepsilon\xi),$$

Proposition 7.10 shows that if $\varepsilon \ne 0$ then $p(x - \varepsilon)$ does not belong to $\mathcal{P}$.

7.1.3 The operators $(1 + D^2)^m$

We introduce some notation. Let $p_m(x) = (1 + x^2)^m$, $\Gamma_m = (2\pi)^{-1}F(1/p_m)$ and put $\lambda_s(x) = (1+|x|)^s e^{-|x|}$. We observe that $\Gamma_m \sim \lambda_{m-1}$, according to Lemma 7.13 below. We define the form

$$Q_m(u) = \operatorname{Re} \int p_m(D)u \cdot \overline{u}\Gamma_m \, dx,$$

and the weighted Sobolev norms

$$\|u\|_{m,s} = \left(\sum_{j=0}^{m} \int |u^{(j)}|^2 \lambda_s \, dx\right)^{1/2}.$$

We remark that the subsequent inequality (7.18) shows that $\|u\|_{m,s} \geq c|u(0)|$, if $m \geq 1$ and $s \geq 0$.

The main result of this section is the following proposition. The five lemmas that follow it are needed for the proof.

Proposition 7.12. *The polynomial $(1 + x^2)^m$ belongs to $\mathcal{P}$ if and only if $m = 0, 1, 2, 3$. The following inequalities hold:*

$$\|u\|_{1,0}^2 \sim Q_1(u), \tag{7.12}$$

$$\|u\|_{2,1}^2 \sim Q_2(u), \tag{7.13}$$

$$c^{-1}\|u\|_{3,1}^2 \leq Q_3(u) \leq c\|u\|_{3,2}^2. \tag{7.14}$$

The inequality (7.14) cannot be improved by replacing any of the squared norms by another one of the type $\|u\|_{3,s}^2$.

Lemma 7.13. *The following identities hold:*

$$\Gamma_1(x) = \frac{1}{2}e^{-|x|},$$

$$\Gamma_2(x) = \frac{1}{4}(|x| + 1)e^{-|x|},$$

$$\Gamma_3(x) = \frac{1}{16}(x^2 + 3|x| + 3)e^{-|x|},$$

$$\Gamma_{m+2}(x) = \frac{2m+1}{2(m+1)}\Gamma_{m+1}(x) + \frac{x^2}{4m(m+1)}\Gamma_m(x).$$

Proof. The recursion formula follows from the relation

$$(1/p)'' = -4m(m+1)/p_{m+2} + 2m(2m+1)/p_{m+1}.$$

The formulas for Γ_1 and Γ_2 can be calculated directly. $\qquad\square$

For the next lemma, which will be used for counterexamples, we construct the functions u_t, $t \geq 1$. Let $\phi \in C_0^\infty((0,2))$ be real with $\phi = 1$ in a neighborhood of 1. Define $\phi_t \in C_0^\infty((0, t+1))$ so that $\phi_t = 1$ on $[1, t]$ and

$$\phi_t(x) = \phi(x), \quad \phi_t(x+t) = \phi(x+1), \quad x \in [0,1].$$

Let ω be a fixed real number and put

$$u_t(x) = e^{|x|/2}\phi_t(|x|)\cos(\omega x/2).$$

Then $u_t \in C_0^\infty$ is real and even.

Lemma 7.14. *Let u_t be as above. As $t \to \infty$ we have*

$$Q_m(u_t) = (2^{-3m}/m!)\operatorname{Re}\left(3 + \omega^2 - 2\omega i\right)^m t^m + \mathcal{O}(t^{m-1}). \tag{7.15}$$

Proof. It follows from Lemma (7.13) that $\Gamma_m(x) = r(|x|)e^{-|x|}$, where r is a polynomial of degree $m-1$ having leading coefficient $2^{-m}/(m-1)!$. Since u_t is real and even,

$$Q_m(u_t) = 2\int_0^\infty r(x)e^{-x}u_t(x)p_m(D)u_t(x)\, dx = 2\int_0^\infty r(x)\psi_t(x)\, dx,$$

where ψ_t is introduced in the obvious way. For $x \in [1, t]$ we have

$$\begin{aligned}
\psi_t(x) &= \cos(\omega x/2)p_m(D - i/2)\cos(\omega x/2)\\
&= \cos(\omega x/2)\operatorname{Re}\left(e^{i\omega x/2}p_m((\omega - i)/2)\right)\\
&= 2^{-(2m+1)}\operatorname{Re}\left((3 + \omega^2 - 2\omega i)^m(1 + e^{i\omega x})\right)
\end{aligned}$$

and it follows that

$$\begin{aligned}
Q_m(u_t) = {}&2\int_0^1 \left(r(\xi)\psi_t(x) + r(x+t)\psi_t(x+t)\right) dx\\
&+ 2^{-2m}\operatorname{Re}\left((3 + \omega^2 - 2\omega i)^m\int_1^t r(x)(1 + e^{i\omega x})\, dx\right).
\end{aligned}$$

Since r has degree $m-1$ the boundedness of $\{\psi_t\}$ implies that the first integral is $\mathcal{O}(t^{m-1})$. After one integration by parts also the integral $\int_1^t r(x)e^{i\omega x}\, dx$ is seen to be $\mathcal{O}(t^{m-1})$, so

$$Q_m(u_t) = 2^{-2m}\operatorname{Re}(3 + \omega^2 - 2\omega i)^m\int_1^t r(x)\, dx + \mathcal{O}(t^{m-1}), \quad \text{as } t \to \infty.$$

This gives (7.15). $\qquad\square$

Lemma 7.15. *If $\varepsilon > 0$, $s \in (1, 2)$ and k is a nonnegative integer then there exist $u, v \in C_0^\infty$ such that*

$$Q_3(u) \leq \varepsilon \int |u^{(j)}|^2 \lambda_s \, dx, \qquad 0 \leq j, \tag{7.16}$$

$$Q_3(v) \geq \varepsilon^{-1} \int |v^{(j)}|^2 \lambda_s \, dx, \quad 0 \leq j \leq k. \tag{7.17}$$

Proof. If we take $\omega = \sqrt{3}$ in the definition of u_t, Lemma 7.14 gives $Q_3(u_t) = \mathcal{O}(t^2)$, as $t \to \infty$. On the other hand, for $x \in [1, t]$, a simple calculation shows that

$$|u_t^{(j)}(x)|^2 = e^x \left(1 + \cos(\sqrt{3}x + j2\pi/3)\right)/2,$$

so for large t, the integral in (7.16) majorizes

$$\int_1^t x^s \left(1 + \cos(\sqrt{3}x + j2\pi/3)\right) dx \geq t^{s+1}/3.$$

This proves (7.16).

To prove (7.17) we take $\omega = 0$. Lemma 7.14 then gives $Q_3(u_t) \geq ct^3$, for large t. But, similarly as in the proof of Lemma 7.14, we see that the right-hand side of (7.17), with u_t in place of v, is $\mathcal{O}(t^{s+1})$. $\qquad\square$

The proof of the following simple lemma, which we use to establish equivalent norms in Proposition 7.12, also indicates the idea behind the more nontrivial Lemma 7.17.

Lemma 7.16. *If $a > 0$ then for every $u \in \mathscr{S}$ we have*

$$0 \leq -2|u(0)|^2 + \int e^{-|x|} \left((1 + a)|u|^2 + a^{-1}|u'|^2\right) dx, \tag{7.18}$$

$$0 \leq \int (1 + |x|)e^{-|x|} \left((1 + a)|u|^2 - 2|u'|^2 + a^{-1}|u''|^2\right) dx. \tag{7.19}$$

Proof. We begin by proving the second inequality. Let $v \in \mathscr{S}$ be a real function that is either even or odd (thus $v(0)v'(0) = 0$). By partial integration

$$\int_0^\infty (1 + x)e^{-x} vv'' \, dx = -\int_0^\infty (1 + x)e^{-x}(v')^2 \, dx - \frac{1}{2} \int_0^\infty (1 - x)e^{-x}v^2 \, dx$$

$$\leq \frac{1}{2} \int_0^\infty (1 + x)e^{-x} \left(v^2 - 2(v')^2\right) dx,$$

so for any $a > 0$,

$$0 \leq a \int_0^\infty (1 + x)e^{-x} \left(v + a^{-1}v''\right)^2 dx$$

$$\leq \int_0^\infty (1 + x)e^{-x} \left((1 + a)v^2 - 2(v')^2 + a^{-1}(v'')^2\right) dx. \tag{7.20}$$

Now, if u is real, (7.19) follows from (7.20) and the observation

$$\int \varphi(|x|)u^{(j)}(x)^2\,dx = 2\int_0^\infty \varphi(x)\big(u_0^{(j)}(x)^2 + u_1^{(j)}(x)^2\big)\,dx,$$

where $u = u_0 + u_1$ is the decomposition of u into even and odd functions. The complex case follows immediately. Similarly, the identity

$$\int_0^\infty e^{-x}(av + a^{-1}v')^2\,dx = -v(0)^2 + \int_0^\infty e^{-x}\big((a+1)v^2 + a^{-1}(v')^2\big)\,dx$$

leads to the first inequality.   □

Lemma 7.17. *For every $u \in \mathscr{S}$ it holds that*

$$0 \le 4|u(0)|^2 + \int e^{-|x|}\big(x^2|u|^2 + 6(1 - |x|)|u'|^2$$
$$+ 3|x|(2 - |x|)|u''|^2 + 2x^2|u'''|^2\big)\,dx. \tag{7.21}$$

Proof. This follows, as in the proof of the preceding lemma, from the identity

$$4v(0)^2 + 2\int_0^\infty e^{-x}\big(x^2v^2 + 6(1 - x)(v')^2 + 3x(2 - x)(v'')^2 + 2x^2(v''')^2\big)\,dx$$
$$= \int_0^\infty x^2 e^{-x}\big(3(v'' - v' + v)^2 + (2v''' - 3v'' + 3v' - v)^2\big)\,dx, \tag{7.22}$$

for a real $v \in \mathscr{S}$. To verify (7.22), one can expand the right-hand side and integrate by parts several times.   □

Proof of Proposition 7.12. The range of the argument for $3+\omega^2-2\omega i$ is $[-\pi/6, \pi/6]$ (the endpoints are attained for $\omega = \pm\sqrt{3}$), so if $m \ge 4$ then $(3+\omega^2-2\omega i)^m$ assumes values with negative real part. By Lemma 7.14, p_m does not belong to $\mathcal{P}$, if $m \ge 4$.

We now turn to proving the inequalities. Using the decompositions

$$\frac{p_1(x) + p_1(y)}{p_1(x - y)} = 1 + \frac{1 + 2xy}{p_1(x - y)},$$

$$\frac{p_2(x) + p_2(y)}{p_2(x - y)} = 1 + \frac{4xy}{p_1(x - y)} + \frac{1 + 2x^2y^2}{p_2(x - y)},$$

$$\frac{p_3(x) + p_3(y)}{p_3(x - y)} = 1 + \frac{6xy}{p_1(x - y)} + \frac{9x^2y^2}{p_2(x - y)} + \frac{1 - 3x^2y^2 + 2x^3y^3}{p_3(x - y)},$$

along with Parseval's formula and the formulas for $F(1/p_m) = 2\pi\Gamma_m$ we obtain (similarly as in the proof of Corollary 7.3) the identities

$$2Q_1(u) = |u(0)|^2 + \frac{1}{2} \int e^{-|x|} \left(|u|^2 + 2|u'|^2 \right) dx, \tag{7.23}$$

$$2Q_2(u) = |u(0)|^2 + \frac{1}{4} \int e^{-|x|} \left(8|u'|^2 + (1+|x|)(|u|^2 + 2|u''|^2) \right) dx, \tag{7.24}$$

$$2Q_3(u) = |u(0)|^2 + \frac{1}{16} \int e^{-|x|} \big((x^2 + 3|x| + 3)|u|^2 + 48|u'|^2$$
$$+ 3(-x^2 + 9|x| + 9)|u''|^2 + 2(x^2 + 3|x| + 3)|u'''|^2 \big) dx. \tag{7.25}$$

Now, (7.23) immediately leads to (7.12).

The inequalities in (7.13) follows from (7.24) and a combination of (7.24) and (7.19).

The right-hand side of (7.25) minus the right-hand side of

$$2Q_3(u) \ge \frac{3}{4}|u(0)|^2 + \frac{3}{16} \int e^{-|x|} \big((1+|x|)|u|^2 + 2(7+|x|)|u'|^2$$
$$+ (9 + 7|x|)|u''|^2 + 2(1+|x|)|u'''|^2 \big) dx, \tag{7.26}$$

multiplied by 16 equals the nonnegative expression in (7.21). Thus (7.26) holds. Now, (7.14) follows from (7.26) and (7.25).

The last statement in the proposition is a consequence of Lemma 7.15. $\qquad \square$

Remark 7.18. From the viewpoint of (7.25), the negative term on the right makes it nontrivial that $Q_3(u) \ge 0$. The proof of Lemma 7.15 shows that (7.25) minus the integral $\int |x|^s e^{-|x|} |u^{(j)}(x)|^2 \, dx$, multiplied by any positive number, can be negative if $s > 1$ and $j \ge 0$.

7.1.4 An application to integral equations

Let p be an even polynomial in $\mathcal{P}$. We can then define a norm (which is induced by an inner product) by

$$||\phi||_H^2 = \int |\phi(x)|^2 \, dx + \iint \frac{p(x) + p(y)}{p(x-y)} \phi(x)\overline{\phi(y)} \, dx \, dy, \quad \phi \in \Sigma.$$

Let H be the completion of the space $\mathscr{S}$ in this norm and define the dual space

$$H^* = \{ u \in \mathscr{S}' : ||u||_{H^*} = \sup_{\phi \in \mathscr{S}} |(u, \phi)| / ||\phi||_H < \infty \},$$

where $\mathscr{S}'$ is the space of tempered distributions. We thus have the inclusions $H \subset L^2 \subset H^*$.

Let a be a measurable function with $0 < M^{-1} \le a(x) \le M$ and define the operator K by

$$K\phi(x) = a(x)\phi(x) + \int \frac{p(x) + p(y)}{p(x-y)} \phi(y) \, dy, \quad \phi \in \Sigma.$$

Then K extends to a continuous linear operator from H to H^*. We introduce an inner product $\langle \, , \, \rangle$ in H by $\langle u, v \rangle = (Ku, v)$. (The induced norm is equivalent to the one defined above.)

If f is any fixed member of H^*, there is by the Riesz representation theorem a unique u in H such that $\langle u, v \rangle = (f, v)$, for all $v \in H$. But this means that u is the unique solution to the integral equation

$$Ku = f \in H^*, \quad u \in H.$$

7.2 Weighted positivity of second-order elliptic systems

Let Ω be a domain in $\mathbb{R}^n$ $(n \geqslant 3)$ with smooth boundary and assume $0 \in \Omega$. Consider the second-order elliptic system on Ω defined by

$$
\begin{aligned}
L_i(x, D_x)u \quad &:= \quad \sum_{j=1}^{N} \sum_{\alpha, \beta=1}^{n} -A_{ij}^{\alpha\beta}(x) \frac{\partial^2 u_j}{\partial x_\alpha \partial x_\beta} \\
&=: \quad -A_{ij}^{\alpha\beta}(x) D_{\alpha\beta} u_j \qquad (i = 1, 2, \ldots, N), \qquad (7.27)
\end{aligned}
$$

where as usual repeated indices indicate summation. We assume that $A_{ij}^{\alpha\beta}(x)$ are real-valued, continuous functions on Ω and there exists $\lambda > 0$ such that the strong Legendre condition

$$A_{ij}^{\alpha\beta}(x) \xi_\alpha^i \xi_\beta^j \geqslant \lambda |\xi|^2, \qquad \forall \xi \in \mathbb{R}^{nN}$$

holds uniformly on Ω. Without loss of generality, we may also assume that

$$A_{ij}^{\alpha\beta}(x) = A_{ij}^{\beta\alpha}(x) \qquad (i, j = 1, 2, \ldots, N, \ \alpha, \beta = 1, 2, \ldots, n).$$

Definition 7.19. The operator L is said to be positive with weight

$$\Psi(x) = (\Psi_{ij}(x))_{i,j=1}^N$$

if

$$\int_\Omega Lu \cdot \Psi u \, dx = -\int_\Omega A_{ik}^{\alpha\beta}(x) D_{\alpha\beta} u_k(x) \cdot u_j(x) \Psi_{ij}(x) \, dx \geqslant 0 \qquad (7.28)$$

for all real-valued vector functions $u = (u_i)_{i=1}^N$, $u_i \in C_0^\infty(\Omega)$.

Remark 7.20. The positivity of $L(x, D_x)$ actually reduces to the positivity of $L(0, D_x)$ (with the same weight). Indeed, if $u = (u_i)_{i=1}^N$ is a smooth vector function that is supported near the origin (say, in a δ-ball B_δ) and $u_\epsilon(x) = u(\epsilon^{-1}x)$, then

$$
\begin{aligned}
\int_\Omega Lu_\epsilon \cdot \Psi u_\epsilon \, dx &= -\int_{B_{\epsilon\delta}} A_{ik}^{\alpha\beta}(x) D_{\alpha\beta} u_{\epsilon k}(x) \cdot u_{\epsilon j}(x) \Psi_{ij}(x) \, dx \\
&= -\epsilon^{-n} \int_{B_{\epsilon\delta}} A_{ik}^{\alpha\beta}(x) (D_{\alpha\beta} u_k)(\epsilon^{-1}x) \cdot u_j(\epsilon^{-1}x) \Psi_{ij}(\epsilon^{-1}x) \, dx \\
&= -\int_{B_\delta} A_{ik}^{\alpha\beta}(\epsilon y) D_{\alpha\beta} u_k(y) \cdot u_j(y) \Psi_{ij}(y) \, dy \qquad (x = \epsilon y).
\end{aligned}
$$

Since the integrand in the last integral is bounded by

$$r^{2-n}\|A_{ik}^{\alpha\beta}\|_{L^\infty(B_\delta)}\|u\|_{C^2}^2\|\Psi_{ij}\|_{L^\infty(S^{n-1})},$$

which is clearly in $L^1(B_\delta)$, the dominated convergence theorem and the continuity of $A_{ij}^{\alpha\beta}$ implies that

$$\lim_{\epsilon\to0^+}\int_\Omega Lu_\epsilon\cdot\Psi u_\epsilon\,dx = -\int_{B_\delta}A_{ik}^{\alpha\beta}(0)D_{\alpha\beta}u_k(y)\cdot u_j(y)\Psi_{ij}(y)\,dy$$

$$=\int_\Omega L(0,D_x)u\cdot\Psi u\,dx.$$

Hence the positivity of L is in effect a local property at the origin.

Guo Luo and Maz'ya [62] proved that if the weighted positivity of general second-order elliptic system holds with respect to a weight Ψ which is smooth in $\mathbb{R}^n\setminus\{0\}$ and positive homogeneous of degree $2-n$, then Ψ has to be the fundamental solution of $L^T(0,D_x)$ multiplied by a semipositive definite constant matrix. Their result is reproduced in this section.

Theorem 7.21. *Suppose L is an elliptic operator as defined in (7.27) and Ψ satisfies (7.29). If L is positive with weight Ψ (and so is $L(0,D_x)$), then $L^T(0,D_x)\Psi = \delta M$ where δ is the Dirac delta function, $L^T(0,D_x)$ is the formal adjoint of $L(0,D_x)$,*

$$L_i^T(0,D_x)u := -A_{ji}^{\alpha\beta}(0)D_{\alpha\beta}u_j \qquad (i=1,2,\dots,N),$$

and $M\in\mathbb{R}^{N\times N}$ is a symmetric, semi-positive definite matrix. Furthermore,

$$\sum_{i,\alpha,\beta}A_{ip}^{\alpha\beta}(r\omega)\xi_\alpha\xi_\beta\Psi_{ip}(\omega)\geqslant 0, \qquad \forall\xi\in\mathbb{R}^n \qquad (p=1,2,\dots,N)$$

for all $r>0$, $\omega\in S^{n-1}$ such that $r\omega\in\Omega$. That is to say, the $n\times n$ matrix $(\sum_i A_{ip}^{\alpha\beta}(r\omega)\Psi_{ip}(\omega))_{\alpha,\beta=1}^n$ is pointwise semi-positive definite.

Remark 7.22. Several extensions of the above result are possible. First, in this theorem we considered only real coefficient elliptic operators and real-valued test functions. It is then natural to ask whether the same result holds for complex cases. Second, it is interesting to ask whether the set of operators that are positive in the sense of (7.1) is "open" in some suitable topology. In other words, we wonder whether a "small" perturbation of a positive operator still leaves the operator positive.

In this section we give the proof of Theorem 7.21. Without loss of generality, for the first part of the theorem we may assume $\Omega=\mathbb{R}^n$ and L is a constant coefficient elliptic operator.

First some preliminaries.

Let $\mathscr{H}_k$ denote the linear space of homogeneous polynomials of degree k that are harmonic; they are the so-called *solid spherical harmonics of degree k*.

The space of restrictions of $\mathscr{H}_k$ to the unit sphere, H_k, are the so-called *surface spherical harmonics of degree* k. It is well known that each $f \in L^2(S^{n-1})$ admits the decomposition

$$f(\omega) = \sum_{k=0}^{\infty} Y_k(\omega), \qquad Y_k \in H_k,$$

where the series converges in the L^2 sense. Since H_k can be shown to be mutually orthogonal (see, for example, Stein [96]), Parseval's identity

$$\int_{S^{n-1}} f(\omega)g(\omega)\, d\sigma = \sum_{k=0}^{\infty} \int_{S^{n-1}} Y_k(\omega) Z_k(\omega)\, d\sigma$$

holds for all $f, g \in L^2(S^{n-1})$ where

$$f(\omega) = \sum_{k=0}^{\infty} Y_k(\omega), \qquad g(\omega) = \sum_{k=0}^{\infty} Z_k(\omega).$$

7.2.1   Support at the origin

The first observation we make is that, in order for L to be positive with weight Ψ, $L^T \Psi$ has to be supported at the origin. We make the following assumption on the weight Ψ:

$$\Psi_{ij} \in C^{\infty}(\mathbb{R}^n \backslash \{0\}), \qquad\qquad (i, j = 1, 2, \ldots, N)$$
$$\Psi(x) = |x|^{2-n} \Psi\left(\frac{x}{|x|}\right) =: r^{2-n} \Psi(\omega), \qquad (7.29)$$

where $r = |x|$ and $\omega = x/|x|$.

Proposition 7.23. *Suppose L is a constant coefficient elliptic operator as defined in (7.27) and Ψ satisfies (7.29). If L is positive with weight Ψ, then $L^T \Psi$ is supported at the origin.*

We start the proof of this proposition by observing some elementary properties of the matrix Ψ.

Lemma 7.24. *Suppose $\Psi = (\Psi_{ij})_{i,j=1}^{N}$ satisfies (7.29). Then*

$$D_\alpha \Psi_{ij}(x) = r^{1-n} \Psi_{ij}^{\alpha}(\omega) \qquad (i, j = 1, 2, \ldots, N),$$
$$D_{\alpha\beta} \Psi_{ij}(x) = r^{-n} \Psi_{ij}^{\alpha\beta}(\omega) \qquad (\alpha, \beta = 1, 2, \ldots, n),$$

where $\Psi_{ij}^{\alpha}, \Psi_{ij}^{\alpha\beta} \in C^{\infty}(S^{n-1})$ and $\displaystyle\int_{S^{n-1}} \Psi_{ij}^{\alpha\beta}(\omega)\, d\sigma = 0$.

Proof. According to (7.29),

$$
\begin{aligned}
D_\alpha \Psi_{ij}(x) &= D_\alpha\left(r^{2-n}\Psi_{ij}(\omega)\right) \\
&= (2-n)r^{1-n}\Psi_{ij}(\omega) \cdot \frac{x_\alpha}{r} + r^{1-n}(D_\beta\Psi_{ij})(\omega)\left(\delta_{\alpha\beta} - \frac{x_\alpha}{r} \cdot \frac{x_\beta}{r}\right) \\
&= r^{1-n}\left[(2-n)\omega_\alpha\Psi_{ij}(\omega) + (D_\alpha\Psi_{ij})(\omega) - \omega_\alpha\omega_\beta(D_\beta\Psi_{ij})(\omega)\right] \\
&=: r^{1-n}\Psi_{ij}^\alpha(\omega),
\end{aligned}
$$

where

$$
\Psi_{ij}^\alpha(\omega) = (2-n)\omega_\alpha\Psi_{ij}(\omega) + (D_\alpha\Psi_{ij})(\omega) - \omega_\alpha\omega_\beta(D_\beta\Psi_{ij})(\omega).
$$

Similarly one can show that

$$
D_{\alpha\beta}\Psi_{ij}(x) = D_\alpha\left(r^{1-n}\Psi_{ij}^\beta(\omega)\right) = r^{-n}\Psi_{ij}^{\alpha\beta}(\omega).
$$

To prove the last statement, we integrate the above identity on $B_2\backslash B_1$ and obtain

$$
\int_{B_2\backslash B_1} D_\alpha\left(r^{1-n}\Psi_{ij}^\beta(\omega)\right)\,dx = \int_{B_2\backslash B_1} r^{-n}\Psi_{ij}^{\alpha\beta}(\omega)\,dx.
$$

Note that

$$
\begin{aligned}
\int_{B_2\backslash B_1} D_\alpha\left(r^{1-n}\Psi_{ij}^\beta(\omega)\right)\,dx &= \int_{\partial(B_2\backslash B_1)} r^{1-n}\Psi_{ij}^\beta(\omega)\nu_\alpha\,d\sigma \\
&= \int_{\partial B_2} r^{1-n}\Psi_{ij}^\beta(\omega)\omega_\alpha\,d\sigma - \int_{\partial B_1} r^{1-n}\Psi_{ij}^\beta(\omega)\omega_\alpha\,d\sigma \\
&= \int_{S^{n-1}} \Psi_{ij}^\beta(\omega)\omega_\alpha\,d\sigma - \int_{S^{n-1}} \Psi_{ij}^\beta(\omega)\omega_\alpha\,d\sigma \\
&= 0,
\end{aligned}
$$

and

$$
\begin{aligned}
\int_{B_2\backslash B_1} r^{-n}\Psi_{ij}^{\alpha\beta}(\omega)\,dx &= \int_1^2 r^{-1}\,dr \int_{S^{n-1}} \Psi_{ij}^{\alpha\beta}(\omega)\,d\sigma \\
&= \log 2 \int_{S^{n-1}} \Psi_{ij}^{\alpha\beta}(\omega)\,d\sigma.
\end{aligned}
$$

So the result follows. □

Since the proof of Proposition 7.23 is long, we break it up into two lemmas.

Lemma 7.25. *Under the assumptions of Proposition 7.23, if L is positive with weight Ψ, then $(L^T\Psi)_{pp}$ $(p = 1, 2, \ldots, N)$ is supported at the origin.*

Proof. Step 1. By definition, we wish to show that

$$\sum_{i,\alpha,\beta} A_{ip}^{\alpha\beta} D_{\alpha\beta} \Psi_{ip} = 0 \quad \text{on} \quad \mathbb{R}^n \backslash \{0\} \qquad (p = 1, 2, \ldots, N).$$

Taking $u = (u_i)_{i=1}^N$ where

$$u_i = \begin{cases} 0, & i \neq p \\ v, & i = p \end{cases}, \qquad v \in C_0^\infty(\mathbb{R}^n \backslash \{0\}),$$

we have

$$\int Lu \cdot \Psi u \, dx = - \int A_{ik}^{\alpha\beta} D_{\alpha\beta} u_k \cdot u_j \Psi_{ij} \, dx$$

$$= - \int \sum_{i,\alpha,\beta} A_{ip}^{\alpha\beta} D_{\alpha\beta} v \cdot v \Psi_{ip} \, dx$$

$$= \int \sum_{i,\alpha,\beta} A_{ip}^{\alpha\beta} D_\alpha v D_\beta v \cdot \Psi_{ip} \, dx + \int \sum_{i,\alpha,\beta} A_{ip}^{\alpha\beta} D_\alpha v \cdot v D_\beta \Psi_{ip} \, dx$$

$$=: I_1 + I_2.$$

Step 2. By assumption (7.29), it is easy to see that

$$|I_1| \leqslant C \int r^{2-n} |Dv|^2 \, dx. \tag{7.30}$$

As for I_2, we observe $D_\alpha v \cdot v = \frac{1}{2} D_\alpha(v^2)$, so integrating by parts once more gives

$$I_2 = -\frac{1}{2} \int \sum_{i,\alpha,\beta} A_{ip}^{\alpha\beta} v^2 D_{\alpha\beta} \Psi_{ip} \, dx. \tag{7.31}$$

Now assume

$$\sum_{i,\alpha,\beta} A_{ip}^{\alpha\beta} D_{\alpha\beta} \Psi_{ip} \not\equiv 0 \quad \text{on} \quad \mathbb{R}^n \backslash \{0\}.$$

By Lemma 7.24,

$$D_{\alpha\beta} \Psi_{ip} = r^{-n} \Psi_{ip}^{\alpha\beta}(\omega),$$

so we may write

$$\sum_{i,\alpha,\beta} A_{ip}^{\alpha\beta} D_{\alpha\beta} \Psi_{ip} = \sum_{i,\alpha,\beta} A_{ip}^{\alpha\beta} r^{-n} \Psi_{ip}^{\alpha\beta}(\omega) =: r^{-n} \Psi_{pp}''(\omega),$$

where

$$\Psi_{pp}''(\omega) = \sum_{i,\alpha,\beta} A_{ip}^{\alpha\beta} \Psi_{ip}^{\alpha\beta}(\omega) \not\equiv 0, \qquad \int_{S^{n-1}} \Psi_{pp}''(\omega) \, d\sigma = 0.$$

Substituting this into (7.31) and switching to spherical coordinates, we have

$$I_2 = -\frac{1}{2} \int_0^\infty r^{-1} \, dr \int_{S^{n-1}} v^2 \Psi''_{pp}(\omega) \, d\sigma. \tag{7.32}$$

Step 3. Let

$$\Psi''_{pp}(\omega) = \sum_{k=m}^\infty Y_k(\omega), \qquad Y_k \in H_k$$

where $Y_m \not\equiv 0$. Note that $m \geqslant 1$ since

$$\int_{S^{n-1}} \Psi''_{pp}(\omega) \, d\sigma = 0.$$

Now take $v(x) = \zeta(r)\Omega(\omega)$ where

$$\zeta \in C_0^\infty(0, \infty) \qquad \text{is to be determined later,}$$
$$\Omega(\omega) = \epsilon^{-1} + Y_m(\omega), \qquad \epsilon > 0.$$

Substituting this into (7.32), applying Parseval's identity and recalling that $m \geqslant 1$, we have

$$I_2 = -\frac{1}{2} \int_0^\infty r^{-1} \zeta^2(r) \, dr \int_{S^{n-1}} \left(\epsilon^{-1} + Y_m(\omega) \right)^2 \sum_{k=m}^\infty Y_k(\omega) \, d\sigma$$

$$\leqslant -\frac{1}{2} \int_0^\infty r^{-1} \zeta^2(r) \, dr \left(2\epsilon^{-1} \int_{S^{n-1}} Y_m^2(\omega) \, d\sigma + C \right).$$

This implies, for small ϵ, that

$$I_2 \leqslant -C_0 \epsilon^{-1} \int_0^\infty r^{-1} \zeta^2(r) \, dr.$$

On the other hand, we note that (7.30) implies that

$$|I_1| \leqslant C \int r^{2-n} \left[(\zeta'(r))^2 \Omega^2(\omega) + r^{-2}\zeta^2(r)|\nabla_\sigma \Omega(\omega)|^2 \right] dx$$

$$= C \int_0^\infty r(\zeta'(r))^2 \, dr \int_{S^{n-1}} \Omega^2(\omega) \, d\sigma$$

$$+ C \int_0^\infty r^{-1}\zeta^2(r) \, dr \int_{S^{n-1}} |\nabla_\sigma Y_m(\omega)|^2 \, d\sigma$$

$$=: I_{11} + I_{12},$$

where ∇_σ is the spherical part of the gradient D.

Step 4. We first choose ϵ small enough so that

$$C \int_{S^{n-1}} |\nabla_\sigma Y_m(\omega)|^2 \, d\sigma < C_0(2\epsilon)^{-1},$$

where C is the constant appearing in (7.30). For this fixed ϵ, we have

$$I_{12} < C_0(2\epsilon)^{-1} \int_0^\infty r^{-1}\zeta^2(r)\, dr, \qquad \forall \zeta \in C_0^\infty(0, \infty).$$

Next, we appeal to Lemma 7.26 below and choose ζ so that

$$I_{11} < C_0(2\epsilon)^{-1} \int_0^\infty r^{-1}\zeta^2(r)\, dr.$$

This shows that

$$I_1 + I_2 < 0$$

and gives us the desired contradiction. $\qquad\qquad\square$

Lemma 7.26. *For any given $C > 0$, there exists $\zeta \in C_0^\infty(0, \infty)$ so that*

$$\int_0^\infty r^{-1}\zeta^2(r)\, dr \geqslant C \int_0^\infty r(\zeta'(r))^2\, dr.$$

Proof. Take $\varphi \in C^\infty(\mathbb{R})$ such that

$$\varphi(r) = \begin{cases} 0, & r \leqslant 0 \\ 1, & r \geqslant 1 \end{cases}, \qquad 0 \leqslant \varphi \leqslant 1.$$

For $0 < \delta < \frac{1}{4}$, define

$$\zeta_\delta(r) = \begin{cases} \varphi(\delta^{-1}r - 1), & 0 \leqslant r < 1 \\ \varphi(-r + 2), & r \geqslant 1 \end{cases}.$$

Clearly $\zeta_\delta \in C_0^\infty(0, \infty)$. Now

$$\int_0^\infty r^{-1}\zeta_\delta^2(r)\, dr \geqslant \int_{2\delta}^1 r^{-1}\, dr = \log\frac{1}{2\delta},$$

$$\int_0^\infty r(\zeta_\delta'(r))^2\, dr = \int_\delta^{2\delta} r(\zeta_\delta'(r))^2\, dr + \int_1^2 r(\zeta_\delta'(r))^2\, dr$$

$$\leqslant \delta^{-2}\|\varphi'\|_\infty^2 \int_\delta^{2\delta} r\, dr + \|\varphi'\|_\infty^2 \int_1^2 r\, dr$$

$$\leqslant C\|\varphi'\|_\infty^2.$$

So the result follows by choosing δ sufficiently small. $\qquad\qquad\square$

While Lemma 7.25 proves the statement of Proposition 7.23 for diagonal elements of $L^T\Psi$, the next one takes care of the off-diagonal elements.

Lemma 7.27. *Under the assumptions of Proposition 7.23, if L is positive with weight Ψ, then $(L^T \Psi)_{pq}$ $(p, q = 1, 2, \ldots, N, p \neq q)$ is supported at the origin.*

Proof. Step 1. By definition, we wish to show that

$$A_{ip}^{\alpha\beta} D_{\alpha\beta} \Psi_{iq} = 0 \quad \text{on} \quad \mathbb{R}^n \backslash \{0\} \qquad (p, q = 1, 2, \ldots, N, p \neq q).$$

Taking $u = (u_i)_{i=1}^N$ where

$$u_i = \begin{cases} 0, & i \neq p, q \\ v, & i = p \\ w, & i = q \end{cases}, \qquad v, w \in C_0^\infty(\mathbb{R}^n \backslash \{0\}),$$

we have

$$\int Lu \cdot \Psi u \, dx = -\int A_{ik}^{\alpha\beta} D_{\alpha\beta} u_k \cdot u_j \Psi_{ij} \, dx$$

$$= \int A_{ik}^{\alpha\beta} D_\alpha u_k D_\beta u_j \cdot \Psi_{ij} \, dx + \int A_{ik}^{\alpha\beta} D_\alpha u_k \cdot u_j D_\beta \Psi_{ij} \, dx$$

$$=: I_1 + I_2.$$

Step 2. By assumption (7.29) and Cauchy's inequality, it is easy to see that

$$|I_1| \leqslant C \int r^{2-n} |Du|^2 \, dx \leqslant C \int r^{2-n} \left(|Dv|^2 + |Dw|^2 \right) dx. \tag{7.33}$$

As for I_2, it follows from Lemma 7.25 that

$$\int A_{ip}^{\alpha\beta} D_\alpha v \cdot v D_\beta \Psi_{ip} \, dx = \int A_{iq}^{\alpha\beta} D_\alpha w \cdot w D_\beta \Psi_{iq} \, dx = 0.$$

So

$$I_2 = \int A_{ip}^{\alpha\beta} D_\alpha v \cdot w D_\beta \Psi_{iq} \, dx + \int A_{iq}^{\alpha\beta} D_\alpha w \cdot v D_\beta \Psi_{ip} \, dx$$

$$= -\int A_{ip}^{\alpha\beta} v \left(D_\alpha w D_\beta \Psi_{iq} + w D_{\alpha\beta} \Psi_{iq} \right) dx + \int A_{iq}^{\alpha\beta} D_\alpha w \cdot v D_\beta \Psi_{ip} \, dx$$

$$= -\int A_{ip}^{\alpha\beta} v w D_{\alpha\beta} \Psi_{iq} \, dx + \int v D_\alpha w \left(A_{iq}^{\alpha\beta} D_\beta \Psi_{ip} - A_{ip}^{\alpha\beta} D_\beta \Psi_{iq} \right) dx$$

$$= -\int r^{-n} v w \Psi_{pq}''(\omega) \, dx + \int r^{1-n} v D_\alpha w \left(\Psi_{\alpha qp}'(\omega) - \Psi_{\alpha pq}'(\omega) \right) dx \tag{7.34}$$

$$=: I_{21} + I_{22}.$$

In the derivation of (7.34) we have used Lemma 7.24 again, where

$$\Psi_{\alpha pq}'(\omega) = A_{ip}^{\alpha\beta} \Psi_{iq}^\beta(\omega),$$

$$\Psi_{pq}''(\omega) = A_{ip}^{\alpha\beta} \Psi_{iq}^{\alpha\beta}(\omega), \qquad \int_{S^{n-1}} \Psi_{pq}''(\omega) \, d\sigma = 0.$$

Now assume

$$A_{ip}^{\alpha\beta} D_{\alpha\beta} \Psi_{iq} \not\equiv 0 \quad \text{on} \quad \mathbb{R}^n \backslash \{0\}.$$

Then we have

$$\Psi''_{pq}(\omega) \not\equiv 0.$$

Step 3. Let

$$\Psi''_{pq}(\omega) = \sum_{k=m}^{\infty} Y_k(\omega), \qquad Y_k \in H_k$$

where $Y_m \not\equiv 0$. Note that $m \geqslant 1$ since

$$\int_{S^{n-1}} \Psi''_{pq}(\omega) \, d\sigma = 0.$$

Now take

$$v(x) = \zeta(r) Y_m(\omega),$$
$$w(x) = \epsilon^{-1} \zeta(r), \qquad \epsilon > 0,$$

where $\zeta \in C_0^\infty(0, \infty)$ is to be determined later. Substituting this into (7.34), applying Parseval's identity and recalling that $m \geqslant 1$, we have

$$I_{21} = -\int_0^\infty r^{-1} \zeta^2(r) \, dr \int_{S^{n-1}} \epsilon^{-1} Y_m(\omega) \sum_{k=m}^{\infty} Y_k(\omega) \, d\sigma$$

$$= -\epsilon^{-1} \int_0^\infty r^{-1} \zeta^2(r) \, dr \int_{S^{n-1}} Y_m^2(\omega) \, d\sigma$$

$$= -C_0 \epsilon^{-1} \int_0^\infty r^{-1} \zeta^2(r) \, dr,$$

$$I_{22} = \epsilon^{-1} \int_0^\infty \zeta(r) \zeta'(r) \, dr \int_{S^{n-1}} \omega_\alpha Y_m(\omega) \left(\Psi'_{\alpha qp}(\omega) - \Psi'_{\alpha pq}(\omega) \right) d\sigma$$

$$= 0.$$

So

$$I_2 = -C_0 \epsilon^{-1} \int_0^\infty r^{-1} \zeta^2(r) \, dr.$$

On the other hand, (7.33) implies that

$$|I_1| \leqslant C \int r^{2-n} \left[(\zeta'(r))^2 Y_m^2(\omega) + r^{-2} \zeta^2(r) |\nabla_\sigma Y_m(\omega)|^2 + \epsilon^{-2} (\zeta'(r))^2 \right] dx$$

$$= C \int_0^\infty r (\zeta'(r))^2 \, dr \int_{S^{n-1}} \left(Y_m^2(\omega) + \epsilon^{-2} \right) d\sigma$$

$$+ C \int_0^\infty r^{-1} \zeta^2(r) \, dr \int_{S^{n-1}} |\nabla_\sigma Y_m(\omega)|^2 \, d\sigma.$$

Now we may proceed as in Lemma 7.25 and choose ϵ, ζ appropriately to derive the desired contradiction. □

Now Proposition 7.23 is a direct consequence of Lemma 7.25 and Lemma 7.27.

7.2.2 Positive Definiteness of $L^T \Psi$

By Proposition 7.23, we can write $L^T \Psi$ as

$$L^T \Psi = \delta M,$$

where δ is the Dirac delta function and M is a real $N \times N$ matrix. Now we show M is symmetric and semi-positive definite.

Proposition 7.28. *Suppose L is a constant coefficient elliptic operator as defined in (7.27) and Ψ satisfies (7.29). If L is positive with weight Ψ, then $L^T \Psi = \delta M$ where δ is the Dirac delta function and $M \in \mathbb{R}^{N \times N}$ is a symmetric, semi-positive definite matrix.*

We start the proof of this proposition by writing M explicitly in terms of $A_{ij}^{\alpha\beta}$ and Ψ_{ij}.

Lemma 7.29. *Under the assumptions of Proposition 7.28, if L is positive with weight Ψ, then $L^T \Psi = \delta M$ where $M \in \mathbb{R}^{N \times N}$,*

$$M_{pq} = - \int_{S^{n-1}} A_{ip}^{\alpha\beta} \omega_\alpha \Psi_{iq}^\beta(\omega)\, d\sigma$$

$$= - \int_{S^{n-1}} \omega_\alpha \Psi'_{\alpha pq}(\omega)\, d\sigma \qquad (p, q = 1, 2, \ldots, N).$$

Here $\Psi_{iq}^\beta(\omega), \Psi'_{\alpha pq}(\omega)$ are as defined in Lemma 7.24 and Lemma 7.27.

Proof. By Lemma 7.24 and Proposition 7.23, for any $u \in C_0^\infty(\mathbb{R}^n)$,

$$M_{pq} u(0) = \left\langle (L^T \Psi)_{pq}, u \right\rangle = \left\langle -A_{ip}^{\alpha\beta} D_{\alpha\beta} \Psi_{iq}, u \right\rangle$$

$$= - \int A_{ip}^{\alpha\beta} \Psi_{iq} D_{\alpha\beta} u\, dx$$

$$= \int A_{ip}^{\alpha\beta} D_\beta \Psi_{iq} D_\alpha u\, dx$$

$$= \lim_{\epsilon \to 0} \int_{\mathbb{R}^n \setminus B_\epsilon} A_{ip}^{\alpha\beta} D_\beta \Psi_{iq} D_\alpha u\, dx$$

$$= \lim_{\epsilon \to 0} \left(\int_{\partial B_\epsilon} A_{ip}^{\alpha\beta} D_\beta \Psi_{iq} \cdot u \nu_\alpha\, d\sigma - \int_{\mathbb{R}^n \setminus B_\epsilon} A_{ip}^{\alpha\beta} D_{\alpha\beta} \Psi_{iq} \cdot u\, dx \right)$$

$$= - \lim_{\epsilon \to 0} \int_{\partial B_\epsilon} A_{ip}^{\alpha\beta} r^{1-n} \Psi_{iq}^\beta(\omega) \cdot u \omega_\alpha\, d\sigma$$

$$= - \lim_{\epsilon \to 0} \int_{S^{n-1}} A_{ip}^{\alpha\beta} \Psi_{iq}^\beta(\omega) u(\epsilon\omega) \omega_\alpha\, d\sigma$$

$$= -u(0) \int_{S^{n-1}} A_{ip}^{\alpha\beta} \omega_\alpha \Psi_{iq}^\beta(\omega)\, d\sigma.$$

So the result follows. □

As before, we break up the proof of Proposition 7.28 into two lemmas.

Lemma 7.30. *Under the assumptions of Proposition 7.28, if L is positive with weight Ψ, then $L^T\Psi = \delta M$ where $M \in \mathbb{R}^{N\times N}$ is symmetric.*

Proof. Step 1. By definition, we wish to show that

$$M_{pq} = M_{qp} \qquad (p, q = 1, 2, \ldots, N,\ p \neq q).$$

As in the proof of Lemma 7.27, we take $u = (u_i)_{i=1}^N$, where

$$u_i = \begin{cases} 0, & i \neq p, q \\ v, & i = p \\ w, & i = q \end{cases}, \qquad v, w \in C_0^\infty(\mathbb{R}^n\backslash\{0\}),$$

and obtain

$$\int Lu \cdot \Psi u\, dx = \int A_{ik}^{\alpha\beta} D_\alpha u_k D_\beta u_j \cdot \Psi_{ij}\, dx + \int A_{ik}^{\alpha\beta} D_\alpha u_k \cdot u_j D_\beta \Psi_{ij}\, dx$$
$$=: I_1 + I_2.$$

Step 2. As before, we have

$$|I_1| \leqslant C \int r^{2-n}\left(|Dv|^2 + |Dw|^2\right) dx \tag{7.35}$$

and

$$I_2 = -\int r^{-n} vw \Psi_{pq}''(\omega)\, dx + \int r^{1-n} v D_\alpha w\left(\Psi_{\alpha qp}'(\omega) - \Psi_{\alpha pq}'(\omega)\right) dx.$$

Note that

$$\Psi_{pq}''(\omega) \equiv 0$$

by Proposition 7.23, so

$$I_2 = \int r^{1-n} v D_\alpha w\left(\Psi_{\alpha qp}'(\omega) - \Psi_{\alpha pq}'(\omega)\right) dx. \tag{7.36}$$

Step 3. Now take

$$v(x) = \zeta(r),$$
$$w(x) = \eta(r) := -\epsilon\, \mathrm{sgn}(M_{pq} - M_{qp}) \int_0^r \rho^{-1}\zeta(\rho)\, d\rho, \qquad \epsilon > 0,$$

where $\zeta \in C_0^\infty(0, \infty)$ is to be determined later. Substituting this into (7.36), switching to spherical coordinates and applying Lemma 7.29, we have

$$I_2 = \int_0^\infty \zeta(r)\eta'(r)\,dr \int_{S^{n-1}} \omega_\alpha \left(\Psi'_{\alpha qp}(\omega) - \Psi'_{\alpha pq}(\omega) \right) d\sigma$$

$$= -\epsilon\,\mathrm{sgn}(M_{pq} - M_{qp}) \cdot (M_{pq} - M_{qp}) \int_0^\infty r^{-1}\zeta^2(r)\,dr$$

$$= -\epsilon|M_{pq} - M_{qp}| \int_0^\infty r^{-1}\zeta^2(r)\,dr.$$

On the other hand, (7.35) implies that

$$|I_1| \leqslant C \int r^{2-n}\left[(\zeta'(r))^2 + (\eta'(r))^2 \right] dx$$

$$= C \left[\int_0^\infty r(\zeta'(r))^2\,dr + \epsilon^2 \int_0^\infty r^{-1}\zeta^2(r)\,dr \right]$$

$$=: I_{11} + I_{12}.$$

Step 4. Assume $M_{pq} - M_{qp} \neq 0$. We first choose ϵ small enough so that

$$C\epsilon < \frac{1}{2}|M_{pq} - M_{qp}|,$$

where C is the constant appearing in (7.35). For this fixed ϵ, we have

$$I_{12} < \frac{\epsilon}{2}|M_{pq} - M_{qp}| \int_0^\infty r^{-1}\zeta^2(r)\,dr, \qquad \forall \zeta \in C_0^\infty(0, \infty).$$

Next, we appeal to Lemma 7.31 below and choose ζ so that

$$I_{11} < \frac{\epsilon}{2}|M_{pq} - M_{qp}| \int_0^\infty r^{-1}\zeta^2(r)\,dr.$$

This shows that

$$I_1 + I_2 < 0$$

and gives us the desired contradiction. $\square$

Lemma 7.31. *For any given $C > 0$, there exists $\zeta \in C_0^\infty(0, \infty)$ so that*

$$\int_0^\infty r^{-1}\zeta^2(r)\,dr \geqslant C \int_0^\infty r(\zeta'(r))^2\,dr$$

and

$$r^{-1}\zeta(r) = \eta'(r) \quad \text{for some } \eta \in C_0^\infty(0, \infty).$$

Proof. We first note that for any $\zeta \in C_0^\infty(0, \infty)$,

$$r^{-1}\zeta(r) = \eta'(r) \quad \text{for some } \eta \in C_0^\infty(0, \infty)$$

if and only if

$$\int_0^\infty r^{-1}\zeta(r)\, dr = 0.$$

Let $\varphi \in C^\infty(\mathbb{R})$ be as given in Lemma 7.26. For $0 < \delta < \frac{1}{4}$ and $R > \frac{5}{4}$, define

$$\zeta_{\delta,R}(r) = \begin{cases} \varphi(\delta^{-1}r - 1), & 0 \leqslant r < \frac{3}{4} \\ 2\varphi\left(-2r + \frac{5}{2}\right) - 1, & \frac{3}{4} \leqslant r < R \\ \varphi(R^{-1}r - 1) - 1, & r \geqslant R \end{cases}.$$

Clearly $\zeta_{\delta,R} \in C_0^\infty(0, \infty)$. For each δ small, we may choose $R = R_\delta$ so that

$$\int_0^\infty \zeta_{\delta,R_\delta}(r)\, dr = 0.$$

This is always possible since the integral above changes continuously with R and

$$\int_0^\infty r^{-1}\zeta_{\delta,5/4}(r)\, dr > 0 \quad \text{if } \delta \text{ is sufficiently small,}$$

$$\int_0^\infty r^{-1}\zeta_{\delta,R}(r)\, dr \to -\infty \quad \text{as } R \to \infty.$$

Now

$$\int_0^\infty r^{-1}\zeta_{\delta,R_\delta}^2(r)\, dr \geqslant \int_{2\delta}^{3/4} r^{-1}\, dr = \log \frac{3}{8\delta},$$

$$\int_0^\infty r(\zeta_{\delta,R_\delta}'(r))^2\, dr = \left[\int_\delta^{2\delta} + \int_{3/4}^{5/4} + \int_{R_\delta}^{2R_\delta}\right] r(\zeta_{\delta,R_\delta}'(r))^2\, dr$$

$$\leqslant \delta^{-2}\|\varphi'\|_\infty^2 \int_\delta^{2\delta} r\, dr + 16\|\varphi'\|_\infty^2 \int_{3/4}^{5/4} r\, dr + R_\delta^{-2}\|\varphi'\|_\infty^2 \int_{R_\delta}^{2R_\delta} r\, dr$$

$$\leqslant C\|\varphi'\|_\infty^2.$$

So the result follows by choosing δ sufficiently small. $\qquad\square$

Now we show M is semi-positive definite.

Lemma 7.32. *Under the assumptions of Proposition 7.28, if L is positive with weight Ψ, then $L^T\Psi = \delta M$ where $M \in \mathbb{R}^{N \times N}$ is semi-positive definite.*

Proof. Step 1. Take $u = (u_i)_{i=1}^N$ where $u_i \in C_0^\infty(\mathbb{R}^n)$ (note that u_i does not necessarily vanish near the origin). As before we have

$$\int Lu \cdot \Psi u \, dx = \int A_{ik}^{\alpha\beta} D_\alpha u_k D_\beta u_j \cdot \Psi_{ij} \, dx + \int A_{ik}^{\alpha\beta} D_\alpha u_k \cdot u_j D_\beta \Psi_{ij} \, dx$$

$$=: I_1 + I_2.$$

Step 2. Clearly

$$|I_1| \leqslant C \int r^{2-n} |Du|^2 \, dx. \tag{7.37}$$

As for I_2, we write

$$I_2 = \int \sum_{k<j} A_{ik}^{\alpha\beta} D_\alpha u_k \cdot u_j D_\beta \Psi_{ij} \, dx + \int \sum_{k>j} A_{ik}^{\alpha\beta} D_\alpha u_k \cdot u_j D_\beta \Psi_{ij} \, dx$$

$$+ \int \sum_{k=j} A_{ik}^{\alpha\beta} D_\alpha u_k \cdot u_j D_\beta \Psi_{ij} \, dx$$

$$=: I_{21} + I_{22} + I_{23}.$$

Similar to the calculations in Lemma 7.29, we have

$$I_{21} = \sum_{k<j} \int A_{ik}^{\alpha\beta} D_\alpha u_k \cdot u_j D_\beta \Psi_{ij} \, dx$$

$$= -\sum_{k<j} \left(u_k(0) u_j(0) \int_{S^{n-1}} \omega_\alpha \Psi'_{\alpha k j}(\omega) \, d\sigma + \int A_{ik}^{\alpha\beta} u_k D_\alpha u_j D_\beta \Psi_{ij} \, dx \right)$$

$$= \sum_{k<j} u_k(0) u_j(0) M_{kj} - \sum_{j<k} \int A_{ij}^{\alpha\beta} u_j D_\alpha u_k D_\beta \Psi_{ik} \, dx,$$

$$I_{23} = \sum_{k=j} \int A_{ik}^{\alpha\beta} D_\alpha u_k \cdot u_j D_\beta \Psi_{ij} \, dx$$

$$= \frac{1}{2} \sum_{k=j} u_k(0) u_j(0) M_{kj}.$$

Since M is symmetric, this implies that

$$I_2 = \frac{1}{2} \sum_{k,j} u_k(0) u_j(0) M_{kj} + \sum_{k>j} \int u_j D_\alpha u_k \left(A_{ik}^{\alpha\beta} D_\beta \Psi_{ij} - A_{ij}^{\alpha\beta} D_\beta \Psi_{ik} \right) dx$$

$$= \frac{1}{2} u^T(0) M u(0) + \sum_{k>j} \int u_j D_\alpha u_k \left(A_{ik}^{\alpha\beta} D_\beta \Psi_{ij} - A_{ij}^{\alpha\beta} D_\beta \Psi_{ik} \right) dx. \tag{7.38}$$

Step 3. Assume M is not semi-positive definite, then there exists $\xi \in \mathbb{R}^N$ such that

$$\xi^T M \xi < 0.$$

Take

$$u_j(x) = \xi_j \varphi \left(\frac{\log r}{\log \epsilon} - 1 \right), \qquad 0 < \epsilon < 1 \qquad (j = 1, 2, \ldots, N),$$

where $\varphi \in C^\infty(\mathbb{R})$ is as given in Lemma 7.26. Substituting this into (7.38), switching to spherical coordinates and applying Lemma 7.30, we have

$$I_2 = \frac{1}{2} \xi^T M \xi + \sum_{k>j} \xi_j \xi_k (M_{jk} - M_{kj}) \int_0^\infty \varphi \left(\frac{\log r}{\log \epsilon} - 1 \right) \left[\varphi \left(\frac{\log r}{\log \epsilon} - 1 \right) \right]' dr$$

$$= \frac{1}{2} \xi^T M \xi.$$

On the other hand

$$|Du|^2 = \sum_i |Du_i|^2 = \sum_i \left| \frac{\xi_i}{\log \epsilon} \varphi' \left(\frac{\log r}{\log \epsilon} - 1 \right) \frac{\omega}{r} \right|^2$$

$$= \frac{|\xi|^2}{r^2 \log^2 \epsilon} \left[\varphi' \left(\frac{\log r}{\log \epsilon} - 1 \right) \right]^2,$$

so (7.37) implies that

$$|I_1| \leqslant C \int \frac{|\xi|^2}{r^n \log^2 \epsilon} \left[\varphi' \left(\frac{\log r}{\log \epsilon} - 1 \right) \right]^2 dx$$

$$\leqslant \frac{C|\xi|^2}{\log^2 \epsilon} \|\varphi'\|_\infty^2 \int_{\epsilon^2}^\epsilon \frac{1}{r} dr$$

$$= \frac{C|\xi|^2}{|\log \epsilon|} \|\varphi'\|_\infty^2.$$

This shows that

$$I_1 + I_2 < 0 \qquad \text{if } \epsilon \text{ is sufficiently small}$$

and gives us the desired contradiction. $\qquad\qquad \square$

Now Proposition 7.28 is a direct consequence of Lemma 7.30 and Lemma 7.32.

It is natural to ask whether one can improve the results of Proposition 7.28 by showing that actually $M = I$, the $N \times N$ identity matrix. The following example shows that this is not the case.

Example 7.33. Assume $n \geqslant 3$ and consider $L = -\Delta \cdot I$, where Δ is the Laplacian:

$$\Delta u = D_{\alpha\alpha} u, \qquad \forall u \in C^2(\mathbb{R}^n),$$

and I is the $N \times N$ identity matrix. It is not hard to see that the fundamental matrix of $L^T = L$ is given by $\Phi = \gamma I$, where

$$\gamma(x) = \frac{1}{\omega_n (n-2)} \cdot r^{2-n}, \qquad \omega_n = \int_{S^{n-1}} dx$$

is the fundamental solution of $-\Delta$. For any $M \in \mathbb{R}^{N \times N}$ with M symmetric and semi-positive definite, we have

$$M = P^T \Lambda P$$

where P is orthogonal and Λ is diagonal with non-negative diagonal elements $\lambda_1, \ldots, \lambda_N$. Now for any $u = (u_i)_{i=1}^N$, $u_i \in C_0^\infty(\mathbb{R}^n)$ $(i = 1, 2, \ldots, N)$,

$$\begin{aligned}
\int Lu \cdot (\Phi M) u \, dx &= - \int \gamma (\Delta u)^T M u \, dx \\
&= - \int \gamma (\Delta u)^T P^T \Lambda P u \, dx \\
&= - \int \gamma (\Delta (Pu))^T \Lambda (Pu) \, dx.
\end{aligned}$$

Setting $v = Pu$, we have

$$\begin{aligned}
- \int \gamma (\Delta v)^T \Lambda v \, dx &= \frac{1}{2} \lambda_i v_i^2(0) + \int \lambda_i |Dv_i|^2 \gamma \, dx \\
&\geqslant \min_{i=1,2,\ldots,N} \{\lambda_i\} \left(\frac{1}{2} |v(0)|^2 + \int |Dv|^2 \gamma \, dx \right) \\
&\geqslant 0.
\end{aligned}$$

7.2.3   Pointwise positive definiteness

With judicious choices of the test function u, we now proceed to show the pointwise "positive definiteness" of Ψ.

Proposition 7.34. *Suppose L is an elliptic operator as defined in (7.27) and Ψ satisfies (7.29). If L is positive with weight Ψ, then*

$$\sum_{i,\alpha,\beta} A_{ip}^{\alpha\beta}(r\omega) \xi_\alpha \xi_\beta \Psi_{ip}(\omega) \geqslant 0, \qquad \forall \xi \in \mathbb{R}^n \qquad (p = 1, 2, \ldots, N)$$

for all $r > 0$, $\omega \in S^{n-1}$ such that $r\omega \in \Omega$. That is to say, the $n \times n$ matrix $(\sum_i A_{ip}^{\alpha\beta}(r\omega) \Psi_{ip}(\omega))_{\alpha,\beta=1}^n$ is pointwise semi-positive definite.

Proof. Let $r > 0$, $\omega \in S^{n-1}$ be fixed and $r\omega \in \Omega$. We follow the idea in Maz'ya [71] and take $u = (u_j)_{j=1}^N$, where

$$u_j(x) = \begin{cases} 0, & j \neq p \\ \epsilon^{-n/2} |\xi|^{-1} \eta(\epsilon^{-1}(x - r\omega)) e^{ix \cdot \xi}, & j = p \end{cases},$$
$$\epsilon > 0, \ \xi \in \mathbb{R}^n, \ 0 \not\equiv \eta \in C_0^\infty(\mathbb{R}^n).$$

By definition (with $y = \epsilon^{-1}(x - r\omega)$),

$$\operatorname{Re} \int_\Omega Lu \cdot \Psi \bar{u} \, dx$$

$$= \operatorname{Re} \left\{ -\epsilon^{-n} |\xi|^{-2} \int_\Omega \sum_{j,\alpha,\beta} A_{jp}^{\alpha\beta} D_{\alpha\beta} \left[\eta(y) e^{ix \cdot \xi} \right] \cdot \eta(y) e^{-ix \cdot \xi} \Psi_{jp} \, dx \right\}$$

$$= -\epsilon^{-n} |\xi|^{-2} \int_\Omega \sum_{j,\alpha,\beta} A_{jp}^{\alpha\beta} \left[\epsilon^{-2} \eta''(y) - \xi_\alpha \xi_\beta \eta(y) \right] \eta(y) \Psi_{jp} \, dx$$

$$\geqslant 0.$$

We first let $|\xi| \to \infty$ along a fixed direction and obtain

$$\epsilon^{-n} \frac{\xi_\alpha}{|\xi|} \cdot \frac{\xi_\beta}{|\xi|} \int_\Omega \sum_{j,\alpha,\beta} A_{jp}^{\alpha\beta} \eta^2(y) \Psi_{jp} \, dx \geqslant 0.$$

By substituting $y = \epsilon^{-1}(x - r\omega)$ for x and letting $\epsilon \to 0$, we then conclude that

$$\sum_{j,\alpha,\beta} A_{jp}^{\alpha\beta}(r\omega) \frac{\xi_\alpha}{|\xi|} \cdot \frac{\xi_\beta}{|\xi|} r^{2-n} \Psi_{jp}(\omega) \int \eta^2 \, dx \geqslant 0,$$

which was to be shown. $\qquad\qquad\qquad\qquad\qquad\qquad\qquad\square$

This completes the proof of Theorem 7.21.

7.3 Weighted positivity of the 3D Lamé system

In the previous section we have seen that – under the assumption that Ψ is smooth and positive homogeneous of order $2 - n$ – the weighted integral inequalities (7.28) holds only if Ψ is the fundamental matrix of L, possibly multiplied by a semi-positive definite constant matrix.

A question that arises naturally is under what conditions are elliptic systems indeed positive definite with such weights. Although it is difficult to answer this question in general, we study, as a special case, the 3D Lamé system

$$Lu = -\Delta u - \alpha \operatorname{grad} \operatorname{div} u, \qquad u = (u_1, u_2, u_3)^T.$$

In this section we give some sufficient conditions for its weighted positive definiteness and show that some restrictions on the elastic constants are inevitable. These results are due to Guo Luo and Maz'ya [63].

In the particular case of the Lamé system, condition (7.28) can be written as

$$\int_{\mathbb{R}^3} (Lu)^T \Psi u \, dx = -\int_{\mathbb{R}^3} \left[D_{kk} u_i(x) + \alpha D_{ki} u_k(x) \right] u_j(x) \Psi_{ij}(x) \, dx \geqslant 0$$

for all real-valued, smooth vector functions $u = (u_i)_{i=1}^3$, $u_i \in C_0^\infty(\mathbb{R}^3 \setminus \{0\})$. Here $\Psi(x) = (\Psi_{ij}(x))_{i,j=1}^3$ and, as usual, D denotes the gradient $(D_1, D_2, D_3)^T$ and Du is the Jacobian matrix of u.

Remark 7.35. The 3D Lamé system satisfies the strong elliptic condition if and only if $\alpha > -1$, and we will make this assumption throughout this section.

The fundamental matrix of the 3D Lamé system is given by $\Phi = (\Phi_{ij})_{i,j=1}^3$, where

$$\Phi_{ij}(x) = c_\alpha r^{-1} \left(\delta_{ij} + \frac{\alpha}{\alpha + 2} \omega_i \omega_j \right) \qquad (i, j = 1, 2, 3), \tag{7.39}$$

$$c_\alpha = \frac{\alpha + 2}{8\pi(\alpha + 1)} > 0.$$

As usual, δ_{ij} is the Kronecker delta, $r = |x|$ and $\omega_i = x_i/|x|$.

The main result we shall prove in this section is the following:

Theorem 7.36. *The 3D Lamé system L is positive definite with weight Φ when $\alpha_- < \alpha < \alpha_+$, where $\alpha_- \approx -0.194$ and $\alpha_+ \approx 1.524$. It is not positive definite with weight Φ when $\alpha < \alpha_-^{(c)} \approx -0.902$ or $\alpha > \alpha_+^{(c)} \approx 39.450$.*

The proof of this theorem is given in the next section.

7.3.1 Proof of Theorem 7.36

We start the proof of Theorem 7.36 by rewriting the integral

$$\int_{\mathbb{R}^3} (Lu)^T \Phi u \, dx = - \int_{\mathbb{R}^3} (D_{kk} u_i + \alpha D_{ki} u_k) u_j \Phi_{ij} \, dx$$

into a more revealing form. In the following, we shall write $\int f \, dx$ instead of $\int_{\mathbb{R}^3} f \, dx$, and by u_{ii}^2 we always mean $\sum_{i=1}^3 u_{ii}^2$; to express $(\sum_{i=1}^3 u_{ii})^2$ we will write $u_{ii} u_{jj}$ instead. Furthermore, we always assume $u_i \in C_0^\infty(\mathbb{R}^3)$ unless otherwise stated.

Lemma 7.37.

$$\int (Lu)^T \Phi u \, dx = \frac{1}{2} |u(0)|^2 + \mathscr{B}(u, u) \tag{7.40}$$

where

$$\mathscr{B}(u, u) = \frac{\alpha}{2} \int (u_j D_k u_k - u_k D_k u_j) D_i \Phi_{ij} \, dx$$

$$+ \int (D_k u_i D_k u_j + \alpha D_k u_k D_i u_j) \Phi_{ij} \, dx.$$

Proof. By definition,

$$\int (Lu)^T \Phi u \, dx = -\int D_{kk} u_i \cdot u_j \Phi_{ij} \, dx - \alpha \int D_{ki} u_k \cdot u_j \Phi_{ij} \, dx$$

$$=: I_1 + I_2.$$

Since the fundamental matrix Φ is symmetric and satisfies the equation

$$-D_{kk}\Phi_{ij} - \alpha D_{ki}\Phi_{kj} = \delta_{ij}\delta(x),$$

where $\delta(x)$ is the Dirac delta distribution with mass concentrated at 0, we have

$$I_1 = -\frac{1}{2}\int D_{kk} u_i \cdot u_j \Phi_{ij} \, dx - \frac{1}{2}\int D_{kk} u_j \cdot u_i \Phi_{ij} \, dx$$

$$= -\frac{1}{2}\int \Big[D_{kk}(u_i u_j) - 2 D_k u_i D_k u_j \Big] \Phi_{ij} \, dx$$

$$= -\frac{1}{2}\int u_i u_j D_{kk}\Phi_{ij} \, dx + \int D_k u_i D_k u_j \cdot \Phi_{ij} \, dx,$$

where the first integral in the last line can be written as

$$-\frac{1}{2}\int u_i u_j D_{kk}\Phi_{ij} \, dx = \frac{1}{2}\int u_i u_j \Big[\delta_{ij}\delta(x) + \alpha D_{ki}\Phi_{kj} \Big] \, dx$$

$$= \frac{1}{2}|u(0)|^2 - \frac{\alpha}{2}\int (D_i u_i \cdot u_j + u_i D_i u_j) D_k \Phi_{kj} \, dx$$

$$= \frac{1}{2}|u(0)|^2 - \frac{\alpha}{2}\int (D_k u_k \cdot u_j + u_k D_k u_j) D_i \Phi_{ij} \, dx.$$

Now a simple integration by parts yields

$$I_2 = \alpha \int D_k u_k \big(D_i u_j \cdot \Phi_{ij} + u_j D_i \Phi_{ij} \big) \, dx,$$

and the lemma follows by adding up the results. $\qquad\square$

Remark 7.38. With $\Phi(x)$ replaced by $\Phi_y(x) := \Phi(x - y)$, we have

$$\int (Lu)^T \Phi_y u \, dx = \int (Lu_y)^T \Phi u_y \, dx \qquad (u_y(x) = u(x + y))$$

$$= \frac{1}{2}|u_y(0)|^2 + \mathscr{B}(u_y, u_y) =: \frac{1}{2}|u(y)|^2 + \mathscr{B}_y(u, u),$$

where

$$\mathscr{B}_y(u, u) = \frac{\alpha}{2}\int (u_j D_k u_k - u_k D_k u_j) D_i \Phi_{y,ij} \, dx$$

$$+ \int (D_k u_i D_k u_j + \alpha D_k u_k D_i u_j) \Phi_{y,ij} \, dx.$$

To proceed, we introduce the following decomposition for $C_0^\infty(\mathbb{R}^3)$ functions:

$$f(x) = \bar{f}(r) + g(x), \qquad \bar{f} \in C_0^\infty[0, \infty), \ g \in C_0^\infty(\mathbb{R}^3),$$

where

$$\bar{f}(r) = \frac{1}{4\pi} \int_{S^2} f(r\omega) \, d\sigma.$$

Note that

$$\int_{S^2} g(r\omega) \, d\sigma = 0, \qquad \forall r \geqslant 0,$$

so we may think of $\bar{f}$ as the "0th order harmonics" of the function f. We shall show below in Lemma 7.39 that all 0th order harmonics in (7.40) are cancelled out, so it is possible to control u by Du.

Lemma 7.39. *With the decomposition*

$$u_i(x) = \bar{u}_i(r) + v_i(x) \qquad (i = 1, 2, 3) \tag{7.41}$$

where

$$\begin{cases} \bar{u}_i(r) = \dfrac{1}{4\pi} \displaystyle\int_{S^2} u_i(r\omega) \, d\sigma \\ \displaystyle\int_{S^2} v_i(r\omega) \, d\sigma = 0 \end{cases} \qquad \forall r \geqslant 0 \qquad (i = 1, 2, 3),$$

we have

$$\int (Lu)^T \Phi u \, dx = \frac{1}{2} |u(0)|^2 + \mathscr{B}^*(u, u)$$

where

$$\mathscr{B}^*(u, u) = \frac{\alpha}{2} \int \left(v_j D_k v_k - v_k D_k v_j \right) D_i \Phi_{ij} \, dx \tag{7.42}$$

$$+ \int \left(D_k u_i D_k u_j + \alpha D_k u_k D_i u_j \right) \Phi_{ij} \, dx.$$

Proof. By Lemma 7.37, it is enough to show

$$\int \left(u_j D_k u_k - u_k D_k u_j \right) D_i \Phi_{ij} \, dx = \int \left(v_j D_k v_k - v_k D_k v_j \right) D_i \Phi_{ij} \, dx.$$

Since

$$\int \left(u_j D_k u_k - u_k D_k u_j \right) D_i \Phi_{ij} \, dx$$

$$= \int \left(\bar{u}_j D_k \bar{u}_k - \bar{u}_k D_k \bar{u}_j \right) D_i \Phi_{ij} \, dx + \int \left(\bar{u}_j D_k v_k - \bar{u}_k D_k v_j \right) D_i \Phi_{ij} \, dx$$

$$+ \int \left(v_j D_k \bar{u}_k - v_k D_k \bar{u}_j \right) D_i \Phi_{ij} \, dx + \int \left(v_j D_k v_k - v_k D_k v_j \right) D_i \Phi_{ij} \, dx$$

$$=: I_1 + I_2 + I_3 + I_4,$$

it suffices to show $I_1 = I_2 = I_3 = 0$. Now

$$
\begin{aligned}
D_i\Phi_{ij} &= D_i\left[c_\alpha r^{-1}\left(\delta_{ij} + \frac{\alpha}{\alpha+2}\omega_i\omega_j\right)\right] \\
&= -c_\alpha r^{-2}\omega_i\delta_{ij} + \frac{c_\alpha\alpha}{\alpha+2}r^{-2}\left[-\omega_i^2\omega_j + (\delta_{ii} - \omega_i^2)\omega_j + (\delta_{ji} - \omega_j\omega_i)\omega_i\right] \\
&= -c_\alpha r^{-2}\omega_j + \frac{c_\alpha\alpha}{\alpha+2}r^{-2}\omega_j =: d_\alpha r^{-2}\omega_j,
\end{aligned}
\tag{7.43}
$$

where

$$
d_\alpha = \frac{-2c_\alpha}{\alpha+2} = \frac{-1}{4\pi(\alpha+1)}.
$$

We have

$$
\begin{aligned}
I_1 &= d_\alpha \int r^{-2}\omega_j\left(\bar{u}_j D_r\bar{u}_k \cdot \omega_k - \bar{u}_k D_r\bar{u}_j \cdot \omega_k\right)dx \qquad (D_r = \partial/\partial r) \\
&= d_\alpha \int r^{-2}\left(\bar{u}_j D_r\bar{u}_k \cdot \omega_j\omega_k - \bar{u}_k D_r\bar{u}_j \cdot \omega_k\omega_j\right)dx = 0, \\
I_3 &= d_\alpha \int r^{-2}\left(v_j D_r\bar{u}_k \cdot \omega_j\omega_k - v_k D_r\bar{u}_j \cdot \omega_k\omega_j\right)dx = 0.
\end{aligned}
$$

As for I_2, we obtain

$$
\begin{aligned}
I_2 &= d_\alpha \int r^{-2}\left(\bar{u}_j D_k v_k \cdot \omega_j - \bar{u}_k D_k v_j \cdot \omega_j\right)dx \\
&= d_\alpha \int r^{-2}\left(\bar{u}_j D_k v_k \cdot \omega_j - \bar{u}_j D_j v_k \cdot \omega_k\right)dx \\
&= -\lim_{\epsilon\to 0^+} d_\alpha \int_{S^2}\left[\bar{u}_j(\epsilon)v_k(\epsilon\omega)\omega_j\omega_k - \bar{u}_j(\epsilon)v_k(\epsilon\omega)\omega_j\omega_k\right]d\sigma \\
&\quad - \lim_{\epsilon\to 0^+} d_\alpha \int_{\mathbb{R}^3\setminus B_\epsilon}\left\{v_k r^{-3}\left[-2\bar{u}_j\omega_j\omega_k + rD_r\bar{u}_j \cdot \omega_j\omega_k + \bar{u}_j(\delta_{jk} - \omega_j\omega_k)\right]\right. \\
&\qquad \left. - v_k r^{-3}\left[-2\bar{u}_j\omega_j\omega_k + rD_r\bar{u}_j \cdot \omega_j\omega_k + \bar{u}_j(\delta_{kj} - \omega_k\omega_j)\right]\right\}dx = 0.
\end{aligned}
$$

The result follows. $\qquad\qquad\qquad\qquad\qquad\qquad\qquad\qquad\qquad\qquad\qquad\square$

Remark 7.40. With $\Phi(x)$ replaced by $\Phi_y(x) := \Phi(x-y)$ and (7.41) replaced by

$$
u_i(x) = \bar{u}_i(r_y) + v_i(x) \qquad (i = 1,2,3),
$$

where $r_y = |x-y|$ and

$$
\begin{cases}
\bar{u}_i(r_y) = \dfrac{1}{4\pi}\displaystyle\int_{S^2} u_i(y + r_y\omega)\,d\sigma \\
\displaystyle\int_{S^2} v_i(y + r_y\omega)\,d\sigma = 0
\end{cases}
\qquad \forall r_y \geqslant 0 \qquad (i = 1,2,3),
$$

we have

$$\int (Lu)^T \Phi_y u \, dx = \frac{1}{2}|u(y)|^2 + \mathscr{B}_y^*(u, u)$$

where

$$\mathscr{B}_y^*(u, u) = \frac{\alpha}{2} \int \left(v_j D_k v_k - v_k D_k v_j \right) D_i \Phi_{y,ij} \, dx$$

$$+ \int \left(D_k u_i D_k u_j + \alpha D_k u_k D_i u_j \right) \Phi_{y,ij} \, dx.$$

In the next lemma, we use the definition of Φ and derive an explicit expression for the bilinear form $\mathscr{B}^*(u, u)$ defined in (7.42).

Lemma 7.41.

$$\mathscr{B}^*(u, u) = c_\alpha \int \left\{ \frac{\alpha}{\alpha + 2} r^{-2} \left[v_k (D_k v) \cdot \omega - (\operatorname{div} v)(v \cdot \omega) \right] \right.$$

$$+ r^{-1} \left[|D_r \bar{u}|^2 + \alpha \frac{2\alpha + 3}{\alpha + 2} (D_r \bar{u}_i)^2 \omega_i^2 + |Dv|^2 + \alpha (\operatorname{div} v)^2 \right.$$

$$+ \frac{\alpha}{\alpha + 2} |(D_k v) \cdot \omega|^2 + \frac{\alpha^2}{\alpha + 2} (\operatorname{div} v)[\omega_i (D_i v) \cdot \omega]$$

$$\left. \left. + \alpha \frac{3\alpha + 4}{\alpha + 2} (D_r \bar{u} \cdot \omega)(\operatorname{div} v) + \alpha (D_r \bar{u} \cdot \omega)[\omega_i (D_i v) \cdot \omega] \right] \right\} dx.$$

Before proving this lemma, we need a simple yet important observation that will be useful in the following computation.

Lemma 7.42. *Let* $g \in C_0^\infty(\mathbb{R}^3)$ *be such that*

$$\int_{S^2} g(r\omega) \, d\sigma = 0, \qquad \forall r \geqslant 0.$$

Then

$$\begin{cases} \displaystyle\int \int f(r) g(x) \, dx = 0 \\ \displaystyle\int \int r^{-1} Df(r) \cdot Dg(x) \, dx = 0 \end{cases} \qquad \forall f \in C_0^\infty[0, \infty).$$

Proof. By switching to polar coordinates, we easily see that

$$\int f(r) g(x) \, dx = \int_0^\infty r^2 f(r) \, dr \int_{S^2} g(r\omega) \, d\sigma = 0.$$

On the other hand,

$$\int r^{-1} Df(r) \cdot Dg(x)\, dx = \int r^{-1} D_r f D_i g \cdot w_i\, dx$$

$$= -\int g\left[-r^{-2}(D_r f)w_i^2 + r^{-1}(D_{rr} f)w_i^2 + r^{-2} D_r f(\delta_{ii} - w_i^2)\right] dx$$

$$= -\int g\left(r^{-2} D_r f + r^{-1} D_{rr} f\right) dx = 0,$$

where the last equality follows by switching to polar coordinates. □

Proof of Lemma 7.41. By definition,

$$\mathscr{B}^*(u, u) = \frac{\alpha}{2} \int \left(v_j D_k v_k - v_k D_k v_j\right) D_i \Phi_{ij}\, dx$$

$$+ \int \left(D_k u_i D_k u_j + \alpha D_k u_k D_i u_j\right)\Phi_{ij}\, dx =: I_1 + I_2.$$

We have shown in Lemma 7.39 that (see (7.43))

$$I_1 = 2^{-1}\alpha d_\alpha \int r^{-2} w_j\left(v_j D_k v_k - v_k D_k v_j\right) dx$$

$$= \frac{c_\alpha \alpha}{\alpha + 2} \int r^{-2}\left[v_k (D_k v) \cdot w - (\operatorname{div} v)(v \cdot w)\right] dx.$$

On the other hand,

$$I_2 = c_\alpha \int r^{-1} D_k u_i D_k u_i\, dx + \frac{c_\alpha \alpha}{\alpha + 2} \int r^{-1} D_k u_i D_k u_j \cdot w_i w_j\, dx$$

$$+ c_\alpha \alpha \int r^{-1} D_k u_k D_i u_i\, dx + \frac{c_\alpha \alpha^2}{\alpha + 2} \int r^{-1} D_k u_k D_i u_j \cdot w_i w_j\, dx$$

$$=: I_3 + I_4 + I_5 + I_6.$$

Substituting $u_i = \bar{u}_i + v_i$ into I_3 and using Lemma 7.42 yields

$$I_3 = c_\alpha \int r^{-1}\left(D_r \bar{u}_i D_r \bar{u}_i \cdot w_k^2 + D_k v_i D_k v_i\right) dx + 2c_\alpha \int r^{-1} D_k \bar{u}_i D_k v_i\, dx$$

$$= c_\alpha \int r^{-1}\left(|D_r \bar{u}|^2 + |Dv|^2\right) dx.$$

Next,

$$I_5 = c_\alpha \alpha \int r^{-1}\left(D_r \bar{u}_k D_r \bar{u}_i \cdot w_k w_i + 2D_i v_i D_r \bar{u}_k \cdot w_k + D_k v_k D_i v_i\right) dx.$$

Note that for $k \neq i$,

$$\int r^{-1} D_r \bar{u}_k D_r \bar{u}_i \cdot w_k w_i\, dx = \int_0^\infty r D_r \bar{u}_k D_r \bar{u}_i\, dr \int_{S^2} w_k w_i\, d\sigma = 0,$$

and therefore

$$I_5 = c_\alpha \alpha \int r^{-1} \left[(D_r \bar{u}_i)^2 \omega_i^2 + 2(\operatorname{div} v)(D_r \bar{u} \cdot \omega) + (\operatorname{div} v)^2 \right] dx.$$

As for I_4,

$$
\begin{aligned}
I_4 &= \frac{c_\alpha \alpha}{\alpha+2} \int r^{-1} D_k(\bar{u}_i + v_i) D_k(\bar{u}_j + v_j) \cdot \omega_i \omega_j \, dx \\
&= \frac{c_\alpha \alpha}{\alpha+2} \int r^{-1} \big(D_r \bar{u}_i D_r \bar{u}_j \cdot \omega_i \omega_j \omega_k^2 + D_r \bar{u}_i D_k v_j \cdot \omega_i \omega_j \omega_k \\
&\qquad\qquad + D_k v_i D_r \bar{u}_j \cdot \omega_i \omega_j \omega_k + D_k v_i D_k v_j \cdot \omega_i \omega_j \big) \, dx \\
&= \frac{c_\alpha \alpha}{\alpha+2} \int r^{-1} \left[(D_r \bar{u}_i)^2 \omega_i^2 + 2(D_r \bar{u} \cdot \omega)[\omega_k(D_k v) \cdot \omega] + |D_k v \cdot \omega|^2 \right] dx.
\end{aligned}
$$

Similarly,

$$
\begin{aligned}
I_6 &= \frac{c_\alpha \alpha^2}{\alpha+2} \int r^{-1} D_k(\bar{u}_k + v_k) D_i(\bar{u}_j + v_j) \cdot \omega_i \omega_j \, dx \\
&= \frac{c_\alpha \alpha^2}{\alpha+2} \int r^{-1} \big(D_r \bar{u}_k D_r \bar{u}_j \cdot \omega_i^2 \omega_j \omega_k + D_r \bar{u}_k D_i v_j \cdot \omega_i \omega_j \omega_k \\
&\qquad\qquad + D_r \bar{u}_j D_k v_k \cdot \omega_i^2 \omega_j + D_k v_k D_i v_j \cdot \omega_i \omega_j \big) \, dx \\
&= \frac{c_\alpha \alpha^2}{\alpha+2} \int r^{-1} \Big[(D_r \bar{u}_j)^2 \omega_j^2 + (D_r \bar{u} \cdot \omega)[\omega_i(D_i v) \cdot \omega] \\
&\qquad\qquad + (D_r \bar{u} \cdot \omega)(\operatorname{div} v) + (\operatorname{div} v)[\omega_i(D_i v) \cdot \omega] \Big] \, dx.
\end{aligned}
$$

The lemma follows by adding up all these integrals. $\square$

With the help of Lemma 7.41, we now complete the proof of Theorem 7.36.

Proof of Theorem 7.36. By Lemma 7.39 and 7.41,

$$-c_\alpha^{-1} \int (Lu)^T \Phi u \, dx = \frac{1}{2} c_\alpha^{-1} |u(0)|^2 + I_1 + I_2 + I_3,$$

where

$$
\begin{aligned}
I_1 &= \int r^{-1} \bigg[|D_r \bar{u}|^2 + \alpha \frac{2\alpha+3}{\alpha+2} (D_r \bar{u}_i)^2 \omega_i^2 + |Dv|^2 \\
&\qquad\qquad + \alpha(\operatorname{div} v)^2 + \frac{\alpha}{\alpha+2} |(D_k v) \cdot \omega|^2 \bigg] \, dx,
\end{aligned}
$$

$$
\begin{aligned}
I_2 &= \int r^{-1} \bigg[\frac{\alpha^2}{\alpha+2} (\operatorname{div} v)[\omega_i(D_i v) \cdot \omega] + \alpha \frac{3\alpha+4}{\alpha+2} (D_r \bar{u} \cdot \omega)(\operatorname{div} v) \\
&\qquad\qquad + \alpha(D_r \bar{u} \cdot \omega)[\omega_i(D_i v) \cdot \omega] \bigg] \, dx,
\end{aligned}
$$

$$I_3 = \int \frac{\alpha}{\alpha+2} r^{-2} \big[v_k(D_k v) \cdot \omega - (\operatorname{div} v)(v \cdot \omega) \big] \, dx.$$

Consider first the case $\alpha \geqslant 0$. By switching to polar coordinates, we have

$$I_1 \geqslant \int r^{-1}\left[|D_r\bar{u}|^2 + \alpha\frac{2\alpha+3}{\alpha+2}(D_r\bar{u}_i)^2\omega_i^2 + |Dv|^2 + \alpha(\mathrm{div}\,v)^2\right]dx$$

$$= \int_0^\infty r\left[\left(1 + \frac{\alpha}{3}\cdot\frac{2\alpha+3}{\alpha+2}\right)\|D_r\bar{u}\|_\omega^2 + \|Dv\|_\omega^2 + \alpha\|\,\mathrm{div}\,v\|_\omega^2\right]dr,$$

where we have written $\|\cdot\|_\omega$ for $\|\cdot\|_{L^2(S^2)}$ and used the fact that

$$\int_{S^2}(D_r\bar{u}_i)^2\omega_i^2\,d\sigma = \frac{4\pi}{3}\sum_{i=1}^3(D_r\bar{u}_i)^2 = \frac{1}{3}\int_{S^2}|D_r\bar{u}|^2\,d\sigma = \frac{1}{3}\|D_r\bar{u}\|_\omega^2.$$

Next,

$$|I_2| \leqslant \int r^{-1}\left[\frac{\alpha^2}{\alpha+2}|\,\mathrm{div}\,v||Dv| + \alpha\frac{3\alpha+4}{\alpha+2}|D_r\bar{u}\cdot\omega||\,\mathrm{div}\,v| + \alpha|D_r\bar{u}\cdot\omega||Dv|\right]dx$$

$$\leqslant \int_0^\infty r\left[\frac{\alpha^2}{\alpha+2}\|\,\mathrm{div}\,v\|_\omega\|Dv\|_\omega + \frac{\alpha}{\sqrt{3}}\cdot\frac{3\alpha+4}{\alpha+2}\|D_r\bar{u}\|_\omega\|\,\mathrm{div}\,v\|_\omega\right.$$

$$\left.+ \frac{\alpha}{\sqrt{3}}\|D_r\bar{u}\|_\omega\|Dv\|_\omega\right]dr,$$

where we have used

$$\|D_r\bar{u}\cdot\omega\|_\omega^2 = \int_{S^2}D_r\bar{u}_iD_r\bar{u}_j\cdot\omega_i\omega_j\,d\sigma$$

$$= D_r\bar{u}_iD_r\bar{u}_j\cdot\frac{4\pi}{3}\delta_{ij} = \frac{4\pi}{3}\sum_{i=1}^3(D_r\bar{u}_i)^2 = \frac{1}{3}\|D_r\bar{u}\|_\omega^2.$$

As for I_3, we note that

$$|I_3| \leqslant \frac{\alpha}{\alpha+2}\int r^{-2}\left(|v||Dv| + |v||\,\mathrm{div}\,v|\right)dx$$

$$\leqslant \frac{\alpha}{\alpha+2}\int_0^\infty \|v\|_\omega\left(\|Dv\|_\omega + \|\,\mathrm{div}\,v\|_\omega\right)dr.$$

Since 2 is the first non-trivial eigenvalue of the Laplace–Beltrami operator on S^2, we have

$$\|v\|_\omega^2 = \int_{S^2}|v(r\omega)|^2\,d\sigma \leqslant \frac{1}{2}\int_{S^2}|D_\omega[v(r\omega)]|^2\,d\sigma$$

$$= \frac{r^2}{2}\int_{S^2}|(D_\omega v)(r\omega)|^2\,d\sigma \leqslant \frac{r^2}{2}\|Dv\|_\omega^2,$$

where D_ω is the gradient operator on S^2. Thus

$$|I_3| \leqslant \frac{1}{\sqrt{2}}\cdot\frac{\alpha}{\alpha+2}\int_0^\infty r\left[\|Dv\|_\omega^2 + \|Dv\|_\omega\|\,\mathrm{div}\,v\|_\omega\right]dr,$$

and by putting all pieces together we obtain

$$I_1 + I_2 + I_3 \geqslant \int_0^\infty r(w^T B_+ w)\, dr,$$

where

$$w = \left(\|D_r \bar{u}\|_\omega, \|Dv\|_\omega, \|\operatorname{div} v\|_\omega\right)^T,$$

$$B_+ = \begin{bmatrix} 1 + \dfrac{\alpha}{3} \cdot \dfrac{2\alpha+3}{\alpha+2} & -\dfrac{\alpha}{2\sqrt{3}} & -\dfrac{\alpha}{2\sqrt{3}} \cdot \dfrac{3\alpha+4}{\alpha+2} \\[2mm] -\dfrac{\alpha}{2\sqrt{3}} & 1 - \dfrac{1}{\sqrt{2}} \cdot \dfrac{\alpha}{\alpha+2} & -\dfrac{\alpha}{2} \cdot \dfrac{\alpha+2^{-1/2}}{\alpha+2} \\[2mm] -\dfrac{\alpha}{2\sqrt{3}} \cdot \dfrac{3\alpha+4}{\alpha+2} & -\dfrac{\alpha}{2} \cdot \dfrac{\alpha+2^{-1/2}}{\alpha+2} & \alpha \end{bmatrix}.$$

Clearly, the weighted positive definiteness of L follows from the positive definiteness of B_+, because the latter implies, for some $c > 0$, that

$$\int_0^\infty r(w^T B_+ w)\, dr \geqslant c \int_0^\infty r|w|^2\, dr$$

$$\geqslant c \int_0^\infty r\left(\|D_r \bar{u}\|_\omega^2 + \|Dv\|_\omega^2\right) dr = c \int r^{-1} |Du|^2\, dx.$$

The positive definiteness of B_+, on the other hand, is equivalent to the positivity of the determinants of all leading principal minors of B_+:

$$p_{+,1}(\alpha) = \frac{2\alpha^2 + 6\alpha + 6}{3(\alpha+2)} > 0, \tag{7.44a}$$

$$p_{+,2}(\alpha) = -\frac{1}{12(\alpha+2)^2}\left[\alpha^4 - 4(1-\sqrt{2})\alpha^3 - 12(3-\sqrt{2})\alpha^2 \right.$$
$$\left. - 12(6-\sqrt{2})\alpha - 48\right] > 0, \tag{7.44b}$$

$$p_{+,3}(\alpha) = -\frac{\alpha}{12(\alpha+2)^3}\left[6\alpha^5 + (23+3\sqrt{2})\alpha^4 + (13+19\sqrt{2})\alpha^3\right.$$
$$\left. - (77-38\sqrt{2})\alpha^2 - (157-24\sqrt{2})\alpha - 96\right] > 0. \tag{7.44c}$$

With the help of computer algebra packages, we find that (7.44) holds for $0 \leqslant \alpha < \alpha_+$, where $\alpha_+ \approx 1.524$ is the largest real root of $p_{+,3}$.

The estimates of I_1, I_2, and I_3 are slightly different when $\alpha < 0$, since now the quadratic term $\alpha\|\operatorname{div} v\|_\omega^2$ in I_1 is negative. This means that it is no longer possible to control the $\|\operatorname{div} v\|_\omega$ terms in I_2, I_3 by $\alpha\|\operatorname{div} v\|_\omega^2$, and in order to obtain positivity we need to bound $\|\operatorname{div} v\|_\omega$ by $\|Dv\|_\omega$ as follows:

$$\|\operatorname{div} v\|_\omega^2 \leqslant 3\|Dv\|_\omega^2.$$

This leads to the following revised estimates:

$$I_1 \geqslant \int_0^\infty r\left[\left(1 + \frac{\alpha}{3}\cdot\frac{2\alpha+3}{\alpha+2}\right)\|D_r\bar{u}\|_\omega^2 + \|Dv\|_\omega^2 + 3\alpha\|Dv\|_\omega^2 + \frac{\alpha}{\alpha+2}\|Dv\|_\omega^2\right]dr,$$

$$|I_2| \leqslant \int_0^\infty r\left[\frac{\sqrt{3}\,\alpha^2}{\alpha+2}\|Dv\|_\omega^2 - \alpha\frac{3\alpha+4}{\alpha+2}\|D_r\bar{u}\|_\omega\|Dv\|_\omega - \frac{\alpha}{\sqrt{3}}\|D_r\bar{u}\|_\omega\|Dv\|_\omega\right]dr,$$

$$|I_3| \leqslant -\frac{1}{\sqrt{2}}\cdot\frac{\alpha}{\alpha+2}\int_0^\infty r\left[\|Dv\|_\omega^2 + \sqrt{3}\,\|Dv\|_\omega^2\right]dr.$$

Hence

$$I_1 + I_2 + I_3 \geqslant \int_0^\infty r\left(w^T B_- w\right)dr,$$

where

$$w = \left(\|D_r\bar{u}\|_\omega, \|Dv\|_\omega\right)^T,$$

$$B_- = \begin{bmatrix} 1 + \dfrac{\alpha}{3}\cdot\dfrac{2\alpha+3}{\alpha+2} & \dfrac{\alpha}{2}\cdot\dfrac{3\alpha+4}{\alpha+2} + \dfrac{\alpha}{2\sqrt{3}} \\ \dfrac{\alpha}{2}\cdot\dfrac{3\alpha+4}{\alpha+2} + \dfrac{\alpha}{2\sqrt{3}} & 1 + 3\alpha + \dfrac{\alpha}{\alpha+2}\left(1 + \dfrac{1+\sqrt{3}}{\sqrt{2}} - \sqrt{3}\,\alpha\right) \end{bmatrix}.$$

The positive definiteness of B_- is equivalent to:

$$p_{-,1}(\alpha) = \frac{2\alpha^2 + 6\alpha + 6}{3(\alpha+2)} > 0, \tag{7.45a}$$

$$p_{-,2}(\alpha) = \frac{1}{6(\alpha+2)^2}\left[-(2+7\sqrt{3})\alpha^4 + 2(15+\sqrt{2}-11\sqrt{3}+\sqrt{6})\alpha^3 \right. \tag{7.45b}$$

$$\left. + 2(57+3\sqrt{2}-10\sqrt{3}+3\sqrt{6})\alpha^2 + 6(20+\sqrt{2}+\sqrt{6})\alpha + 24\right] > 0,$$

and (7.45) holds for $\alpha_- < \alpha < 0$, where $\alpha_- \approx -0.194$ is the smallest real root of $p_{-,2}$.

Now we show that the 3D Lamé system is not positive definite with weight Φ when α is either too close to -1 or too large. By Proposition 7.34, the 3D Lamé system is positive definite with weight Φ only if

$$\sum_{i,\beta,\gamma} A_{ip}^{\beta\gamma}\xi_\beta\xi_\gamma\Phi_{ip}(\omega) \geqslant 0, \qquad \forall\xi\in\mathbb{R}^3,\ \forall\omega\in S^2 \qquad (p=1,2,3),$$

where

$$A_{ij}^{\beta\gamma} = \delta_{ij}\delta_{\beta\gamma} + \frac{\alpha}{2}\left(\delta_{i\beta}\delta_{j\gamma} + \delta_{i\gamma}\delta_{j\beta}\right)$$

and (see equation (7.39))

$$\Phi_{ij}(\omega) = c_\alpha r^{-1}\left(\delta_{ij} + \frac{\alpha}{\alpha+2}\omega_i\omega_j\right) \qquad (i,j=1,2,3).$$

This means, in particular, that the matrix

$$A(\omega;\alpha) := \left(\sum_{i=1}^{3} A_{i1}^{\beta\gamma}\Phi_{i1}(\omega)\right)_{\beta,\gamma=1}^{3}$$

$$= \frac{c_\alpha r^{-1}}{2(\alpha+2)}\begin{bmatrix} 2(\alpha+1)(\alpha+2+\alpha\omega_1^2) & \alpha^2\omega_1\omega_2 & \alpha^2\omega_1\omega_3 \\ \alpha^2\omega_1\omega_2 & 2(\alpha+2+\alpha\omega_1^2) & 0 \\ \alpha^2\omega_1\omega_3 & 0 & 2(\alpha+2+\alpha\omega_1^2) \end{bmatrix}$$

is semi-positive definite for any $\omega \in S^2$ if the 3D Lamé system is positive definite with weight Φ. But $A(\omega;\alpha)$ is semi-positive definite only if the determinant of its leading principal minor

$$d_2(\omega;\alpha) := \det\begin{bmatrix} 2(\alpha+1)(\alpha+2+\alpha\omega_1^2) & \alpha^2\omega_1\omega_2 \\ \alpha^2\omega_1\omega_2 & 2(\alpha+2+\alpha\omega_1^2) \end{bmatrix}$$

$$= 4(\alpha+1)(\alpha+2+\alpha\omega_1^2)^2 - \alpha^4\omega_1^2\omega_2^2$$

is non-negative, and elementary estimate shows that

$$\min_{\omega\in S^2} d_2(\omega;\alpha) \leqslant d_2\big[(2^{-1/2},2^{-1/2},0);\alpha\big]$$

$$= (\alpha+1)(3\alpha+4)^2 - \frac{\alpha^4}{4} =: q(\alpha).$$

It follows that the 3D Lamé system is not positive definite with weight Φ when $q(\alpha) < 0$, which holds for $\alpha < \alpha_-^{(c)} \approx -0.902$ or $\alpha > \alpha_+^{(c)} \approx 39.450$. $\qquad\square$

Remark 7.43. We have in fact shown that, for $\alpha_- < \alpha < \alpha_+$ and some $c > 0$ depending on α,

$$\int (Lu)^T\Phi u\,dx \geqslant \frac{1}{2}|u(0)|^2 + c\int |Du(x)|^2\frac{dx}{|x|}.$$

If we replace $\Phi(x)$ by $\Phi_y(x) := \Phi(x-y)$, then

$$\int (Lu)^T\Phi_y u\,dx = \int [Lu(x+y)]^T\Phi u(x+y)\,dx$$

$$\geqslant \frac{1}{2}|u(y)|^2 + c\int |Du(x+y)|^2\frac{dx}{|x|}$$

$$\geqslant \frac{1}{2}|u(y)|^2 + c\int \frac{|Du(x)|^2}{|x-y|}\,dx.$$

7.4 The polyharmonic operator

7.4.1 The case $n \geqslant 2m$

Here we consider L^2-weighted positivity of the polyharmonic operator. The weight Ψ is the fundamental solution of the operator $(-\Delta)^m$:

$$\Psi(x) = \begin{cases} \gamma |x|^{2m-n} & \text{for } 2m < n \\ \gamma \log \frac{D}{|x|} & \text{for } 2m = n \end{cases}$$

where D is a positive constant and

$$\gamma = \begin{cases} 2^{1-m}[(m-1)!(n-2)(n-4)\ldots(n-2m)\omega_{n-1}]^{-1} & \text{for } 2m < n \\ [2^{m-1}(m-1)!]^{-2}(\omega_{n-1})^{-1} & \text{for } 2m = n, \end{cases}$$

ω_{n-1} being the $(n-1)$-dimensional measure of the unit sphere.

Proposition 7.44. *Let $n \geqslant 2m$ and let*

$$\int_\Omega u(x)(-\Delta)^m u(x)\Psi(x-p)dx \geqslant 0 \tag{7.46}$$

for all $u \in C_0^\infty(\Omega)$ and, at least, for one point $p \in \Omega$. Then

$$n = 2m, 2m+1, 2m+2 \quad \text{for } m > 2$$

and

$$n = 4, 5, 6, 7 \quad \text{for } m = 2.$$

Proof. Assume that $n \geqslant 2m + 3$ for $m > 2$ and $n \geqslant 8$ for $m = 2$. Denote by (r, ω) the spherical coordinates with center $p, r > 0, \omega \in \partial B_1(p)$, and by G the image of Ω under the mapping $x \to (t, \omega), t = -\log r$. Since

$$r^2 \Delta u = r^{2-n}\left(r\frac{\partial}{\partial r}\right)\left(r^{n-2}\left(r\frac{\partial}{\partial r}\right)u\right) + \delta_\omega u,$$

where δ_ω is the Beltrami operator on $\partial B_1(p)$, then

$$\Delta = e^{2t}\left(\frac{\partial^2}{\partial t^2} - (n-2)\frac{\partial}{\partial t} + \delta_\omega\right) = e^{2t}\left\{\left(\frac{\partial}{\partial t} - \frac{n-2}{2}\right)^2 - A\right\}$$

where

$$A = -\delta_\omega + \frac{(n-2)^2}{4}.$$

Hence

$$r^{2m}\Delta^m = \prod_{j=0}^{m-1}\left\{\left(\frac{\partial}{\partial t} - \frac{n-2}{2} + 2j\right)^2 - A\right\}. \tag{7.47}$$

Let u be a function in $C_0^\infty(\Omega)$, which depends only on $|x - p|$. We set $w(t) = u(x)$. Clearly

$$\int_\Omega (-\Delta)^m u(x) \cdot u(x) \Psi(x - p) dx = \int_{\mathbb{R}^1} w(t) P(d/dt) w(t) dt, \qquad (7.48)$$

where

$$P(\lambda) = (-1)^m \gamma \omega_{n-1} \prod_{j=0}^{m-1} (\lambda + 2j)(\lambda - n + 2 + 2j)$$

$$= (-1)^m \gamma \omega_{n-1} \lambda(\lambda - n + 2) \prod_{j=1}^{m-1} (\lambda + 2j)(\lambda - n - 2m + 2 + 2j).$$

Let

$$P(\lambda) = (-1)^m \gamma \omega_{n-1} \lambda^{2m} + \sum_{k=1}^{2m-1} a_k \lambda^k.$$

We have

$$a_2 = (\lambda^{-1} P(\lambda))'\big|_{\lambda=0} = \frac{1}{2 - n} + \sum_{j=1}^{2m-1} \left(\frac{1}{2j} - \frac{1}{n - 2 - 2m + 2j} \right).$$

Hence and by $n \geqslant 2m + 3$,

$$a_2 = \frac{1}{2} - \frac{1}{n - 2} - \frac{1}{n - 2m} + \sum_{j=1}^{m-1} \frac{n - 2 - 2m}{2j(n - 2 - 2m + 2j)}$$

$$\geqslant \frac{1}{2} - \frac{1}{n - 2} - \frac{1}{n - 2m} > 0.$$

We choose a real-valued function $\eta \in C_0^\infty(1, 2)$ normalized by

$$\int_{\mathbb{R}^1} |\eta'(\sigma)|^2 d\sigma = 1$$

and we set $u(x) = \eta(\varepsilon t)$, where ε is so small that $\operatorname{supp} u \subset \Omega$. The quadratic form on the right in (7.48) equals

$$\int_{\mathbb{R}^1} \left(\varepsilon^{2m} \gamma \omega_{n-1} |\eta^{(m)}(\varepsilon t)|^2 + \sum_{k=1}^{m-1} a_{2k}(-1)^k \varepsilon^{2k} |\eta^{(k)}(\varepsilon t)|^2 \right) dt = -a_2 \varepsilon + \mathcal{O}(\varepsilon^3) < 0,$$

which contradicts the assumption (7.46).   □

Now we prove the converse statement. By ∇_k we mean the gradient of order k, i.e., $\nabla_k = \{\partial^\alpha\}$ with $|\alpha| = k$.

Proposition 7.45. *Let $\Psi_p(x) = \Psi(x - p)$, where $p \in \Omega$. If*

$$
\begin{aligned}
n &= 2m, 2m + 1, 2m + 2 && \text{for } m > 2,\\
n &= 4, 5, 6, 7 && \text{for } m = 2,\\
n &= 2, 3, 4, \dots && \text{for } m = 1,
\end{aligned}
$$

then for all $u \in C_0^\infty(\Omega)$

$$
\int_\Omega u(x)(-\Delta)^m u(x)\Psi(x - p)dx \geqslant 2^{-1}u^2(p) + c\sum_{k=1}^m \int_\Omega \frac{|\nabla_k u(x)|^2}{|x - p|^{2(m-k)}} \Psi(x - p)dx
$$

$$(7.49)$$

(in the case $n = 2m$ the constant D in the definition of Ψ is greater than $|x - p|$ for all $x \in \operatorname{supp} u$).

Proof. We preserve the notation introduced in the proof of Proposition 7.44. We note first that (7.49) becomes identity when $m = 1$. The subsequent proof will be divided into four parts.

(i) *The case* $n = 2m + 2$. By (7.47),

$$
r^{-2m}\Delta^m = \prod_{j=0}^{m-1}\left(\frac{\partial}{\partial t} - m + 2j - A^{1/2}\right)\prod_{j=0}^{m-1}\left(\frac{\partial}{\partial t} - m + 2j + A^{1/2}\right)
$$

where $A = -\delta_\omega + m^2$ and $A^{1/2}$ is defined by using spherical harmonics. By setting $k = m - j$ in the second product, we rewrite the right-hand side as

$$
\prod_{j=0}^{m-1}\left(\frac{\partial}{\partial t} - m + 2j - A^{1/2}\right)\prod_{k=1}^m\left(\frac{\partial}{\partial t} + m - 2k + A^{1/2}\right).
$$

This can be represented in the form

$$
\left(\frac{\partial}{\partial t} - m - A^{1/2}\right)\left(\frac{\partial}{\partial t} - m + A^{1/2}\right)\prod_{j=1}^{m-1}\left(\frac{\partial^2}{\partial t^2} - B_j^2\right),
$$

where $B_j = A^{1/2} + m - 2j$. Therefore

$$
\begin{aligned}
2^m\Delta^m &= \left(\frac{\partial^2}{\partial t^2} + \delta_\omega - 2m\frac{\partial}{\partial t}\right)\prod_{j=1}^{m-1}\left(\frac{\partial^2}{\partial t^2} - B_j^2\right)\\
&= \left(\frac{\partial^2}{\partial t^2} + \delta_\omega\right)\prod_{j=1}^{m-1}\left(\frac{\partial^2}{\partial t^2} - B_j^2\right)\\
&\quad + (-1)^m 2m\frac{\partial}{\partial t}\sum_{j=0}^{m-1}\left(-\frac{\partial^2}{\partial t^2}\right)^{m-j-1}\sum_{k_1<\cdots<k_j} B_{k_1}^2 \dots B_{k_j}^2.
\end{aligned}
$$

We extend u by zero outside Ω and introduce the function w defined by $w(t, \omega) = u(x)$. We write the left-hand side of (7.49) in the form $\gamma(I_1 + I_2)$, where γ is the constant in the definition of Ψ,

$$I_1 = 2m \int_G \frac{\partial}{\partial t} \sum_{j=0}^{m-1} \left(-\frac{\partial^2}{\partial t^2}\right)^{m-j-1} \sum_{k_1 < \cdots < k_j} B_{k_1}^2 \ldots B_{k_j}^2 w \cdot w \, dt \, d\omega$$

and

$$I_2 = (-1)^m \int_G \left(\frac{\partial^2}{\partial t^2} + \delta_\omega\right) \prod_{j=1}^{m-1} \left(\frac{\partial^2}{\partial t^2} - B_j^2\right) w \cdot w \, dt \, d\omega.$$

Since the operators B_j are symmetric, it follows that

$$I_1 = m \sum_{j=0}^{m-1} \sum_{k_1 < \cdots < k_j} \int_{\mathbb{R}^1} \frac{\partial}{\partial t} \int_{\partial B_1} \left(\frac{\partial^{m-j-1}}{\partial t^{m-j-1}} B_{k_1} \ldots B_{k_j} w\right)^2 d\omega \, dt$$

$$= m \sum_{j=0}^{m-1} \sum_{k_1 < \cdots < k_j} \int_{\partial B_1} \left| \left(\frac{\partial^{m-j-1}}{\partial t^{m-j-1}} B_{k_1} \ldots B_{k_j} w\right) (+\infty, \omega)\right|^2 d\omega.$$

By $u \in C^\infty(\Omega)$, we have $w(t, \omega) = u(p) + \mathcal{O}(e^{-t})$ as $t \to +\infty$, and this can be differentiated. Therefore, all the terms with $j < m - 1$ are equal to zero and we find

$$I_1 = m \int_{\partial B_1} |(B_1 \ldots B_{m-1} w)(+\infty, \omega)|^2 d\omega = mu^2(p) \int_{\partial B_1} |B_1 \ldots B_{m-1} 1|^2 d\omega.$$

By $B_j = (-\delta_\omega + m^2)^{1/2} + m - 2j$, we have

$$I_1 = 4^{m-1} m [(m-1)!]^2 \omega_{2m+1}.$$

Since in the case $n = 2m + 2$

$$\gamma^{-1} = 2^{2m-1} m [(m-1)!]^2 \omega_{2m+1},$$

we conclude that

$$I_1 = (2\gamma)^{-1} u^2(p). \tag{7.50}$$

We now wish to obtain the lower bound for I_2. Let $\widetilde{w}$ denote the Fourier transform of w with respect to t. Then

$$I_2 = \int_{\partial B_1} \int_{\mathbb{R}^1} (\lambda^2 - \delta_\omega) \prod_{j=1}^{m-1} (\lambda^2 + B_j^2) \widetilde{w}(\lambda, \omega) \overline{\widetilde{w}(\lambda, \omega)} \, d\lambda \, d\omega.$$

Clearly,

$$B_j \geqslant (m^2 - \delta_\omega)^{1/2} - m + 2 \geqslant 2m^{-1}(m^2 - \delta_\omega)^{1/2},$$

and

$$\lambda^2 + B_j^2 \geq 4m^{-2}(\lambda^2 + 1 - \delta_\omega),$$

the operators being compared with respect to their quadratic forms. Thus

$$\left(\frac{m}{2}\right)^{2m-2} I_2 \geq \int_{\partial B_1} \int_{\mathbb{R}^1} (\lambda^2 - \delta_\omega)(\lambda^2 + 1 - \delta_\omega)^{m-1} \widetilde{w}(\lambda, \omega) \overline{\widetilde{w}(\lambda, \omega)} \, d\lambda \, d\omega$$

$$\geq c \left(\left\| \frac{\partial w}{\partial t} \right\|_{H^{m-1}(G)}^2 + \|\nabla_\omega w\|_{H^{m-1}(G)}^2 \right),$$

where H^{m-1} is the Sobolev space. This is equivalent to the inequality

$$I_2 \geq c \int_\Omega \sum_{k=1}^m \frac{|\nabla_k u(x)|^2}{|x - p|^{n-2k}} \, dx,$$

which along with (7.50) completes the proof for $n = 2m + 2$.

(ii) *The case $n = 2m+1$.* We shall treat this case by descent from $n = 2m+2$ to $n = 2m + 1$. Let $z = (x, s)$, where $x \in \Omega$, $s \in \mathbb{R}^1$, and $q = (p, 0)$, where $p \in \Omega$, $0 \in \mathbb{R}^1$. We introduce a cut-off function $\eta \in C_0^\infty(-2, 2)$ which satisfies $\eta(s) = 1$ for $|s| \leq 1$ and $0 \leq \eta \leq 1$ on $\mathbb{R}^1$. Let

$$U_\varepsilon(z) = u(x)\eta(\varepsilon s)$$

and let $\Psi^{(n)}$ denote the fundamental solution of $(-\Delta)^m$ in $\mathbb{R}^n$.

By integrating

$$(-\Delta_z)^m \Psi^{(n+1)}(z, q) = \delta(z - q),$$

with respect to $s \in \mathbb{R}^1$ we have

$$\Psi^{(n)}(x, y) = \int_{\mathbb{R}^1} \Psi^{(k+1)}(z, q) \, ds. \tag{7.51}$$

From part (i) of the present proof we obtain

$$\int_{\Omega \times \mathbb{R}^1} (-\Delta_z)^m U_\varepsilon(z) \cdot U_\varepsilon(z) \Psi^{(k+1)}(z - q) \, dz$$

$$\geq \frac{1}{2} U_\varepsilon^2(q) + c \int_{\Omega \times \mathbb{R}^1} \sum_{k=1}^m \frac{|\nabla_k U_\varepsilon(z)|^2}{|z - q|^{2(m+1-k)}} \, dx.$$

By letting $\varepsilon \to 0$, we find

$$\int_{\Omega \times \mathbb{R}^1} (-\Delta_x)^m u(x) \cdot u(x) \Psi^{(n+1)}(z - q) \, ds \, dx$$

$$\geq \frac{1}{2} u^2(p) + c \int_{\Omega \times \mathbb{R}^1} \sum_{k=1}^m \frac{|\nabla_k u(x)|^2}{|z - q|^{2(m+1-k)}} \, ds \, dx.$$

The result follows from (7.51).

(iii) *The case $m = 2, n = 7$.* By (7.47),

$$30\omega_6 \int_\Omega \Delta^2 u(x) \cdot u(x) \Psi(x-p)\, dx = \int_G (w_{tt} - 5w_t + \delta_\omega w)(w_{tt} + w_t - 6w + \delta_\omega w)\, dt\, d\omega \,.$$

Since $w(t, \omega) = u(p) + \mathcal{O}(e^{-t})$ as $t \to +\infty$, the last integral equals

$$\int_G \left(w_{tt}^2 - 5w_t^2 - 6w_{tt}w + 2w_{tt}\delta_\omega w + (\delta_\omega w)^2 - 6w\delta_\omega w \right) dt\, d\omega + 15\omega_6 u^2(p) \,.$$

After integrating by parts we rewrite this in the form

$$\int_G \left(w_{tt}^2 + (\delta_\omega w)^2 + 2(\nabla_\omega w_t)^2 + 6(\nabla_\omega w)^2 + w_t^2 \right) dt\, d\omega + 15\omega_6 u^2(p) \,.$$

Using the variables (r, ω), we obtain that the left-hand side exceeds

$$c \int_\Omega \left(\frac{(\Delta u(x))^2}{|x-p|^3} + \frac{|\nabla u(x)|^2}{|x-p|} \right) dx + 15\omega_6 u^2(p) \,.$$

Since

$$|\nabla_2 u|^2 - (\Delta u)^2 = \Delta((\nabla u)^2) - \frac{\partial^2}{\partial x_i \partial x_j}\left(\frac{\partial u}{\partial x_i}\frac{\partial u}{\partial x_j} \right),$$

it follows that

$$\int_\Omega \frac{(\nabla_2 u(x))^2}{|x-p|^3}\, dx \leqslant \int_\Omega \frac{(\Delta u(x))^2}{|x-p|}\, dx + c \int_\Omega \frac{(\nabla u(x))^2}{|x-p|}\, dx \,,$$

which completes the proof.

(iv) *The case $n = 2m$.* By (7.47),

$$r^{2m}\Delta^m = \prod_{j=0}^{m-1} \left\{ \left(\frac{\partial}{\partial t} - m + 1 + 2j \right)^2 - (m-1)^2 + \delta_\omega \right\}$$

$$= \prod_{j=0}^{m-1} \left(\frac{\partial}{\partial t} - m + 1 + 2j - \mathcal{E}^{1/2} \right) \prod_{j=0}^{m-1} \left(\frac{\partial}{\partial t} - m + 1 + 2j + \mathcal{E}^{1/2} \right),$$

where $\mathcal{E} = -\delta_\omega + (m-1)^2$. We introduce $k = m - 1 - j$ in the second product and obtain

$$r^{2m}\Delta^m = \prod_{j=0}^{m-1} \left(\frac{\partial^2}{\partial t^2} - \mathcal{F}_j^2 \right),$$

where $\mathcal{F}_j = m - 1 - 2j + \mathcal{E}^{1/2}$. Hence

$$\int_\Omega (-\Delta)^m u(x) \cdot u(x) \Psi(x-p)\, dx = \gamma \int_G \prod_{j=0}^{m-1} \left(-\frac{\partial^2}{\partial t^2} + \mathcal{F}_j^2 \right) w \cdot (\ell+t)\, w\, dt\, d\omega \quad (7.52)$$

where $\ell = \log D$. Since $w(t, \omega) = u(p) + \mathcal{O}(e^{-t})$ and

$$\prod_{j=0}^{m-1} \left(-\frac{\partial^2}{\partial t^2} + \mathcal{F}_j^2 \right) = \sum_{j=0}^m \left(-\frac{\partial^2}{\partial t^2} \right)^{m-j} \sum_{k_1 < \cdots < k_j} \mathcal{F}_{k_1}^2 \cdots \mathcal{F}_{k_j}^2,$$

the right-hand side in (7.52) can be rewritten as

$$\gamma \int_G \sum_{j=0}^m \sum_{k_1 < \cdots < k_j} \left(\frac{\partial}{\partial t} \right)^{m-j} \mathcal{F}_{k_1} \cdots \mathcal{F}_{k_j} w \left(\frac{\partial}{\partial t} \right)^{m-j} ((\ell + t) \mathcal{F}_{k_1} \cdots \mathcal{F}_{k_j} w) dt \, d\omega$$

$$= \gamma \int_G \sum_{j=0}^m \sum_{k_1 < \cdots < k_j} \left(\left(\frac{\partial}{\partial t} \right)^{m-j} \mathcal{F}_{k_1} \cdots \mathcal{F}_{k_j} w \right)^2 (\ell + t) dt \, d\omega$$

$$+ \frac{\gamma}{2} \int_G \sum_{j=0}^{m-1} \sum_{k_1 < \cdots < k_j} (m - j) \frac{\partial}{\partial t} \left(\left(\frac{\partial}{\partial t} \right)^{m-1-j} \mathcal{F}_{k_1} \cdots \mathcal{F}_{k_j} w \right)^2 dt \, d\omega.$$

The second integral in the right-hand side equals

$$\lim_{t \to +\infty} \int_{\partial B_1(p)} \sum_{j=0}^{m-1} \sum_{k_1 < \cdots < k_j} (m - j) \left(\left(\frac{\partial}{\partial t} \right)^{m-1-j} \mathcal{F}_{k_1} \cdots \mathcal{F}_{k_j} w \right)^2 d\omega$$

$$= \lim_{t \to +\infty} \int_{\partial B_1(p)} \sum_{k_1 < \cdots < k_{m-1}} (\mathcal{F}_{k_1} \cdots \mathcal{F}_{k_{m-1}} e)^2 d\omega$$

and since $\mathcal{F}_{m-1}(t, \omega) = \mathcal{O}(e^{-t})$ the last expression is equal to

$$\lim_{t \to +\infty} \int_{\partial B_1(p)} (\mathcal{F}_0 \cdots \mathcal{F}_{m-2} w)^2 d\omega = (2^{m-1} (m - 1)!)^2 \omega_{n-1} u^2(p).$$

Hence

$$\int_\Omega (-\Delta)^m u(x) \cdot u(x) \Psi(x - p) \, dx$$

$$= \frac{1}{2} u^2(p) + \gamma \int_G (\ell + t) \sum_{j=0}^{m-1} \sum_{k_1 < \cdots < k_j} \left(\left(\frac{\partial}{\partial t} \right)^{m-1-j} \mathcal{F}_{k_1} \cdots \mathcal{F}_{k_j} w \right)^2 dt \, d\omega.$$

Since $\mathcal{F}_{m-1} \geqslant c(-\delta)^{1/2}$ and $\mathcal{F}_k \geqslant c(-\delta + 1)^{1/2}$ for $k < m - 1$, the last integral majorizes

$$c \int_G (\ell + t) \sum_{1 \leqslant \mu + \nu \leqslant m-1} \left(\left(\frac{\partial}{\partial t} \right)^\mu (-\delta)^{\nu/2} w \right)^2 dt \, d\omega$$

$$\geqslant c \int_\Omega \log \frac{D}{|x - p|} \sum_{k=1}^m \frac{|\nabla_k u(x)|^2}{|x - p|^{2(m-k)}} dx,$$

which completes the proof. $\qquad \square$

7.4.2   Accretivity of the biharmonic operator in $\mathbb{R}^3$

In this section we show that one can have L^2-positivity also for weights which are not power weights, as previously done. In fact we shall show that the biharmonic operator is L^2-positive with respect to certain a weight g.

Let (r, ω) be the spherical coordinates in $\mathbb{R}^3$, i.e., $r = |x| \in (0, \infty)$ and $\omega = x/|x| \in S^2$, the unit sphere. Now let $t = \log r^{-1}$. By $\varkappa$ we denote the mappings $\mathbb{R}^3 \ni x \to (t, \omega) \in \mathbb{R} \times S^2$.

The symbols δ_ω and ∇_ω refer, respectively, to the Laplace–Beltrami operator and the gradient on S^2.

Theorem 7.46. *Let ω be an open set in $\mathbb{R}^3$, $u \in C_0^\infty(\Omega)$ and $v = e^t(u \circ \varkappa^{-1})$. Then*

$$\int_{\mathbb{R}^3} u(x)\, \Delta^2 u(x)\, |x|^{-1} G(\log |x|^{-1})\, dx \tag{7.53}$$
$$= \int_{\mathbb{R}} \int_{S^2} \Big[(\delta_\omega v)^2 G + 2(\partial_t \nabla_\omega v)^2 G + (\partial_t^2 v)^2 G - (\nabla_\omega v)^2 (\partial_t^2 G + \partial_t G + 2G)$$
$$- (\partial_t v)^2 (2\partial_t^2 G + 3\partial_t G - G) + \frac{1}{2} v^2 (\partial_t^4 G + 2\partial_t^3 G - \partial_t^2 G - 2\partial_t G) \Big]\, d\omega\, dt,$$

for every function G on $\mathbb{R}$ such that both sides of (7.53) are well defined.

Proof. In the system of coordinates (t, ω) the 3-dimensional Laplacian can be written as

$$\Delta = e^{2t} \Lambda(\partial_t, \delta_\omega), \quad \text{where} \quad \Lambda(\partial_t, \delta_\omega) = \partial_t^2 - \partial_t + \delta_\omega.$$

Then passing to the coordinates (t, ω), we have

$$\int_{\mathbb{R}^3} u(x)\, \Delta^2 u(x)\, |x|^{-1} G(\log |x|^{-1})\, dx$$
$$= \int_{\mathbb{R}^3} \Delta u(x)\, \Delta \big(u(x)\, |x|^{-1} G(\log |x|^{-1}) \big)\, dx$$
$$= \int_{\mathbb{R}} \int_{S^2} \Lambda(\partial_t - 1, \delta_\omega) v\, \Lambda(\partial_t, \delta_\omega)(vG)\, d\omega\, dt$$
$$= \int_{\mathbb{R}} \int_{S^2} (\partial_t^2 v - 3\partial_t v + 2v + \delta_\omega v)(\partial_t^2(vG) - \partial_t(vG) + G\delta_\omega v)\, d\omega\, dt$$
$$\int_{\mathbb{R}} \int_{S^2} (\partial_t^2 v - 3\partial_t v + 2v + \delta_\omega v)$$
$$\times (G\delta_\omega v + G\partial_t^2 v + (2\partial_t G - G)\partial_t v + (\partial_t^2 G - \partial_t G)v)\, d\omega\, dt$$
$$= \int_{\mathbb{R}} \int_{S^2} \Big(\big((\delta_\omega v)^2 + 2\delta_\omega v \partial_t^2 v + (\partial_t^2 v)^2 \big) G + (v\delta_\omega v + v\partial_t^2 v)(\partial_t^2 G - \partial_t G + 2G)$$
$$+ (\partial_\omega v \partial_t v + \partial_t^2 v \partial_t v)(2\partial_t G - 4G) + (\partial_t v)^2(-6\partial_t G + 3G)$$
$$+ v\partial_t v(-3\partial_t^2 G + 7\partial_t G - 2G) + v^2(2\partial_t^2 G - 2\partial_t G) \Big)\, d\omega\, dt.$$

This, in turn, is equal to

$$
\int_{\mathbb{R}} \int_{S^2} \Big(g(\partial_\omega v)^2 - 2G\delta_\omega \partial_t v \partial_t v + G(\partial_t^2 v)^2
$$
$$
+ (\nabla_\omega v)^2 (-\partial_t^2 G - (\partial_t^2 G - \partial_t G + 2G) + (\partial_t^2 G - 2\partial_t G))
$$
$$
+ (\partial_t v)^2 \left(-(\partial_t^2 G - \partial_t G + 2G) + (-\partial_t^2 G + 2\partial_t G) + (-6\partial_t G + 3G) \right)
$$
$$
+ v\partial_t v \left(-(\partial_t^2 G - \partial_t^2 G + 2\partial_t G) + (-3\partial_t^2 G + 7\partial_t G - 2G) \right)
$$
$$
+ v^2 (2\partial_t^2 G - 2\partial_t G) \Big) d\omega dt,
$$

and integrating by parts once again we obtain (7.53).   □

In order to single out the term with v^2 in (7.53) we shall need the following auxiliary result.

Lemma 7.47. *Consider the equation*

$$
\frac{d^4 g}{dt^4} + 2\frac{d^3 g}{dt^3} - \frac{d^2 g}{dt^2} - 2\frac{dg}{dt} = \delta, \tag{7.54}
$$

where δ stands for the Dirac delta function. A unique solution to (7.54) which is bounded and vanishes at $+\infty$ is given by

$$
g(t) = -\frac{1}{6} \begin{cases} e^t - 3, & t < 0, \\ e^{-2t} - 3e^{-t}, & t > 0. \end{cases} \tag{7.55}
$$

Proof. Since the equation (7.54) is equivalent to

$$
\frac{d}{dt}\left(\frac{d}{dt} + 2\right)\left(\frac{d}{dt} + 1\right)\left(\frac{d}{dt} - 1\right) g = \delta,
$$

a bounded solution of (7.54) vanishing at $+\infty$ must have the form

$$
g(t) = \begin{cases} ae^t + b, & t < 0, \\ ce^{-2t} + de^{-t}, & t > 0, \end{cases}
$$

for some constants a, b, c, d. Once this is established, we find the system of coefficients so that $\partial_t^k g$ is continuous for $k = 0, 1, 2$ and $\lim_{t\to 0+} \partial_t^3 g(t) - \lim_{t\to 0-} \partial_t^3 g(t) = 1$.   □

With Lemma 7.47 at hand, a suitable choice of the function G yields the positivity of the left-hand side of 7.53. The details are as follows.

Theorem 7.48. *Let Ω be a bounded domain in $\mathbb{R}^3$, $0 \in \mathbb{R}^3 \setminus \Omega$, $u \in C_0^\infty(\Omega)$ and $v = e^t(u \circ \varkappa^{-1})$. Then for every $\xi \in \Omega$ and $\tau = \log|\xi|^{-1}$ we have*

$$
\frac{1}{2} \int_{S^2} v^2(\delta, \omega) \, d\omega \leqslant \int_{\mathbb{R}^3} u(x) \, \Delta^2 u(x) \, |x|^{-1} g(\log|x|^{-1}) \, dx
$$

where g is given by (7.55).

Proof. Representing v as a series of spherical harmonics and noting that the eigenvalues of the Laplace–Beltrami operator on the unit sphere are $k(k+1)$, $k = 0, 1, \ldots$, we arrive at the inequality

$$\int_{S^2} |\delta_\omega v|^2 d\omega \geqslant 2 \int_{S^2} |\nabla_\omega v|^2 d\omega. \tag{7.56}$$

Now, let us take $G(t) = g(t - \tau)$, $t \in \mathbb{R}$. Since $g \geqslant 0$, the combination of Lemma 7.47, (7.53) and (7.56) allows to obtain the estimate

$$\int_{\mathbb{R}^3} \int_{\mathbb{R}^3} u(x) \, \Delta^2 u(x) \, |x|^{-1} g(\log |x|^{-1}) \, dx$$

$$\geqslant \int_{\mathbb{R}} \int_{S^2} \Big[- (\nabla_\omega v)^2 \left(\partial_t^2 g(t - \tau) + \partial_t g(t - \tau) \right)$$

$$- (\partial_t v)^2 \left(2\partial_t^2 g(t - \tau) + 3\partial_t g(t - \tau) - g(t - \tau) \right) \Big] d\omega dt + \frac{1}{2} \int_{S^2} v^2(\tau, \omega) \, d\omega \,.$$

Thus, the matters are reduced to showing that

$$\partial_t^2 g + \partial_t g \leqslant 0 \quad \text{and} \quad 2\partial_t^2 g + 3\partial_t g - g \leqslant 0.$$

Indeed we compute

$$\partial_t g(t) = -\frac{1}{6} \begin{cases} e^t, & t < 0, \\ -2e^{-2t} + 3e^{-t}, & t > 0, \end{cases}$$

and

$$\partial_t^2 g(t) = -\frac{1}{6} \begin{cases} e^t, & t < 0, \\ 4e^{-2t} - 3e^{-t}, & t > 0, \end{cases}$$

which gives

$$\partial_t^2 g(t) + \partial_t g(t) = -\frac{1}{3} \begin{cases} e^t, & t < 0, \\ e^{-2t}, & t > 0, \end{cases}$$

and

$$2\partial_t^2 g(t) + 3\partial_t g(t) - g(t) = -\frac{1}{6} \begin{cases} 4e^t + 3, & t < 0, \\ e^{-2t} + 6e^{-t}, & t > 0. \end{cases}$$

Clearly, both functions above are non-positive. $\qquad \square$

7.5 Weighted positivity for real positive powers of the Laplacian

7.5.1 Notation and preliminaries

Here we let $\mathscr{S} = \mathscr{S}(\mathbb{R}^n)$ be the Schwartz class of complex-valued functions. Let $\mathscr{L}$ be the class of functions in $\mathscr{S}$ with all moments equal to 0 and $\hat{\mathscr{L}}$ the space of

Fourier transforms of these functions, that is

$$\mathscr{L} = \{u \in \mathscr{S} : \partial^\alpha \hat{u}(0) = 0, \quad |\alpha| \in \mathbb{N}\}, \quad \hat{\mathscr{L}} = \{\hat{u} : u \in \mathscr{L}\}.$$

Here, α is a multi-index, $\mathbb{N} = \{0, 1, \dots\}$ and

$$\hat{u}(\xi) = \int e^{-ix\cdot\xi} u(x) \, dx.$$

Whenever we omit writing out the domain of integration, we mean integration over $\mathbb{R}^n$. The Fourier–Laplace transform $\hat{u}(\xi)$, with a complex ξ will be used only in $\mathbb{R}^1$ in cases where the integral is absolutely convergent.

For $\mu \in \mathbb{R}$ and u a sufficiently good function (like $u \in \mathscr{S}$ if $\mu > -n/2$, $u \in \mathscr{L}$ otherwise), we define $(-\Delta)^\mu$ by

$$((-\Delta)^\mu u)^\wedge(\xi) = |\xi|^{2\mu} \hat{u}(\xi).$$

Notice that $(-\Delta)^\mu$ is bijective on $\mathscr{L}$.

We define the lth order gradient by $\nabla_l u = \{(l!/\alpha!)^{1/2} \partial^\alpha u\}_{|\alpha|=l}$.

The gamma-function Γ is analytic in the complex plane, except for simple poles at $0, -1, \dots$. We write $|\Gamma(-m)| = \infty$ when $m \in \mathbb{N}$. The asymptotic formula

$$\Gamma(\alpha + z)/\Gamma(\beta + z) = z^{\alpha-\beta}(1 + \Omega(1/|z|)), \quad |z| \to \infty \qquad (7.57)$$

for $\alpha, \beta \in \mathbb{R}$ and $|\arg(z)| < \pi$, will be useful. We write $(\lambda)_m = \Gamma(\lambda + m)/\Gamma(\lambda)$.

The psi-function is defined by $\psi = \Gamma'/\Gamma$. When working with this function, we will only need the formula

$$\psi(z) - \psi(w) = \sum_{m=0}^{\infty} \left(\frac{1}{m+w} - \frac{1}{m+z} \right).$$

We define power weights,

$$\Gamma_\lambda(x) = c_\lambda |x|^{2\lambda-n}, \quad \text{with } c_\lambda = 2^{-2\lambda} \pi^{-n/2} \Gamma(n/2 - \lambda)\Gamma(\lambda)^{-1}.$$

For $\lambda \in -\mathbb{N}$, we interpret $c_\lambda = 0$. If $x \neq 0$, the function $\lambda \mapsto \Gamma_\lambda(x)$ is analytic except for simple poles at $n/2 + \mathbb{N}$. If $0 < \lambda < n/2$, we have $\widehat{\Gamma_\lambda}(x) = |x|^{-2\lambda}$ in the sense of distributions.

Let $\{S_{j,k}\}_{k=1}^{d_{n,j}}$ be an orthogonal base (with respect to the scalar product in $L^2(S^{n-1})$) for the space of all spherical harmonic functions having degree j. Then,

$$d_{n,j} = \begin{cases} \binom{n+j-1}{j} - \binom{n+j-3}{j-2}, & j \geq 2, \\ n, & j = 1, \\ 1, & j = 0. \end{cases}$$

In case of $\mathbb{R}^1$, $d_{n,j} = 0$ for $j \geq 2$ and we only define the two functions:

$$S_{0,1}(1) = S_{0,1}(-1) = S_{1,1}(1) = -S_{1,1}(-1) = 1/\sqrt{2}.$$

S_j will mean any normalized spherical harmonic function of degree j. We write $S_j(x) = |x|^j S_j(x')$, where $x = |x| x'$. Let $\omega_{n-1} = 2\pi^{n/2}\Gamma(n/2)^{-1}$ denote the area of S^{n-1}.

The letter c denotes a finite positive constant, which value we allow to change within a series of inequalities.

7.5.2 Diagonalization

This section contains the basic facts about the diagonalization of certain quadratic forms.

Let $\sigma, \tau, \eta \in \mathbb{R}$, $\gamma = l+s$, where $l \in \mathbb{N}$ and $0 \leq s \leq 1$. We put $\lambda = \sigma+\tau+\gamma+\eta$, and assume throughout this section that $\lambda < n/2$. We define the quadratic form

$$I_\eta^{\sigma,\tau,\gamma}(f) = \iint \frac{|x|^{2\sigma}|y|^{2\tau}(2x \cdot y)^l(|x|^{2s} + |y|^{2s} - |x - y|^{2s})}{|x - y|^{2\lambda}} f(x)\overline{f(y)}\,dx\,dy,$$

where f is a sufficiently good function. By changing the function f, one of the parameters σ, τ, η may be regarded as redundant, but it will be convenient to include all of them. For simplicity we assume that $f \in \mathscr{S}$ if the above kernel belongs to $L^1_{\mathrm{loc}}(\mathbb{R}^{2n})$ and $f \in \mathscr{L}$ otherwise. Then the double-integral will be always absolutely convergent.

The following lemma can be proved by assuming $\tau \leq \sigma$, integrating the modulus of the kernel over the set $\{|y| \leq |x| \leq 1\}$ and changing variables according to $x = rx'$, $y = try'$.

Lemma 7.49. *The kernel of $I_\eta^{\sigma,\tau,\gamma}$ belongs to $L^1_{\mathrm{loc}}(\mathbb{R}^{2n})$ if and only if $\eta < n$ and $2\min(\sigma,\tau) + l + \min(1,2s) > -n$.*

We shall use the following decomposition of f:

$$f(x) = |x|^{-n}\sum_{j=0}^{\infty}\sum_{k=1}^{d_{n,j}} f_{j,k}(\log|x|)S_{j,k}(x'), \quad x = |x|x',$$

where the functions $S_{j,k}$ are described in the previous section. In case $n = 1$, this is just a decomposition into even and odd parts. In the sequel, we write the double sum as $\sum_{j,k}$.

We collect some basic facts about the functions $f_{j,k}$ in the following lemma. As the statements are easily checked we omit the proof.

Lemma 7.50. *Let f be decomposed as above. If $f \in \mathscr{S}$ then $\widehat{f_{j,k}}$ is analytic above the line $t - in$, $t \in \mathbb{R}$ in the complex plane. If $f \in \mathscr{L}$ then $\widehat{f_{j,k}}$ is entire.*

The following holds when the $\widehat{f_{j,k}}$ are addressed in the appropriate region as above:

(i) *The function $\xi \mapsto \widehat{f_{j,k}}(\xi + i\mu)$ belongs to $\mathscr{S}$.*

(ii) *The transformation $f \mapsto |x|^{\mu} f$ corresponds to $\widehat{f_{j,k}}(\xi) \mapsto \widehat{f_{j,k}}(\xi + i\mu)$.*

(iii) *$(-\Delta)^{\mu/2} \overline{S_{j,k}(\partial)} \widehat{f}(0) = i^j \widehat{f_{j,k}}(i(\mu + j))$.*

(iv) *$|x|^{\mu} f \in \mathscr{L}$ iff $\widehat{f_{j,k}}(i(\mu + j + m)) = 0$, for $m = 0, 1, \ldots$.*

In order to diagonalize the form $I_{\eta}^{\sigma,\tau,\gamma}$ we need to introduce functions $\Phi_{\eta,j}^{\sigma,\tau,\gamma}$ and functionals $A_{\eta,j}^{\sigma,\tau,\gamma}$. For each of these we again associate the number $\lambda = \sigma + \tau + \gamma + \eta < n/2$.

First define

$$\Phi_{0,j}^{\sigma,\tau,0}(\xi) = \frac{\omega_{n-1}}{2} \sum_{m=0}^{\infty} a_{j,m}^{\lambda} \left(\frac{1}{\sigma + m + z} + \frac{1}{\tau + m + \overline{z}} \right), \quad z = \frac{j + i\xi}{2}, \qquad (7.58)$$

where $a_{j,m}^{\lambda} = (\lambda)_{j+m}(\lambda + 1 - n/2)_m / (n/2)_{j+m} m!$. These coefficients behave like $a_{j,m}^{\lambda} = \mathcal{O}(m^{2\lambda-n})$ for large m, so the restriction on λ guarantees that the series converges uniformly in ξ. By formula 1.4(3) in Bateman [5] vol. 1, this can be written in a closed form,

$$\Phi_{0,j}^{\sigma,\tau,0}(\xi) = \frac{\pi^n 2^{2\lambda} c_{\lambda} \Gamma(\sigma + z) \Gamma(\tau + \overline{z})}{\Gamma(n/2 - \tau + z) \Gamma(n/2 - \sigma + \overline{z})}. \qquad (7.59)$$

Both (7.58) and (7.59) will be useful in the sequel.

We extend the definition by means of the formulas

$$\begin{aligned}
\Phi_{\eta,j}^{\sigma,\tau,0} &= \Phi_{0,j}^{\sigma+\eta/2,\tau+\eta/2,0}, \\
\Phi_{\eta,j}^{\sigma,\tau,l+s} &= \Phi_{\eta,j}^{\sigma+s,\tau,l} + \Phi_{\eta,j}^{\sigma,\tau+s,l} - \Phi_{\eta,j}^{\sigma,\tau,l}, \quad 0 \le s \le 1.
\end{aligned} \qquad (7.60)$$

From (7.57) and the recursion formula (7.60), we obtain the asymptotic formula,

$$\Phi_{\eta,j}^{\sigma,\tau,\gamma}(\xi) = 2^{l+\lceil s \rceil + 2\lambda} c_{\lambda} \pi^n |z|^{2\lambda-n} (1 + \omega(1)), \quad |z| \to \infty, \qquad (7.61)$$

where $\lceil s \rceil$ is the smallest integer greater than or equal to s and where $\omega(1) \to 0$ when $|z| \to \infty$.

Let us interpret $\operatorname{Re} \Phi_{\eta,j}^{\sigma,\tau,\gamma}(0)$ as $\lim_{\xi \to 0} \operatorname{Re} \Phi_{\eta,j}^{\sigma,\tau,\gamma}(\xi)$. Then all $\operatorname{Re} \Phi$ become continuous as functions of ξ, as is seen in (7.58).

We proceed now with the definition of the Q's. First introduce the auxiliary quantity

$$Q_j^{\sigma,\tau}(\phi) = \frac{\omega_{n-1}}{2} \sum_{m=0}^{\infty} (1 - \operatorname{sgn}(b_m)) a_{j,m}^{\lambda} \phi(ib_m) \overline{\phi(-ib_m)},$$

where $b_m = 2(\sigma + m) + j$. Next we define

$$Q_{0,j}^{\sigma,\tau,0}(\phi) = Q_j^{\sigma,\tau}(\phi) + \overline{Q_j^{\tau,\sigma}(\phi)}.$$

Finally we extend the definition in exactly the same manner as was done with the functions Φ.

If $\gamma > 0$, some terms in the Q will cancel (see Lemma 7.52). This is essential in the next lemma when we apply it to functions that are less regular than the definition may suggest is needed.

Let us finally introduce

$$Q_\eta^{\sigma,\tau,\gamma}(f) = \sum_{j,k} A_{\eta,j}^{\sigma,\tau,\gamma}(T_{i\eta}\widehat{f_{j,k}}), \tag{7.62}$$

$$R_\eta^{\sigma,\tau,\gamma}(f) = \frac{1}{2\pi} \sum_{j,k} \int_{-\infty}^{\infty} \Phi_{\eta,j}^{\sigma,\tau,\gamma} |T_{i\eta}\widehat{f_{j,k}}|^2 \, d\xi, \tag{7.63}$$

where $T_{i\eta}\widehat{f_{j,k}}(\xi) = \widehat{f_{j,k}}(\xi - i\eta)$ and the integral is to be taken as a principal value if the imaginary part of $\Phi_{\eta,j}^{\sigma,\tau,\gamma}$ is singular at 0.

We are now ready to formulate the main lemma concerning the diagonalization.

Lemma 7.51. *Let f be as in the definition of $I_\eta^{\sigma,\tau,\gamma}(f)$. Then*

$$I_\eta^{\sigma,\tau,\gamma}(f) = Q_\eta^{\sigma,\tau,\gamma}(f) + R_\eta^{\sigma,\tau,\gamma}(f). \tag{7.64}$$

Proof. First let $f \in \hat{\mathscr{L}}$, even if the kernel of the I is in $L_{\text{loc}}^1(\mathbb{R}^{2n})$. Then, since $I_\eta^{\sigma,\tau,\gamma}(f) = I_0^{\sigma+\eta/2,\tau+\eta/2,\gamma}(|x|^{-\eta}f)$ and since multiplying f by $|x|^{-\eta}$ amounts to applying $T_{i\eta}$ to all $\widehat{f_{j,k}}$'s (Lemma 7.50 (ii)), we may assume that $\eta = 0$. Also, since I, Q and R satisfies the same recursion formula (7.60), we can let $\gamma = 0$. (For $f \in \hat{\mathscr{L}}$, all occurring terms will be well defined.)

We introduce new variables by

$$x = e^s x', \quad y = e^t y', \quad p = t - s, \quad \nu = x' \cdot y'.$$

Let $K^{\sigma,\tau,\gamma}(p,\nu)$ be the kernel of $I_0^{\sigma,\tau,\gamma}$ in those variables. (For later reference, we keep γ arbitrary for a while.) We may define functions $K_j^{\sigma,\tau,\gamma}$ by

$$\int_{S^{n-1}} K(p,\nu)S_j(x')\,dx' = K_j(p)S_j(y'), \tag{7.65}$$

where we omit writing out the parameters. Passing to the variables s and t and using the orthogonality of the $S_{j,k}$'s, we find

$$I_0^{\sigma,\tau,\gamma}(f) = \sum_{j,k}(K_j * f_{j,k}, f_{j,k}), \tag{7.66}$$

where $(,)$ is the $L^2(\mathbb{R}^1)$ scalar product.

We complete the proof only for $n \geq 3$. Then by the Funk–Hecke theorem (see Bateman [5] vol. 2),

$$K_j(p) = A_j \int_{-1}^{1} K(p,\nu) C_j^{n/2-1}(\nu)(1-\nu^2)^{(n-3)/2} \, d\nu, \tag{7.67}$$

where $A_j = (4\pi)^{n/2-1}\Gamma(n/2-1)j!/(j+n-3)!$ and C_j^μ is a Gegenbauer polynomial. For $\gamma = 0$, we have

$$
\begin{aligned}
K(p,\nu) &= (1 - 2e^{-|p|}\nu + e^{-2|p|})^{-\lambda}(e^{-2\sigma p}\chi_+(p) + e^{2\tau p}\chi_-(p)) \\
&= \sum_{\kappa=0}^{\infty} C_\kappa^\lambda(\nu)\big(e^{-(2\sigma+\kappa)p}\chi_+(p) + e^{(2\tau+\kappa)p}\chi_-(p)\big),
\end{aligned} \tag{7.68}
$$

where χ_+ and χ_- are the characteristic functions of $\mathbb{R}_+$ and $\mathbb{R}_-$ respectively.

For $p \neq 0$, we are allowed to integrate termwise in (7.67). Doing so and applying the formula

$$A_j \int_{-1}^{1} C_\kappa^\lambda(\nu) C_j^{n/2-1}(\nu)(1-\nu^2)^{(n-3)/2} \, d\nu = \begin{cases} \omega_{n-1} a_{j,m}^\lambda, & \text{if } \kappa = 2m + j, \\ 0, & \text{otherwise,} \end{cases}$$

(where j, κ, m are nonnegative integers, see formula 2.21.18.15 in vol. 2 of Prudnikov, Brychkov and Marichev [88]) we obtain

$$K_j(p) = \sum_{m=0}^{\infty} a_{j,m}^\lambda \big(e^{-(2\sigma+2m+j)p}\chi_+(p) + e^{(2\tau+2m+j)p}\chi_-(p)\big).$$

The result now follows from (7.66) and the following handy consequence of Parseval's formula,

$$(K_+^a * \phi, \phi) = \overline{(K_-^a * \phi, \phi)} = \varepsilon\hat\phi(ia)\overline{\hat\phi(-ia)} + \frac{1}{2\pi} \int_{-\infty}^{\infty} \frac{1}{a+i\xi}|\hat\phi(\xi)|^2 \, d\xi,$$

where $K_+^a(p) = e^{-ap}\chi_+(p)$, $K_-^a(p) = e^{ap}\chi_-(p)$, $\varepsilon = (1 - \operatorname{sgn}(a))/2$ and ϕ is a sufficiently good function. If $a = 0$ the integral is understood as a principal value. (This case follows by taking a limit, $a \to +0$ or $a \to -0$.)

Having completed the proof for all $f \in \mathcal{L}$, we assume now that the kernel is locally in L^1 and let f be merely in $\mathcal{S}$. Let $\zeta_\varepsilon(|x|) = \zeta(|x|/\varepsilon)$, where ζ is a smooth function that vanishes together with all its derivatives at 0 and tends to 1 at ∞. Then (7.64) holds with $\zeta_\varepsilon f$ in place of f and we only need to see that both sides tend to the original expressions when $\varepsilon \to 0$. For the left-hand side this is clear by the assumption on the kernel.

As for the R, we have in (7.63), with $\varphi_\varepsilon(t) = \zeta_\varepsilon(e^t)$,

$$\widehat{T_{i\eta} f_{j,k} \varphi_\varepsilon}(\xi) = \frac{1}{2\pi} \int_{-\infty}^{\infty} \widehat{f_{j,k}}(\xi - i\eta - \rho)\widehat{\varphi_\varepsilon}(\rho) \, d\rho,$$

in place of $T_{in}\widehat{f_{j,k}}(\xi)$. By Lemma 7.49, $\eta < n$, so Lemma 7.50 (i) implies that this convolution tends uniformly to $T_{in}\widehat{f_{j,k}}(\xi)$ as $\varepsilon \to 0$.

Similarly, $Q_\eta^{\sigma,\tau,\gamma}(\zeta_\varepsilon f) \to Q_\eta^{\sigma,\tau,\gamma}(f)$ because Lemma 7.49 together with Lemma 7.52 below shows that the $\widehat{f_{j,k}}$'s in (7.62) only become addressed at points above $-in$ on the imaginary axis. $\qquad\square$

Lemma 7.52. $Q_\eta^{\sigma,\tau,\gamma}(f) = 0$ *for all* f *if and only if*

$$\eta + l + 2\min(\sigma,\tau) + \min(1,2s) > 0.$$

If $Q_\eta^{\sigma,\tau,\gamma}$ does not vanish identically then the point

$$i(l + 2\min(\sigma,\tau) + \min(1,2s))$$

is the lowest one on the imaginary axis that assigns to a function $\widehat{f_{j,k}}$ in (7.62).

Proof. Let $\eta = 0$ and K, K_j be as in the proof of Lemma 7.51. It is clear from that proof that $Q = 0$ if and only if $K_j(p) \to 0$, as $|p| \to \infty$, for all j.

Using the formulas

$$K^{\sigma,\tau,l+s} = (2\nu)^l K^{\sigma+l/2,\tau+l/2,s}$$
$$= (2\nu)^l \left(K^{\sigma+l/2+s,\tau+l/2,0} + K^{\sigma+l/2,\tau+l/2+s,0} - K^{\sigma+l/2,\tau+l/2,0}\right)$$

and then (7.68) (which concerns only the case $\gamma = 0$), we obtain the asymptotic formula

$$K^{\sigma,\tau,\gamma}(p,\nu) = (2\nu)^l e^{-l|p|}\left(e^{-2\sigma p}\chi_+(p) + e^{2\tau p}\chi_-(p)\right)$$
$$\times \left(e^{-2sp} + 2s\nu e^{-|p|} + \Omega(e^{-(2s+1)|p|} + e^{-2|p|})\right), \quad \text{as } |p| \to \infty.$$

(We also substituted for the Gegenbauer polynomials, $C_0^\lambda(\nu) = 1$ and $C_1^\lambda(\nu) - C_1^{\lambda-s}(\nu) = 2s\nu$.) Now we see from (7.65) that all $K_j(p) \to 0$ iff $K(p,\nu) \to 0$ for all ν, which in turn happens if and only if the condition in the statement is satisfied.

Similarly, the second statement follows from the fact that $e^{-\mu|p|}K_j(p) \to 0$ for all j as $|p| \to \infty$ if and only if $\mu < \eta + l + 2\min(\sigma,\tau) + \min(1,2s)$. $\qquad\square$

In the next lemma, we see that the values of $\Phi_{\eta,j}^{\sigma,\tau,\gamma}$ can be "reached" through sequences of good functions.

Lemma 7.53. *Let $\alpha_k \in \mathbb{R}$, $k = 1, 2, \ldots, k_0$. For fixed j, η and $\xi_0 \neq 0$, there are functions $f_\varepsilon \in \hat{\mathscr{L}}$ with $|x|^{\alpha_k} f_\varepsilon \in \mathscr{L}$ which are even or odd according as j is even or odd, such that*

$$R_\eta^{\sigma,\tau,\gamma}(f_\varepsilon) \to \Phi_{\eta,j}^{\sigma,\tau,\gamma}(\xi_0), \tag{7.69}$$

$$R_\eta^{\sigma,\tau,\gamma}(\mathrm{Re}\, f_\varepsilon) \to \mathrm{Re}\, \Phi_{\eta,j}^{\sigma,\tau,\gamma}(\xi_0), \tag{7.70}$$

as $0 < \varepsilon \to 0$.

Proof. Let S_j be a real normalized surface harmonic function of degree j. Omitting writing the index ε, we shall take

$$f(x) = c|x|^{-n} f_j(\log|x|) S_j(x'), \tag{7.71}$$

for appropriately chosen f_j and $c(\varepsilon) > 0$.

Put

$$\hat{g}_j(\xi) = \exp\left(-((\xi - \xi_0 + i\eta)/2\varepsilon)^2\right).$$

Then g_j correspond via (7.71), with $c = 1$, to the function

$$g(x) = \frac{\varepsilon}{\sqrt{\pi}} |x|^{i\xi_0 - n - \eta - (\varepsilon \log|x|)^2} S_j(x').$$

Clearly $g \in \hat{\mathscr{L}}$. Now put

$$h(\xi) = \prod_{k=1}^{k_0} \left(\cosh(2\pi(\xi + i\eta)) - \cos(2\pi\alpha_k)\right).$$

Then $\hat{f}_j = h\hat{g}_j$ is a linear-combination of functions of the type $\hat{g}_j$, with different ξ_0. Hence also $f \in \mathscr{L}$.

Since, on the real line, $T_{i\eta}\hat{f}_j(\xi)$ concentrates to the point $\xi = \xi_0$, when $\varepsilon \to 0$ and since $T_{i\eta}h(\xi_0) \neq 0$, we may choose $c(\varepsilon)$ so that (7.69) holds. Similarly, (7.70) follows after noticing that conjugating f or $\Phi_{\eta,j}^{\sigma,\tau,\gamma}(\xi_0)$ only amounts to changing the sign of ξ_0.

Finally, the fact that $h(\xi) = 0$ at points $\xi = i(m - \eta \pm \alpha_k)$, m integer implies that $|x|^{-\eta \pm \alpha_k} f \in \mathscr{L}$ thanks to Lemma 7.50 (iv). $\square$

From Lemmas 7.51 and 7.53, formula (7.61) and the continuity of $\mathrm{Re}\,\Phi_{\eta,j}^{\sigma,\tau,\gamma}$, we immediately obtain the following corollary which contains a large family of inequalities.

Corollary 7.54. *Let $\sigma + \tau + \gamma + \eta < n/2$ be fixed parameters such that $\mathrm{Re}\,\Phi_{\eta,j}^{\sigma,\tau,\gamma} > 0$. If $\sigma' + \tau' + \gamma' \leq \sigma + \tau + \gamma$ then*

$$\left|\mathrm{Re}\left(I_\eta^{\sigma',\tau',\gamma'}(f) - Q_\eta^{\sigma',\tau',\gamma'}(f)\right)\right| \leq C\,\mathrm{Re}\left(I_\eta^{\sigma,\tau,\gamma}(f) - Q_\eta^{\sigma,\tau,\gamma}(f)\right),$$

where the best constant is given by

$$C = \sup |\mathrm{Re}\,\Phi_{\eta,j}^{\sigma',\tau',\gamma'}(\xi)|(\mathrm{Re}\,\Phi_{\eta,j}^{\sigma,\tau,\gamma}(\xi))^{-1} < \infty$$

and where the supremum is taken over all $\xi \in \mathbb{R}$, $j = 0, 1$ if $n = 1$ and $j = 0, 1, \ldots$ if $n \geq 2$.

If we include the appropriate α_k's in the f_ε in Lemma 7.53, we see that the constant is best possible also among the real functions in $\mathscr{L} \cap \hat{\mathscr{L}}$ for which the Q vanishes on both sides of the inequality.

Let us now introduce the quadratic form that is dual to $I_\eta^{\sigma,\tau,\gamma}$. We recall that $\lambda = \sigma + \tau + \gamma + \eta < n/2$ and $\gamma = l + s$, where $l \in \mathbb{N}$ and $0 \le s \le 1$. For $0 \le s < 1$, we define the expression

$$
J_\eta^{\sigma,\tau,\gamma}(u) = \begin{cases} 2^l \displaystyle\int (-\Delta)^\sigma \nabla_l u \cdot ((-\Delta)^\tau \nabla_l \bar{u}) \Gamma_\lambda \, dx, & s = 0, \\[2ex] 2^l A_s \displaystyle\iint \frac{(\Delta_y (-\Delta)^\sigma \nabla_l u) \cdot (\Delta_y (-\Delta)^\tau \nabla_l \bar{u})}{|y|^{n+2s}} \Gamma_\lambda \, dx \, dy, & s \ne 0. \end{cases}
$$

where $(\Delta_y v)(x) = v(x) - v(x-y)$, $A_s^{-1} = \int (1 - \cos(x_1)) |x|^{-n-2s} \, dx$ and $\Gamma_\lambda(x)$ is defined in the previous section. We shall only deal with the form J in situations where the occurring integral or double-integral is absolutely convergent.

Lemma 7.55. *Let $\sigma, \tau, \gamma \ge 0$ and $0 < \sigma + \tau + \gamma + \eta < n/2$. Then*

$$
I_\eta^{\sigma,\tau,\gamma}(f) = J_\eta^{\sigma,\tau,\gamma}(\hat{f}), \tag{7.72}
$$

for all $f \in \mathscr{S}$.

Proof. If this has been proved for $f \in \hat{\mathscr{L}}$, the case of $f \in \mathscr{S}$ follows by approximation. Namely if $\varphi_\varepsilon(x) = \varphi(x/\varepsilon)$, where $\varphi \in C_0^\infty$ equals 1 in a neighborhood of the origin, we have

$$
I_\eta^{\sigma,\tau,\gamma}(f - \varphi_\varepsilon f) = J_\eta^{\sigma,\tau,\gamma}(\hat{f} - (2\pi)^{-n} \widehat{\varphi_\varepsilon} * \hat{f})
$$

and when $\varepsilon \to 0$, each side converges to the corresponding side of (7.72). We omit the details.

Now, let $f \in \hat{\mathscr{L}}$ and consider first $\gamma = 0$. This case follows from Parseval's formula, for instance if we regard the integration in x on the left as a Riesz potential:

$$
I_\eta^{\sigma,\tau,0}(f) = (2\pi)^n c_\lambda \int (-\Delta)^{\lambda - n/2} (|y|^{2\sigma} f) |y|^{2\tau} \overline{f} \, dy = J_\eta^{\sigma,\tau,0}(\hat{f}).
$$

Next, distributing the terms from $(2x \cdot y)^l$ onto $f(x)\overline{f(y)}$ we have

$$
I_\eta^{\sigma,\tau,l}(f) = 2^l \sum_{|\alpha|=l} \frac{l!}{\alpha!} I_{\eta+l}^{\sigma,\tau,0}(x^\alpha f).
$$

For J the same formula holds, but with ∂^α in place of x^α. Hence we are done for $\gamma = l$.

It is clear that if $\lambda > s$ then $I_\eta^{\sigma,\tau,\gamma}$ satisfies the recursion formula (7.60). To see that the same is true for J, we apply the representation

$$v(-\Delta)^s\overline{w} + \overline{w}(-\Delta)^s v - (-\Delta)^s(v\overline{w}) = A_s \int \frac{(\Delta_y v)\overline{(\Delta_y w)}}{|y|^{n+2s}}\, dy, \quad 0 < s < 1,$$

for $v, w \in \mathscr{S}$ (see Lemma 1 in Eilertsen [23]), together with the formula $(-\Delta)^s\Gamma_\lambda = \Gamma_{\lambda-s}$, which is valid in the sense of distributions.

Finally, the case $0 < \lambda \le s$ follows by analyticity in the parameter η. $\qquad\square$

Other instances of the identity (7.72) may be obtained by performing analytic continuation in some of the parameters, with or without imposing additional requirements on f.

7.5.3 The weighted positivity of $(-\Delta)^\lambda$

We say that the operator $(-\Delta)^\lambda$, $0 < \lambda < n/2$ is positive with weight Γ_λ provided there exist a $c > 0$ with

$$\int ((-\Delta)^\lambda u)u\Gamma_\lambda\, dx \ge c \int \left((-\Delta)^{\lambda/2}u\right)^2\Gamma_\lambda\, dx, \tag{7.73}$$

for all real $u \in \mathscr{L} \cap \hat{\mathscr{L}}$. By Lemmas 7.55, 7.51 and 7.53, this is equivalent to $\operatorname{Re}\Phi_{0,j}^{\lambda,0,0}(\xi) > 0$, for all $j = 0, 1, \ldots$, and $\xi \in \mathbb{R}$, where

$$\Phi_{0,j}^{\lambda,0,0}(\xi) = \frac{\pi^n 2^{2\lambda} c_\lambda \Gamma(\lambda + z)\Gamma(\overline{z})}{\Gamma(n/2 + z)\Gamma(n/2 - \lambda + \overline{z})}, \quad z = \frac{j + i\xi}{2}.$$

The continuity of the real part of this function implies that $(-\Delta)^\lambda$ has the positivity property for λ in an open subset of $(0, n/2)$ and that (7.73) holds with $c = 0$ precisely for λ in the closure of this subset. The main object of this section is to characterize these sets.

Since $S_{0,1}$ is a constant with modulus equal to $1/\sqrt{\omega_{n-1}}$ and $a_{0,0}^\lambda = 1$, we find

$$Q_0^{\lambda,0,0}(f) = \frac{\omega_{n-1}}{2}a_{0,0}^\lambda|\widehat{f_{0,1}}(0)|^2 = \frac{1}{2}|S_{0,1}(\partial)\widehat{f}(0)|^2 = \frac{1}{2}u(0)^2,$$

where $u = \widehat{f}$. The second step is an application of Lemma 7.50 (iii). Now Corollary 7.54 shows that (7.73) implies

$$\int ((-\Delta)^\lambda u)u\Gamma_\lambda\, dx \tag{7.74}$$

$$\ge \frac{1}{2}u(0)^2 + c\left(\sum_{l=1}^{\lfloor\lambda\rfloor}\int |\nabla_l u|^2\Gamma_l\, dx + \iint \frac{|\nabla_k u(x) - \nabla_k u(y)|^2}{|x-y|^{n+2s}}\Gamma_{k+s}(x)\, dx\, dy\right),$$

for all real $u \in \mathcal{S}$, where $\lfloor \lambda \rfloor$ is the integer part of λ, $0 < s < 1$ and $k + s \le \lambda$. The latter kind of inequality is what was needed in Eilertsen [23] to deduce that a boundary point is regular if it satisfies the Wiener test.

Remark: Actually, (7.73) and (7.74) are equivalent. This follows for example from the fact that $\Phi_{\eta,j}^{\rho,\rho,\gamma} > 0$ for $0 < 2\rho + \gamma + \eta < n/2$.

It turns out that the value $\operatorname{Re} \Phi_{0,0}^{\lambda,0,0}(0)$ plays the crucial role for the positivity. Let us now find an expression for this value. Put

$$g(\xi) = \frac{\Gamma(\lambda + i\xi/2)\Gamma(1 - i\xi/2)}{\Gamma(n/2 + i\xi/2)\Gamma(n/2 - \lambda - i\xi/2)}.$$

Then $\Phi_{0,0}^{\lambda,0,0}(\xi) = i\omega_{n-1}g(\xi)/\xi g(0)$ so

$$\operatorname{Re} \Phi_{0,0}^{\lambda,0,0}(0) = \omega_{n-1}ig(0)^{-1} \lim_{\xi \to 0}(g(\xi) - g(-\xi))/2\xi = \omega_{n-1}ig'(0)/g(0)$$

$$= 2^{-1}\omega_{n-1}f(n,\lambda),$$

where we introduce the function

$$f(n, \lambda) = \psi(n/2) - \psi(n/2 - \lambda) - \psi(\lambda) + \psi(1).$$

We now state some properties of $f(n, \lambda)$.

Lemma 7.56. *Let $0 < \lambda < n/2$. If $n \le 7$ then $f(n, \lambda) > 0$.*

If $n \ge 8$ then $f(n, \lambda)$ has exactly two zeros, λ_n and $n/2 - \lambda_n$. We have $f(n, \lambda) > 0$ if and only if $\lambda \notin [\lambda_n, n/2 - \lambda_n]$.

The numbers λ_n satisfy $2 > \lambda_n \searrow 1$, as $n \to \infty$.

Proof. First notice that $f(n, \lambda) = f(n, n/2 - \lambda)$. Differentiating the expansion

$$f(n, \lambda) = \sum_{m=0}^{\infty} \left(\frac{1}{n/2 - \lambda + m} - \frac{1}{n/2 + m} + \frac{1}{\lambda + m} - \frac{1}{1 + m} \right),$$

we find that the terms in $f_\lambda'(n, \lambda)$ are

$$\frac{(4\lambda - n)(4m + n)}{(n - 2\lambda + 2m)^2(\lambda + m)^2},$$

so for a fixed n, $f(n, \lambda)$ has its minimum when $\lambda = n/4$. Similarly, $f_n'(n, \lambda) < 0$. Now, the first statement follows from the fact that

$$f(7, 7/4) = 2/5 + 4\ln 2 - \pi > 0.$$

For $n \ge 8$ we have

$$f(n, 2) = \frac{10n - 20 - n^2}{(n - 2)(n - 4)} < 0.$$

On the other hand,

$$\lim_{n \to \infty} f(n, \lambda) = \psi(1) - \psi(\lambda)$$

has the same sign as $1 - \lambda$. The last statements follow. $\qquad\square$

Theorem 7.57. *Let $0 < \lambda < n/2$. The inequality (7.73) holds with a positive c if and only if $\lambda \notin [\lambda_n, n/2 - \lambda_n]$.*

If $\lambda = \lambda_n$ or $\lambda = n/2 - \lambda_n$ then the left integral in (7.73) is positive unless $u = 0$.

If $\lambda \in (\lambda_n, n/2 - \lambda_n)$ then the left integral in (7.73) takes on negative values for some u.

Proof. Since $\lambda \mapsto \operatorname{Re} \Phi_{0,0}^{\lambda,0,0}(0)$ is negative in $(\lambda_n, n/2 - \lambda_n)$ and positive outside the closure of this set, the theorem will follow once we prove the implication

$$\operatorname{Re} \Phi_{0,0}^{\lambda,0,0}(0) \geq 0 \Rightarrow \operatorname{Re} \Phi_{0,j}^{\lambda,0,0}(\xi) > 0, \quad \text{for } \xi \neq 0. \tag{7.75}$$

First notice that $\Phi_{0,j}^{\lambda,0,0}(\xi)$ is a product of a positive function and a function that is analytic in the variable $z = (j + i\xi)/2$. Since by (7.61), $\operatorname{Re} \Phi_{0,j}^{\lambda,0,0}(\xi) > 0$ if $|z|$ is large, the maximum principle shows that it suffices to prove (7.75) for $j = 0$.

The fact that $\Phi_{0,j}^{n/2-\lambda,0,0}/\Phi_{0,j}^{\lambda,0,0} > 0$, shows that (any side of) (7.75) is true if and only if it is true with $n/2 - \lambda$ in place of λ. We may therefore restrict ourselves to $\lambda \in [n/4, n/2)$. If $\lambda \geq n/2 - 1$, then $a_{j,m}^{\lambda} \geq 0$ and $a_{j,0}^{\lambda} > 0$ in (7.58) so the right-hand side of (7.75) is true. If, on the other hand, $\lambda \geq n/4$ and the left-hand side of (7.75) is true, then it follows from Lemma 7.56 that $\lambda > n/2 - 2$.

In conclusion, we need only to prove the implication in the special case $\lambda \in (n/2 - 2, n/2 - 1)$ and $j = 0$.

By first using (7.59) together with $\Gamma(z+1) = z\Gamma(z)$ and then (7.58), we find

$$\Phi_{0,0}^{\lambda,0,0}(2\xi) = \frac{\lambda(n/2 - \lambda - 1)}{(\lambda + i\xi)(-i\xi)} \Phi_{0,1}^{\lambda+1/2,1/2,0}(2\xi) = \sum_{m=0}^{\infty} c_m A_m, \tag{7.76}$$

where

$$c_m = 2^{-1}\omega_{n-1}\lambda(n/2 - \lambda - 1)(\lambda + 2m + 2)a_{1,m}^{\lambda+1}$$

and

$$A_m = \frac{1}{(\lambda + i\xi)(-i\xi)(\lambda + 2m + 2)} \left(\frac{1}{\lambda + m + 1 + i\xi} + \frac{1}{m + 1 - i\xi} \right).$$

For those values of λ we consider, $c_m > 0$ (this is the reason for rewriting the Φ in (7.76) before expanding it). If we put

$$b_m = m^2 + (\lambda + 2)m + \lambda + 1 - \lambda^2,$$
$$B_m(\xi) = |(\lambda + i\xi)(\lambda + m + 1 + i\xi)(m + 1 + i\xi)|^{-2},$$

we obtain after simplification,

$$\operatorname{Re} A_m = (b_m + \xi^2) B_m(\xi).$$

Now, fix $\xi \neq 0$ and put $q = B_{m_0}(\xi)/B_{m_0}(0)$, where m_0 is the smallest nonnegative integer with $b_{m_0} \geq 0$. We have

$$\operatorname{Re}\big(\Phi_{0,0}^{\lambda,0,0}(2\xi) - q\Phi_{0,0}^{\lambda,0,0}(0)\big) \geq \sum_{m=0}^{\infty} c_m b_m (B_m(\xi) - qB_m(0)).$$

Since b_m and $B_m(\xi)/B_m(0)$ grow as functions of m, and $c_m, B_m > 0$, all terms (except one that vanishes) are positive. The proof is complete. $\qquad\square$

7.6 L^p-positivity of fractional powers of the Laplacian

In this section we do not consider variable weights. We deal with the L^p-positivity of the fractional powers of the Laplacian $(-\Delta)^\alpha$ $(0 < \alpha < 1)$ for any $p \in (1, \infty)$.

By $\|v\|_{\mathcal{L}^{\alpha,2}}$ we denote the semi-norm

$$\left(\int_{\mathbb{R}^n} \int_{\mathbb{R}^n} |v(x+t) - v(x)|^2 \frac{dx\,dt}{|t|^{n+2\alpha}}\right)^{1/2}.$$

Theorem 7.58. *Let* $0 < \alpha < 1$. *We have, for any* $u \in C_0^\infty(\mathbb{R}^n)$,

$$\int_{\mathbb{R}^n} \langle(-\Delta)^\alpha u, u\rangle |u|^{p-2} dx \geq \frac{2\,c_\alpha}{p\,p'} \||u|^{p/2}\|_{\mathcal{L}^{\alpha,2}(\mathbb{R}^n)}^2,$$

where

$$c_\alpha = -\pi^{-n/2} 4^\alpha \Gamma(\alpha + n/2)/\Gamma(-\alpha) > 0. \tag{7.77}$$

Proof. As proved by Stein (see [94, p. 104], [96, pp. 161–162]), we may write

$$(-\Delta)^\alpha u(x) = -c_\alpha \lim_{\varepsilon \to 0} \int_{|t| \geq \varepsilon} \frac{u(x+t) - u(x)}{|t|^{n+2\alpha}} dt \tag{7.78}$$

for any $u \in C_0^\infty(\mathbb{R}^n)$, the constant c_α being given by (7.77).

From (7.78) it follows that

$$\int_{\mathbb{R}^n} \langle(-\Delta)^\alpha u, v\rangle dx = \frac{c_\alpha}{2} \int_{\mathbb{R}^n} \int_{\mathbb{R}^n} (u(x+t) - u(x))\,(v(x+t) - v(x)) \frac{dx\,dt}{|t|^{n+2\alpha}} \tag{7.79}$$

for any $u, v \in C_0^\infty(\mathbb{R}^n)$. Note that, since u and v have compact supports and

$$|(u(x+t) - u(x))\,(v(x+t) - v(x))|\,|t|^{-n-2\alpha} \leq \|\nabla u\|_\infty \|\nabla v\|_\infty |t|^{2-n-2\alpha},$$

the integral in (7.79) is absolutely convergent.

Given $u \in C_0^\infty(\mathbb{R}^n)$ and $\varepsilon > 0$, define

$$g_\varepsilon(s) = \sqrt{s^2 + \varepsilon^2}, \quad v_\varepsilon(x) = (g_\varepsilon[u(x)])^{p-2}u(x).$$

In view of (7.79) we can write

$$\int_{\mathbb{R}^n} \langle(-\Delta)^\alpha u, v_\varepsilon\rangle\, dx = \frac{c_\alpha}{2}\int_{\mathbb{R}^n}\int_{\mathbb{R}^n}(u(x+t) - u(x))(v_\varepsilon(x+t) - v_\varepsilon(x))\frac{dxdt}{|t|^{n+2\alpha}}.$$
$$(7.80)$$

As $\varepsilon \to 0$, $v_\varepsilon(x)$ tends to $|u(x)|^{p-2}u(x)\chi_u(x)$, where $\chi_u(x) = 1$ if $u(x) \neq 0$ and $\chi_u(x) = 0$ if $u(x) = 0$. Applying Lemma 4.35, we obtain from (7.80)

$$\int_{\mathbb{R}^n} \langle(-\Delta)^\alpha u, u\rangle|u|^{p-2}dx = \frac{c_\alpha}{2}\int_{\mathbb{R}^n}\int_{\mathbb{R}^n}(u(x+t) - u(x))$$
$$\times (|u(x+t)|^{p-2}u(x+t)\chi_u(x+t) - |u(x)|^{p-2}u(x)\chi_u(x))\frac{dxdt}{|t|^{n+2\alpha}}.$$

On the other hand we have

$$(x - y)(|x|^{p-2}x - |y|^{p-2}y) \geq \frac{4}{pp'}(|x|^{p/2} - |y|^{p/2})^2$$

for any $x, y \in \mathbb{R}$. This can be proved in an elementary way determining the infimum of the function of one real variable

$$\frac{(t-1)(|t|^{p-2}t - 1)}{(|t|^{p/2} - 1)^2}.$$

Therefore

$$\int_{\mathbb{R}^n} \langle(-\Delta)^\alpha u, u\rangle|u|^{p-2}dx \geq \frac{2c_\alpha}{pp'}\int_{\mathbb{R}^n}\int_{\mathbb{R}^n}(|u(x+t)|^{p/2} - |u(x)|^{p/2})^2\frac{dxdt}{|t|^{n+2\alpha}}$$
$$= \frac{2c_\alpha}{pp'}\||u|^{p/2}\|^2_{\mathcal{L}^{\alpha,2}(\mathbb{R}^n)}$$

and the theorem is proved. $\square$

7.7 *L²*-positivity for the Stokes system

It is obvious that

$$\int_\Omega (-\Delta u + \nabla p)\, u\, dx \geq 0$$

for any $u \in C_0^\infty(\mathbb{R}^n)$ such that div $u = 0$.

Because of the n-dimensional Hardy inequality

$$\int_{\mathbb{R}^n} |x|^{-2}|u|^2 dx \leqslant \frac{4}{(n-2)^2} \int_{\mathbb{R}^n} |\nabla u|^2 dx, \qquad (7.81)$$

we may obtain a more precise estimate

$$\int_{\mathbb{R}^n} |x|^{-2}|u|^2 dx \leqslant \frac{4}{(n-2)^2} \int_{\mathbb{R}^n} (-\Delta u + \nabla p)\,\overline{u} dx, \qquad (7.82)$$

holding for any $u \in C_0^\infty(\mathbb{R}^n)$ such that $\operatorname{div} u = 0$.

It is well known that inequality (7.81) is sharp. One can ask whether the restriction $\operatorname{div} u = 0$ can improve the constant in (7.82).

In the present section we show that this is the case indeed if $n > 2$ and the vector field u is axisymmetric by proving that the afore-mentioned constant can be replaced by the (smaller) optimal value

$$\frac{4}{(n-2)^2}\left(1 - \frac{8}{(n+2)^2}\right), \qquad (7.83)$$

which, in particular, evaluates to $68/25$ in three dimensions.

In the following theorem we use the same notations for the vector $u = (u_\varrho, u_\vartheta, u_\Phi)$ as in Section 3.4 (see p. 87). Here the condition of axial symmetry means that u depends only on ϱ and ϑ.

Theorem 7.59. *Let $n > 2$, and let u be an axisymmetric divergence-free vector field in $C_0^\infty(\mathbb{R}^n)$ and $p \in C^1(\mathbb{R}^n)$. Then*

$$\int_{\mathbb{R}^n} |x|^{-2}|u|^2 dx \leqslant C_n \int_{\mathbb{R}^n} (-\Delta u + \nabla p)\,\overline{u}\,dx, \qquad (7.84)$$

with the best value of C_n given by (7.83).

Proof. Obviously, proving (7.84) for a divergence-free vector field in $C_0^\infty(\mathbb{R}^n)$ is equivalent to proving

$$\int_{\mathbb{R}^n} |x|^{-2}|u|^2 dx \leqslant C_n \int_{\mathbb{R}^n} |\nabla u|^2\,dx. \qquad (7.85)$$

In the spherical coordinates introduced previously, we have

$$\operatorname{div} u = \varrho^{1-n}\frac{\partial}{\partial \varrho}(\varrho^{n-1}u_\varrho) + \varrho^{-1}(\sin\vartheta)^{2-n}\frac{\partial}{\partial\vartheta}((\sin\vartheta)^{n-2}u_\vartheta)$$

$$+ \sum_{k=1}^{n-3}(\varrho\sin\vartheta\sin\vartheta_{n-3}\ldots\sin\vartheta_{k+1})^{-1}(\sin\vartheta_k)^{-k}\frac{\partial}{\partial\vartheta_k}((\sin\vartheta_k)^k u_{\vartheta_k}) \quad (7.86)$$

$$+ (\varrho\sin\vartheta\sin\vartheta_{n-3}\ldots\sin\vartheta_1)^{-1}\frac{\partial u_\varphi}{\partial\varphi}.$$

Since the components u_φ and u_{ϑ_k}, $k = 1, \ldots, n-3$, depend only on ϱ and ϑ, (7.86) becomes

$$\operatorname{div} u = \varrho^{1-n} \frac{\partial}{\partial \varrho}(\varrho^{n-1} u_\varrho(\varrho, \vartheta)) + \varrho^{-1}(\sin \vartheta)^{2-n} \frac{\partial}{\partial \vartheta}((\sin \vartheta)^{n-2} u_\vartheta(\varrho, \vartheta))$$

$$+ \sum_{k=1}^{n-3} k(\sin \vartheta_{n-3} \ldots \sin \vartheta_{k+1})^{-1} \cot \vartheta_k \frac{u_{\vartheta_k}(\varrho, \vartheta)}{\varrho \sin \vartheta} .$$

By the linear independence of the functions

$$1, \ (\sin \vartheta_{n-3} \ldots \sin \vartheta_{k+1})^{-1} \cot \vartheta_k, \ k = 1, \ldots, n-3,$$

the divergence-free condition is equivalent to the collection of $n-2$ identities

$$\varrho \frac{\partial u}{\partial \varrho} + (n-1) u_\varrho + \left(\frac{\partial}{\partial \vartheta} + (n-2) \cot \vartheta \right) u_\vartheta = 0,$$

$$u_{\vartheta_k} = 0, \ k = 1, \ldots, n-3.$$

If the right-hand side of (7.85) diverges, there is nothing to prove. Otherwise, let us introduce the vector field

$$v(x) = u(x) |x|^{-1+n/2}.$$

The inequality (7.85) becomes

$$\left(\frac{1}{C_n} - \frac{(n-2)^2}{4} \right) \int_{\mathbb{R}^n} |x|^{-n} |v|^2 dx \leqslant \int_{\mathbb{R}^n} |x|^{2-n} |\nabla v|^2 dx.$$

The condition $\operatorname{div} u = 0$ is equivalent to

$$\varrho \operatorname{div} v = \frac{n-2}{2} v_\varrho . \tag{7.87}$$

To simplify the exposition, we assume first that $v_\varphi = 0$. Now, (7.87) can be written as

$$\varrho \frac{\partial v_\varrho}{\partial \varrho} + \frac{n}{2} v_\varrho + \mathcal{D} v_\vartheta = 0,$$

where

$$\mathcal{D} = \frac{\partial}{\partial \vartheta} + (n-2) \cot \vartheta.$$

Note that $\mathcal{D}$ is the adjoint of $-\partial/\partial \vartheta$ with respect to the scalar product

$$\int_0^\pi f(\vartheta) \overline{g(\vartheta)} (\sin \vartheta)^{n-2} d\vartheta.$$

A straightforward, though lengthy calculation yields

$$\varrho^2 |\nabla v|^2 = \varrho^2 \left(\frac{\partial v_\varrho}{\partial \varrho} \right)^2 + \varrho^2 \left(\frac{\partial v_\vartheta}{\partial \varrho} \right)^2 + \left(\frac{\partial v_\varrho}{\partial \vartheta} \right)^2 + \left(\frac{\partial v_\vartheta}{\partial \vartheta} \right)^2$$
$$+ v_\vartheta^2 + (n-1)v_\varrho^2 + (n-2)(\cot \vartheta)^2 v_\vartheta^2 + 2 \left(v_\varrho D v_\vartheta - v_\vartheta \frac{\partial v_\varrho}{\partial \vartheta} \right).$$

Hence, denoting by S^{n-1} the unit sphere,

$$\varrho^2 \int_{S^{n-1}} |\nabla v|^2 ds = \int_{S^{n-1}} \left\{ \varrho^2 \left(\frac{\partial v_\varrho}{\partial \varrho} \right)^2 + \varrho^2 \left(\frac{\partial v_\vartheta}{\partial \varrho} \right)^2 + \left(\frac{\partial v_\varrho}{\partial \vartheta} \right)^2 + \left(\frac{\partial v_\vartheta}{\partial \vartheta} \right)^2 \right.$$
$$\left. + v_\vartheta^2 + (n-1)v_\varrho^2 + (n-2)(\cot \vartheta)^2 v_\vartheta^2 + 4 v_\varrho D v_\vartheta \right\} ds. \quad (7.88)$$

Changing the variable ϱ to $t = \log \varrho$ and applying the Fourier transform with respect to t,

$$v(t, \vartheta) \mapsto w(\lambda, \vartheta),$$

we derive

$$\int_{\mathbb{R}^n} |x|^{2-n} |\nabla v|^2 dx$$
$$= \int_{\mathbb{R}} \int_{S^{n-1}} \left\{ (\lambda^2 + n - 1)|w_\varrho|^2 + (\lambda^2 - n + 3)|w_\vartheta|^2 \right. \quad (7.89)$$
$$\left. + \left| \frac{\partial w_\varrho}{\partial \vartheta} \right|^2 + \left| \frac{\partial w_\vartheta}{\partial \vartheta} \right|^2 + (n-2)(\sin \vartheta)^{-2} |w_\vartheta|^2 + 4 \operatorname{Re}(\overline{w}_\varrho D w_\vartheta) \right\} ds d\lambda$$

and

$$\int_{\mathbb{R}^n} |x|^{-n} |v|^2 dx = \int_{\mathbb{R}} \int_{S^{n-1}} |w|^2 ds d\lambda. \quad (7.90)$$

From (7.87), we obtain

$$w_\varrho = - \frac{D w_\vartheta}{i\lambda + n/2}, \quad (7.91)$$

which implies

$$|w_\varrho|^2 = \frac{4 D w_\vartheta}{4\lambda^2 + n^2} \quad \text{and} \quad \operatorname{Re}(\overline{w}_\varrho D w_\vartheta) = - \frac{2n |D w_\vartheta|^2}{4\lambda^2 + n^2}.$$

Introducing this into (7.89), we arrive at the identity

$$\int_{\mathbb{R}^n} |x|^{2-n} |\nabla v|^2 dx$$
$$= \int_{\mathbb{R}} \int_{S^{n-1}} \left\{ (\lambda^2 + n - 1) \frac{4 |D w_\vartheta|^2}{4\lambda^2 + n^2} + (\lambda^2 - n + 3)|w_\vartheta|^2 + \left| \frac{\partial w_\vartheta}{\partial \vartheta} \right|^2 \right.$$
$$\left. + (n-2)(\sin \vartheta)^{-2} |w_\vartheta|^2 + \frac{4}{4\lambda^2 + n^2} \left| \frac{\partial}{\partial \vartheta} D w_\vartheta \right|^2 - \frac{8n |D w_\vartheta|^2}{4\lambda^2 + n^2} \right\} ds d\lambda.$$

We simplify the right-hand side to obtain

$$\int_{\mathbb{R}^n} |x|^{2-n} |\nabla v|^2 dx = \int_{\mathbb{R}} \left\{ \left(\frac{4(\lambda^2 - n - 1)}{4\lambda^2 + n^2} + 1 \right) |\mathcal{D}w_\vartheta|^2 \right. \tag{7.92}$$
$$\left. + (\lambda^2 - n + 3)|w_\vartheta|^2 + \frac{4}{4\lambda^2 + n^2} \left| \frac{\partial}{\partial \vartheta} \mathcal{D}w_\vartheta \right|^2 \right\} ds d\lambda.$$

On the other hand, by (7.90) and (7.91)

$$\int_{\mathbb{R}^n} |x|^{2-n} |\nabla v|^2 dx = \int_{\mathbb{R}} \int_{S^{n-1}} \left(\frac{4|\mathcal{D}w_\vartheta|^2}{4\lambda^2 + n^2} + |w_\vartheta|^2 \right) ds d\lambda. \tag{7.93}$$

Defining the self-adjoint operator

$$T = -\frac{\partial}{\partial \vartheta} \mathcal{D},$$

or equivalently,

$$T = -\delta_\vartheta + (n - 2)(\sin \vartheta)^{-2},$$

where δ_ϑ is the ϑ part of the Laplace–Beltrami operator on S^{n-1}, we write (7.92) and (7.93) as

$$\int_{\mathbb{R}^n} |x|^{2-n} |\nabla v|^2 dx = \int_{\mathbb{R}} \int_{S^{n-1}} Q(\lambda, w_\vartheta) ds d\lambda \tag{7.94}$$

and

$$\int_{\mathbb{R}^n} |x|^{-n} |v|^2 dx = \int_{\mathbb{R}} \int_{S^{n-1}} q(\lambda, w_\vartheta) ds d\lambda,$$

respectively, where Q and q are sesquilinear forms in w_ϑ, defined by

$$Q(\lambda, w_\vartheta) = \left(\frac{4(\lambda^2 - n - 1)}{4\lambda^2 + n^2} + 1 \right) T w_\vartheta \overline{w_\vartheta} + (\lambda^2 - n + 3)|w_\vartheta|^2 + \frac{4}{4\lambda^2 + n^2} |T w_\vartheta|^2$$

and

$$q(\lambda, w_\vartheta) = \frac{4 T w_\vartheta \overline{w_\vartheta}}{4\lambda^2 + n^2} + |w_\vartheta|^2.$$

The eigenvalues of T are $\alpha_\nu = \nu(\nu + n - 2), \nu \in \mathbb{Z}^+$. Representing w_ϑ as an expansion in eigenfunctions of T, we find

$$\inf_{w_\vartheta} \frac{\int_{\mathbb{R}} \int_{S^{n-1}} Q(\lambda, w_\vartheta) ds d\lambda}{\int_{\mathbb{R}} \int_{S^{n-1}} q(\lambda, w_\vartheta) ds d\lambda} = \inf_{\lambda \in \mathbb{R}} \inf_{\nu \in \mathbb{N}} \frac{\left(\frac{4(\lambda^2 - n - 1)}{4\lambda^2 + n^2} + 1 \right) \alpha_\nu + \lambda^2 - n + 3 + \frac{4\alpha_\nu^2}{4\lambda^2 + n^2}}{\frac{4\alpha_\nu^2}{4\lambda^2 + n^2}}. \tag{7.95}$$

Thus our minimization problem reduces to finding

$$\inf_{x \geqslant 0} \inf_{\nu \in \mathbb{N}} f(x, \alpha_\nu), \tag{7.96}$$

where
$$f(x, \alpha_\nu) = x - n + 3 + \alpha_\nu \left(1 - \frac{16}{4x + 4\alpha_\nu + n^2} \right).$$

The function f being increasing in x, the value (7.96) is equal to

$$\inf_{\nu \in \mathbb{N}} f(0, \alpha_\nu) = \inf_{\nu \in \mathbb{N}} \left(3 - n + \alpha_\nu \left(1 - \frac{16}{4\alpha_\nu + n^2} \right) \right).$$

We have
$$\frac{\partial}{\partial \alpha_\nu} f(0, \alpha_\nu) = 1 - \frac{16}{4\alpha_\nu + n^2}.$$

Noting that
$$4\alpha_\nu + n^2 \geqslant 4(n-1) + n^2 \geqslant 4n\sqrt{n-1} \, ,$$

we see that
$$\frac{\partial}{\partial \alpha_\nu} f(0, \alpha_\nu) \geqslant 1 - \frac{1}{n(n-1)} > 0.$$

Thus the minimum of $f(0, \alpha_\nu)$, is attained at $\alpha_1 = n - 1$ and equals

$$3 - n + (n-1)\left(1 - \frac{16}{4(n-1) + n^2} \right) = \frac{2(n-2)^2}{4(n-1) + n^2}. \tag{7.97}$$

This completes the proof for the case $v_\varphi = 0$.

If we drop the assumption $v_\varphi = 0$, then, to the integrand on the right-hand of (7.88), we should add the terms

$$\varrho^2 \left(\frac{\partial v_\varphi}{\partial \varrho} \right)^2 + \left(\frac{\partial v_\varphi}{\partial \vartheta} \right)^2 + (\sin \vartheta \sin \vartheta_{n-3} \ldots \sin \vartheta_1)^{-2} v_\varphi^2. \tag{7.98}$$

The expression in (7.98) equals

$$\varrho^2 |\nabla(v_\varphi e^{i\varphi})|^2.$$

As a result, the right-hand side of (7.94) is augmented by

$$\int_{\mathbb{R}} \int_{S^{n-1}} R(\lambda, w_\varphi) \, ds d\lambda \, ,$$

where
$$R(\lambda, w_\varphi) = |\lambda|^2 |w_\varphi|^2 + |\nabla_\omega(w_\varphi e^{i\varphi})|^2$$

with $\omega = (\vartheta, \vartheta_{n-3}, \ldots, \varphi)$. Hence,

$$\inf_v \frac{\int_{\mathbb{R}^n} |x|^{2-n} |\nabla v|^2 dx}{\int_{\mathbb{R}^n} |x|^{-n} |v|^2 dx} = \inf_{w_\vartheta, w_\varphi} \frac{\int_{\mathbb{R}} \int_{S^{n-1}} (Q(\lambda, w_\vartheta) + R(\lambda, w_\varphi)) \, ds d\lambda}{\int_{\mathbb{R}} \int_{S^{n-1}} (q(\lambda, w_\vartheta) + |w_\varphi|^2) \, ds d\lambda}.$$

Using the fact that w_ϑ and w_φ are independent, the right-hand side is the minimum of (7.95) and

$$\inf_{w_\varphi} \frac{\int_{\mathbb{R}} \int_{S^{n-1}} R(\lambda, w_\varphi) \, ds d\lambda}{\int_{\mathbb{R}} \int_{S^{n-1}} |w_\varphi|^2 ds d\lambda}. \tag{7.99}$$

Since $w_\varphi e^{i\varphi}$ is orthogonal to one on S^{n-1}, we have

$$\int_{S^{n-1}} |\nabla_\omega(w_\varphi e^{i\varphi})|^2 ds \geqslant (n-1) \int_{S^{n-1}} |w_\varphi|^2 ds.$$

Hence the infimum in (7.99) is at most $n-1$, which exceeds the value in (7.97), and the result follows. □

7.8 Sharp Gårding inequality for a pseudo-differential operator with Lipschitz symbol

In this section we deal with the semi-boundedness of a certain pseudo-differential operator $\sigma(x, D)$. We obtain a variant of the so-called sharp Gårding inequality for it. The interest of this result is that the symbol $\sigma(x, \xi)$ is not continuously differentiable in x but only satisfies a Lipschitz condition (see Comments at the end of the chapter).

Theorem 7.60. *Let*

$$\sigma(x, \xi) = |f(x)| \, e(x, \xi) \, |\xi|,$$

where e is a sufficiently smooth symbol of order zero, $e(x, \xi) > 0$ for $|\xi| \neq 0$, and let f be a function in C^3 such that $x_n f(x) \geqslant 0$. Then the inequality

$$\mathrm{Re}(\sigma u, u) \geqslant c_1 \| \, |f|^{1/2} e^{1/2} u\|^2_{H^{1/2}} - c_2 \|u\|^2_{L^2}$$

holds for all $u \in C_0^\infty$. Here $e^{1/2}$ is a singular integral operator with the symbol $e^{1/2}(x, \xi)$.

Proof. Let $\varphi(x) = x_n^{-1} f(x)$. Since $\varphi(x) \geqslant 0$ and $\varphi \in C^2$, we have $D\varphi^{1/2} \leqslant c$. Using A.P. Calderòn's theorem on commutators [9], we obtain[1]

$$\mathrm{Re}(\sigma u, u) = \mathrm{Re}(|x_n|\varphi^{1/2} e^{1/2} u, (-\Delta)^{1/2} \varphi^{1/2} e^{1/2} u) + \mathcal{O}(\|u\|^2).$$

We introduce the function $v(x) = \varphi^{1/2}(x) e^{1/2}(x, D) u$. It remains to show that

$$\mathrm{Re}(|x_n| v, (-\Delta)^{1/2} v) \geqslant c_1 \| \, |x_n|^{1/2} v\|^2_{H^{1/2}} - c_2 \|v\|_{L^2}. \tag{7.100}$$

---

[1] According to this theorem, a commutator of functions in $C^{0,1}$ and a first-order pseudo-differential operator with a smooth symbol is an operator of a nonpositive order.

If v is extended as a harmonic function on $\mathbb{R}^{n+1}_+ : \{(x_1,\dots,x_{n+1}) : x_{n+1} > 0\}$, the left-hand side in the last inequality takes on the form

$$
-\operatorname{Re}\left(|x_n|v,\frac{\partial v}{\partial x_{n+1}}\right) = \int_{\mathbb{R}^{n+1}_+}(x_n^2 + x_{n+1}^2)^{1/2}|Dv|^2\,dx\,dx_{n+1}
$$
$$
- \int_{\mathbb{R}^{n+1}_+}(x_n^2 + x_{n+1}^2)^{-1/2}|v|^2\,dx\,dx_{n+1}\,. \tag{7.101}
$$

Moreover we have

$$
\int_{\mathbb{R}^{n+1}_+}(x_n^2 + x_{n+1}^2)^{-1/2}|v|^2\,dx\,dx_{n+1} \leqslant c\int_{\mathbb{R}^{n+1}_+}x_{n+1}|Dv|^2\,dx\,dx_{n+1}\,. \tag{7.102}
$$

In fact, set $\varrho^2 = x^n + x_{n+1}^2$ and denote the integral in the left-hand side of (7.102) by J. Integrating by parts, we obtain

$$
J = -2\int_{\mathbb{R}^{n+1}_+} x_{n+1}\varrho^{-1}v\,v_{x_{n+1}}\,dx\,dx_{n+1}
$$
$$
+ \left(\int_{0<2x_{n+1}<\varrho} + \int_{2x_{n+1}>\varrho}\right)x_{n+1}^2\varrho^{-3}v^2\,dx\,dx_{n+1} \tag{7.103}
$$
$$
\leqslant 8\int_{\mathbb{R}^{n+1}_+} x_{n+1}|Dv|^2\,dx\,dx_{n+1} + \frac{3}{4}J + \int_{2x_{n+1}>\varrho}x_{n+1}^2\varrho^{-3}v^2\,dx\,dx_{n+1}\,.
$$

It remains to estimate the last integral. We introduce the cylindrical coordinate (z,ϱ,ϑ): $z \in \mathbb{R}^{n-1}$, $0 < \vartheta < \pi$, $0 < \varrho < \infty$. Then

$$
J_1 := \int_{2x_{n+1}>\varrho}x_{n+1}^2\varrho^{-3}v^2\,dx\,dx_{n+1} = -2\int_{\mathbb{R}^{n-1}}dz\int_{2x_{n+1}>\varrho}v\frac{\partial v}{\partial \varrho}\sin^2\vartheta\,d\varrho\,d\vartheta
$$
$$
\leqslant 2J_1^{1/2}\left(\int_{2x_{n+1}>\varrho}\varrho\,|Dv|^2dx\right)^{1/2}
$$

and therefore

$$
J_1 \leqslant 8\int_{\mathbb{R}^{n+1}_+}x_{n+1}|Dv|^2\,dx\,dx_{n+1}\,.
$$

By this estimate and (7.103), we obtain (7.102).

Combining (7.102) and the identity

$$
2\int_{\mathbb{R}^{n+1}_+}x_{n+1}|Dv|^2\,dx\,dx_{n+1} = \int_{\mathbb{R}^n}|v|^2dx
$$

with (7.101), we obtain estimate (7.100). $\square$

7.9 Comments to Chapter 7

We collect here bibliographical information concerning results of the present chapter.

Let us note first that different kinds of weighted positivity of partial differential and pseudo-differential operators have a number of applications in qualitative theory of elliptic boundary value problems (see, e.g., Eilertsen [23], Guo Luo and Maz'ya [63], Maz'ya [69, 70], Maz'ya and Donchev [74], Maz'ya and Mayboroda [64, 65] *et al.*).

Jaye, Maz'ya and Verbitsky [43], in connection with the Schrödinger operator, obtained a characterization of distributional potentials σ satisfying the semi-boundedness property

$$\langle \sigma, h^2 \rangle \leqslant \int_\Omega |\nabla h|^2 dx, \qquad \forall\, h \in C_0^\infty(\Omega). \tag{7.104}$$

They proved that a real-valued distribution σ satisfies (7.104) if and only if there exists a vector $\Gamma \in [L^2_{\text{loc}}(\Omega)]^n$ such that

$$\sigma \leqslant \text{div}(\Gamma) - |\Gamma|^2$$

in the sense of distributions.

We remark that, by means of these results, one can characterize the L^p dissipativity of the operator

$$Au = \nabla(\mathscr{A}\,\nabla u) + \sigma u$$

in a domain $\Omega \subset \mathbb{R}^n$. Here $\mathscr{A}$ is a matrix with real L^∞ entries and there exist $m, M > 0$ such that

$$m\,|\xi|^2 \leqslant \mathscr{A}(x)\xi \cdot \xi, \quad |\mathscr{A}(x)\xi| \leqslant M\,|\xi|$$

for any $\xi \in \mathbb{R}^n \setminus \{0\}$.

As remarked on p. 44, A is L^p-dissipative if and only if

$$\langle \sigma, h^2 \rangle \leqslant \frac{4}{pp'} \int_\Omega \langle \mathscr{A}\,\nabla h, \nabla h \rangle\, dx, \qquad \forall\, h \in C_0^\infty(\Omega).$$

Thanks to the above Jaye, Maz'ya and Verbitsky result, we can say that A is L^p-dissipative if and only if there exists $\Gamma \in [L^2_{\text{loc}}(\Omega)]^n$ such that

$$\sigma \leqslant \frac{4}{pp'}(\text{div}(\mathscr{A}\,\Gamma) - (\mathscr{A}\,\Gamma) \cdot \Gamma)$$

in the sense of distributions.

In the same paper [43] the authors consider also the characterization of distributional potentials σ satisfying the boundedness property

$$|\langle \sigma, h^2 \rangle| \leqslant \int_{\mathbb{R}^n} |\nabla h|^2 dx, \qquad \forall \, h \in C_0^\infty(\mathbb{R}^n). \tag{7.105}$$

In particular they give a new and simpler proof of the following theorem, proved for the first time by Maz'ya and Verbitsky [78]: the distribution σ satisfies inequality (7.105) if and only if σ can be represented as $\sigma = \operatorname{div}(\Gamma)$, where $\Gamma \in [L_{\text{loc}}^2(\mathbb{R}^n)]^n$ obeys

$$\int_{\mathbb{R}^n} h^2 |\Gamma|^2 dx \leqslant C \int_{\mathbb{R}^n} |\nabla h|^2 dx, \qquad \forall \, h \in C_0^\infty(\mathbb{R}^n).$$

Results given in Sections 7.1 and 7.5 are due to Eilertsen [22] and [24], respectively. In [24] Eilertsen has studied also inequalities for other similar quadratic forms. As a curious example, he considers the inequality

$$\int_{\mathbb{R}^n} \Delta^2 \nabla u \cdot \nabla u \, \frac{dx}{|x|^{n-4}} > 0, \tag{7.106}$$

which is obtained replacing u by ∇u in

$$\int_{\mathbb{R}^n} (\Delta^2 u) \, u \, \frac{dx}{|x|^{n-4}} > 0, \tag{7.107}$$

studied in Section 7.4. While, as proved for the first time in Maz'ya [70], (7.107) holds for $n = 5, 6, 7$ but not for $n \geqslant 8$, the seemingly similar (7.106) holds for $n = 5, 6, \ldots, 13$ but not for $n \geqslant 14$.

In Sections 7.2 and 7.3 we follow Guo Luo and Maz'ya [62, 63].

The material in Subsections 7.4.1 and 7.4.2 can be found in Maz'ya [70] and Mayboroda and Maz'ya [64] respectively. The extension of results in 7.4.2 to any dimensions is quite complicated. The results depend on the parity of m, n and $m - n/2$ and different weights can be employed. We state only two theorems, just in order to give a flavour of results and we refer to Maz'ya and Mayboroda [65] for the proofs and related results.

We start with n odd. Let us denote by $L^{m,n}$ the operator

$$L^{m,n} = (-1)^m \prod_{j=0}^{m-1} (\partial_t - c_j)(\partial_t + c_j + 1),$$

where

$$c_j = 2j - (m - (n-1)/2), \qquad 0 \leqslant j \leqslant m - 1.$$

Theorem 7.61. *Assume that $m \in \mathbb{N}$ and $n \in [3, 2m+1] \cap \mathbb{N}$ is odd. Let Ω be a bounded domain in $\mathbb{R}^n$, $O \in \mathbb{R}^n \setminus \Omega$, $u \in C_0^\infty(\Omega)$ and $v = e^{(m-(n-1)/2)t}(u \circ \varkappa)$. Then for every $\xi \in \Omega$ and $\tau = \log |\xi|^{-1}$,*

$$\int_{S^{n-1}} v^2(\tau, \omega) d\omega \leqslant C \int_{\mathbb{R}^n} (-\Delta)^m u(x) u(x) g(\log |x|^{-1}, \log |\xi|^{-1}) dx,$$

where

$$g(t, \tau) = e^t (C_1 h(t - \tau) + C_2), \quad t, \tau \in \mathbb{R},$$

and h is a unique solution of the equation

$$L^{m,n} h = \delta$$

(δ being the Dirac delta function) which is bounded and vanishes at $+\infty$. Here C, C_1, C_2 are some constants depending on m and n only.

Consider now the case of n even.

Theorem 7.62. *Assume that $m \in \mathbb{N}$ and $n \in [2, 2m] \cap \mathbb{N}$ is even. Let Ω be a bounded domain in $\mathbb{R}^n$, $O \in \mathbb{R}^n \setminus \Omega$, $u \in C_0^\infty(\Omega)$ and $v = e^{(m-n/2)t}(u \circ \varkappa)$. Let R be a positive constant such that the support of u is contained in B_{2R}. Then there exist positive constants C, C', C'', depending on m and n only, such that, for every $\xi \in B_{2R}$ and $\tau = \log |\xi|^{-1}$,*

$$\int_{S^{n-1}} v^2(\tau, \omega) d\omega \leqslant C \int_{\mathbb{R}^n} (-\Delta)^m u(x) u(x) g(\log |x|^{-1}, \log |\xi|^{-1}) dx,$$

where $C_R = \log(4R)$ and g is defined by

$$g(t, \tau) = h(t - \tau) + \mu(C_R + \tau) + C' + C''(C_R + t),$$

where h is a unique solution of the equation

$$\begin{cases} \prod_{j=0}^{m-1} \left(-\partial_t^2 + \left(m - \dfrac{n}{2} - 2j \right)^2 \right) h = \delta, & \text{if } m - n/2 \text{ is even} \\ \prod_{j=0}^{m-1} \left(-\partial_t^2 + \left(m + \dfrac{n}{2} - 2j - 1 \right)^2 \right) h = \delta, & \text{if } m - n/2 \text{ is odd} \end{cases}$$

which vanishes at $+\infty$ and has at most linear growth or decay at $-\infty$. μ is a constant depending on m and n only.

With respect to Section 7.6, we mention a theorem by Kato [46], which states that, if A is an accretive operator in a Hilbert space, then the fractional powers A^α, with $0 < \alpha < 1$, are accretive too. This suggests that the fractional powers $(-\Delta)^\alpha$ ($0 < \alpha < 1$) would be accretive on L^p also for $p \neq 2$. This is in fact true and

Theorem 7.58 gives an explicit lower bound for the relevant form. The material in Section 7.6 seems to be new.

Theorem 7.59, in Section 7.7, is a particular case of more general weighted inequalities obtained by Costin and Maz'ya [15] (see also [73, Th. 1–2, pp. 220–229]).

The so-called "sharp Gårding inequality" appeared in the paper [40] by Hörmander: if a belongs to a certain class of smooth symbols and $\operatorname{Re} a \geqslant 0$, then

$$\operatorname{Re}(a(x,D)u,u) \geqslant -C\|u\|_{H^m}^2, \quad u \in C_0^\infty.$$

The matrix case was treated by Lax and Nirenberg [56]. Further development is due to Melin [79], Beals [6], Fefferman and Phong [28] *et al.* The symbol of a in all these works is sufficiently smooth, at least it belongs to C^1. Theorem 7.60, which is due to Maz'ya [68], contains an improvement of the sharp Gårding inequality for a class of symbols $a(x,\xi)$, whose first derivatives in x may be discontinuous.

Bibliography

[1] AGMON, S., DOUGLIS, A., NIRENBERG, L., Estimates near the boundary for solutions of elliptic partial differential equations satisfying general boundary conditions. I, *Comm. Pure Appl. Math.*, 12, 1959, 623–727.

[2] AGMON, S., DOUGLIS, A., NIRENBERG, L., Estimates near the boundary for solutions of elliptic partial differential equations satisfying general boundary conditions. II. *Comm. Pure Appl. Math.* 17, 1964, 35–92.

[3] AMANN, H., Dual semigroups and second order elliptic boundary value problems, *Israel J. Math.*, 45, 1983, 225–254.

[4] AUSCHER, P., BARTHÉLEMY, L., BÉNILAN, P., OUHABAZ, EL M., Absence de la L^∞-contractivité pour les semi-groupes associés aux opérateurs elliptiques complexes sous forme divergence, *Poten. Anal.*, 12, 2000, 169–189.

[5] BATEMAN, H.: *Higher Transcendental Functions*, (A. Erdelyi, W. Magnus, F. Oberhettinger and F.G. Tricomi, eds.), vol. **1, 2**, McGraw-Hill, New York, 1953.

[6] BEALS, R., Square roots of nonnegative systems and the sharp Gårding inequality, *J. Differential Equations* 24, 1977, 235–239.

[7] BERCHIO, E., On the sign of solutions to some linear parabolic biharmonic equations, *Adv. Differential Equations* 13, 2008, 59–976.

[8] BREZIS, H., STRAUSS, W.A., Semi-linear second order elliptic equations in L^1, *J. Math. Soc. Japan*, 25, 1973, 565–590.

[9] CALDERÒN, A. P., Commutators of singular integral operators, *Proc. Nat. Acad. Sci. U.S.A.* 53, 1965, 1092–1099.

[10] CHILL, R., FAŠANGOVÁ, E., METAFUNE, G., PALLARA, D. The sector of analyticity of the Ornstein–Uhlenbeck semigroup on L^p spaces with respect to invariant measure, J. London Math. Soc. (2), 2005, 703–722.

[11] CIALDEA, A., MAZ'YA, V., Criterion for the L^p-dissipativity of second order differential operators with complex coefficients, *J. Math. Pures Appl.*, 84, 2005, 1067–1100.

[12] CIALDEA, A., MAZ'YA, V., Criteria for the L^p-dissipativity of systems of second order differential equations, *Ricerche Mat.* 55, 2006, 233–265.

[13] CIALDEA, A., MAZ'YA, V., A quasicommutative property of the Poisson and composition operators, *J. Math. Sci. (N.Y.)*, 164, 2010, 415–426.

[14] CIALDEA, A., MAZ'YA, V., L^p-dissipativity of the Lamé operator *Mem. Differential Equations Math. Phys*, to appear.

[15] COSTIN, O., MAZ'YA, V., Sharp Hardy–Leray inequality for axisymmetric divergence-free fields, *Calc. Var. Partial Differ. Equ.* 32, 2008, 523–532.

[16] CYCON, H.L., FROESE, R.G., KIRSCH, W., SIMON, B.: Schrödinger Operators, Springer-Verlag Berlin Heidelberg, 1987

[17] DANERS, D., Heat kernel estimates for operators with boundary conditions, *Math. Nachr.*, 217, 2000, 13–41.

[18] DAVIES, E.B., *One-parameter semigroups*, Academic Press, London-New York, 1980.

[19] DAVIES, E.B., *Heat Kernels and Spectral Theory*, Cambridge University Press, Cambridge, U.K., 1989.

[20] DAVIES, E.B., L^p spectral independence and L^1 analyticity, *J. London Math. Soc.* (2), 52, 1995, 177–184.

[21] DAVIES, E.B., Uniformly elliptic operators with measurable coefficients, *J. Funct. Anal.*, 132, 1995, 141–169.

[22] EILERTSEN, S., On weighted positivity of ordinary differential operators, *J. of Inequal. & Appl.*, 4, 1999, 301–314.

[23] EILERTSEN, S., On weighted positivity and the Wiener regularity of a boundary point for the fractional Laplacian, *Ark. Mat.* 38, 2000, 53–75.

[24] EILERTSEN, S., On weighted fractional integral inequalities, *J. Funct. Anal.*, 185, 2001, 342–366.

[25] ENGEL, K.J., NAGEL, R., *One-parameter semigroups for linear evolution equations. With contributions by S. Brendle, M. Campiti, T. Hahn, G. Metafune, G. Nickel, D. Pallara, C. Perazzoli, A. Rhandi, S. Romanelli and R. Schnaubelt*, Grad. Texts in Math. 194, Springer-Verlag, New York, 2000.

[26] FATTORINI, H.O., *The Cauchy Problem*, Encyclopedia Math. Appl., 18, Addison-Wesley, Reading, Mass., 1983.

[27] FATTORINI, H.O., On the angle of dissipativity of ordinary and partial differential operators, in ZAPATA, G.I. (ed.), *Functional Analysis, Holomorphy and Approximation Theory, II*, Math. Studies, 86, North-Holland, Amsterdam, 1984, 85–111.

[28] FEFFERMAN, C., PHONG, D.H., The uncertainty principle and sharp Gårding inequalities, *Comm. Pure Appl. Math.* 34, 1981, 285–331.

[29] FICHERA, G., *Linear elliptic differential systems and eigenvalue problems*, Lecture Notes in Math., 8, Springer-Verlag, Berlin, 1965.

[30] GAZZOLA, F., On the moments of solutions to linear parabolic equations involving the biharmonic operator, *Discrete Contin. Dyn. Syst.* 33, 2013, 3583–3597.

[31] GOLDSTEIN, J.A., *Semigroups of Linear Operators and Applications*, Oxford University Press, Oxford, 1985

[32] GOLDSTEIN, J.A., *(More-or-Less) Complete Bibliography of Semigroups of Operators through 1984*, Tulane Univ., New Orleans, La., 1985.

[33] GRIGOR'YAN, A., Gaussian upper bounds for the heat kernel on arbitrary manifolds, *J. Diff. Geom.*, 45, 1997, 33–52.

[34] GRISVARD, P., *Elliptic Problems in Nonsmooth Domains*, Pitman Publishing Ltd., London, 1985

[35] GURTIN, M.E., The linear theory of elasticity, in *Mechanics of Solids* – Volume II, C. TRUESDELL (ed.), Springer-Verlag (1984).

[36] HEWITT, E., STROMBERG, K. *Real and Abstract Analysis*, Springer-Verlag Berlin Heidelberg New York, 1969.

[37] HILLE, E., *Functional Analysis and Semi-Groups*, Amer. Math. Soc. Colloq. Pub., vol. 31. AMS, New York, 1948.

[38] HILLE, E., PHILLIPS, R.S., *Functional analysis and semi-groups*, Third printing of the revised edition of 1957. Amer. Math. Soc. Colloq. Publ., Vol. XXXI. AMS, Providence, R.I., 1974.

[39] HÖMBERG, D., KRUMBIEGEL, K., REHBERG, J., Optimal Control of a Parabolic Equation with Dynamic Boundary Condition, *Appl. Math. Optim.*, 67, 2013, 3–31.

[40] HÖRMANDER, L., Pseudo-differential operators and non-elliptic boundary problems, *Ann. of Math.* (2) 83, 1966, 129–209.

[41] HÖRMANDER, L., *The Analysis of Linear Partial Differential Operators I*, Springer-Verlag Berlin Heidelberg, 1983, 1990.

[42] HÖRMANDER, L., *The Analysis of Linear Partial Differential Operators III*, Springer-Verlag, Berlin Heidelberg, 1985.

[43] JAYE, B.J., MAZ'YA, V., VERBITSKY, I.E., Existence and regularity of positive solutions of elliptic equations of Schrödinger type. *J. Anal. Math.* 118, 2012, 577–621.

[44] JAYE, B., MAZ'YA, V., VERBITSKY, I., Quasilinear elliptic equations and weighted Sobolev–Poincaré inequalities with distributional weights, *Adv. Math.* 232, 2013, 513–542.

[45] KARRMANN, S., Gaussian estimates for second order operators with unbounded coefficients, *J. Math. Anal. Appl.*, 258, 2001, 320–348.

[46] KATO, T., Fractional powers of dissipative operators, *J. Math. Soc. Japan*, 13, 1961, 246–274.

[47] KATO, T., *Perturbation Theory for Linear Operators*, 2nd ed., Springer-Verlag, 1976.

[48] KOVALENKO, V., SEMENOV, YU., C_0-semigroups in $L^p(\mathbb{R}^d)$ and $\hat{C}(\mathbb{R}^d)$ spaces generated by the differential expression $\Delta + b \cdot \nabla$, *Theory Probab. Appl.*, 35, 1990, 443–453.

[49] KRESIN, G., Sharp constants and maximum principles for elliptic and parabolic systems with continuous boundary data. In: The Maz'ya Anniversary Collection, Vol. 1: On Maz'ya's work in functional analysis, partial differential equations and applications, *Oper. Theory Adv. Appl.*, **109**, Birkhäuser, Basel, 249–306 (1999)

[50] KRESIN, G., MAZ'YA, V., Criteria for validity of the maximum modulus principle for solutions of linear parabolic systems, *Ark. Mat.*, 32, 1994, 121–155.

[51] KRESIN, G., MAZ'YA, V., *Maximum Principles and Sharp Constants for Solutions of Elliptic and Parabolic Systems*, Mathematical Surveys and Monographs, 183. AMS, Providence, 2012.

[52] LADYŽHENSKAYA, O.A., *The mathematical theory of viscous incompressible flow*, Gordon and Breach, New York (1969).

[53] LADYŽHENSKAYA, O.A., SOLONNIKOV, V.A., URAL'CEVA, N.N., *Linear and Quasilinear Equations of Parabolic Type* Amer. Math. Soc. Providence, R.I., 1968.

[54] LANGER, M., L^p-contractivity of semigroups generated by parabolic matrix differential operators, in *The Maz'ya Anniversary Collection*, Vol. 1: *On Maz'ya's work in functional analysis, partial differential equations and applications*, Birkhäuser, Basel, 1999, 307–330.

[55] LANGER, M., MAZ'YA, V., On L^p-contractivity of semigroups generated by linear partial differential operators, *J. of Funct. Anal.*, 164, 1999, 73–109.

[56] LAX, P.D., NIRENBERG, L., On stability for difference schemes: A sharp form of Gårding's inequality, *Comm. Pure Appl. Math.*, 19, 1966 473–492.

[57] LISKEVICH, V.A., On C_0-semigroups generated by elliptic second order differential expressions on L^p-spaces, *Differential Integral Equations*, 9, 1996, 811–826.

[58] LISKEVICH, V.A., PERELMUTER M.A., Analyticity of Submarkovian Semigroups, *Proc. Amer. Math. Soc.* 123, 1995, 1097–1104.

[59] LISKEVICH, V.A., SEMENOV, YU.A, Some problems on Markov semigroups. In: Demuth, M. (ed.) et al., *Schrödinger operators, Markov semigroups, wavelet analysis, operator algebras.* Berlin: Akademie Verlag. Math. Top. 11, 1996, 163–217.

[60] LISKEVICH, V.A., SOBOL, Z., VOGT, H., On the L_p-theory of C_0 semigroups associated with second order elliptic operators. II, *J. Funct. Anal.*, 193, 2002, 55–76.

[61] LUMER, G., PHILLIPS, R.S.: Dissipative operators in a Banach space, *Pacific J. Math.* 11, 1961, 679–698.

[62] LUO, G., MAZ'YA, V., Weighted positivity of second order elliptic systems, *Potential Anal.* 27, 2007, 251–270.

[63] LUO, G., MAZ'YA, V., Wiener type regularity of a boundary point for the 3D Lamé system, *Potential Anal.* 32, 2010, 133–151.

[64] MAYBORODA, S., MAZ'YA, V., Boundedness of the gradient of a solution and Wiener test of order one for the biharmonic equation, *Invent. Math.* 175 2009, 287–334.

[65] MAYBORODA, S., MAZ'YA, V., Regularity of solutions to the polyharmonic equation in general domains, *Invent. Math.*, to appear, DOI 10.1007/s00222-013-0464-1

[66] MAZ'YA, V., The negative spectrum of the higher-dimensional Schrödinger operator, (Russian), *Dokl. Akad. Nauk SSSR*, 144, 1962, 721–722.

[67] MAZ'YA, V., On the theory of the multidimensional Schrödinger operator, (Russian), *Izv. Akad. Nauk SSSR Ser. Mat.*, 28, 1964, 1145–1172.

[68] MAZ'YA, V., On a degenerating problem with directional derivative, *Math. USSR Sb.* 16 429, 1972, 429–469.

[69] MAZ'YA, V., Behaviour of solutions to the Dirichlet problem for the biharmonic operator at a boundary point. In: Equadiff IV (Proc. Czechoslovak Conf. Differential Equations and their Applications, Prague, 1977), *Lecture Notes in Math.*, 703, Springer, Berlin, 1979, 250–262.

[70] MAZ'YA, V., On the Wiener-type regularity of a boundary point for the polyharmonic operator, *Appl. Anal.* 71, 1999, 149–165.

[71] MAZ'YA, V., The Wiener test for higher order elliptic equations, *Duke Math. J.* 115(3), 2002, 479–512.

[72] MAZ'YA, V., Analytic criteria in the qualitative spectral analysis of the Schrödinger operator, *Proc. Sympos. Pure Math.*, 76.1, 2007, 257–288.

[73] MAZ'YA, V., *Sobolev Spaces*, 2nd Ed., Springer-Verlag, 2011.

[74] MAZ'YA, V., DONCHEV, T., Regularity in the sense of Wiener of a boundary point for a polyharmonic operator. (Russian) *C. R. Acad. Bulgare Sci.* 36, 1983, 177–179. English translation *Amer. Math. Soc. Transl.*, 137, 2, 1987, 53–55.

[75] MAZ'YA, V., PLAMENEVSKII, B., On the asymptotics of the fundamental solutions of elliptic boundary value problems in regions with conical points, (in Russian), *Probl. Mat. Anal.* 7, 1979, 100–145; Engl. transl.: *Sel. Math. Sov.* 4, 1985, 363–397.

[76] MAZ'YA, V., SHAPOSHNIKOVA, T., *Theory of Sobolev multipliers. With applications to differential and integral operators*, Grundlehren Math. Wiss., 337, Springer-Verlag, Berlin, 2009.

[77] MAZ'YA, V., SOBOLEVSKIĬ, P., On the generating operators of semigroups (Russian), *Uspekhi Mat. Nauk*, 17, 1962, 151–154.

[78] MAZ'YA, V., VERBITSKY, I.E., The Schrödinger operator on the energy space: boundedness and compactness criteria. *Acta Math.* 188, 2002, 263–302.

[79] MELIN, A., Lower bounds for pseudo-differential operators. *Ark. Mat.* 9, 1971, 117–140.

[80] METAFUNE, G., PALLARA, D., PRÜSS, J., SCHNAUBELT, R., L^p-theory for elliptic operators on $\mathbb{R}^d$ with singular coefficients, *Z. Anal. Anwendungen*, 24, 2005, 497–521.

[81] NITTKA, R., Projections onto convex sets and L^p-quasi-contractivity of semi-groups, *Arch. Math.* 98, 2012, 341–353.

[82] OKAZAWA, N., Sectorialness of second order elliptic operators in divergence form, *Proc. Amer. Math. Soc.*, 113, 1991, 701–706.

[83] OSTERMANN, A., SCHRATZ, K., Stability of exponential operator splitting methods for noncontractive semigroups, *SIAM J. Numer. Anal.*, 51, 2013, 191–203.

[84] OUHABAZ, E.M., Gaussian estimates and holomorphy of semigroups, *Proc. Amer. Math. Soc.* 123, 1995, 1465–1474.

[85] OUHABAZ, E.M., Gaussian upper bounds for heat kernels of second-order elliptic operators with complex coefficients on arbitrary domains, *J. Operator Theory*, 51, 2004, 335–360.

[86] OUHABAZ, E.M., *Analysis of heat equations on domains*, London Math. Soc. Monogr. Ser., 31, Princeton Univ. Press, Princeton, 2005.

[87] PAZY, A., *Semigroups of Linear Operators and Applications to Partial Differential Equations*, Applied Mathematical Sciences, 44, Springer-Verlag, 1983.

[88] PRUDNIKOV, A.P., BRYCHKOV, YU.A., MARICHEV, O.I., *Integrals and Series, Vol. 1, 2: Elementary Functions, Special Functions*, Nauka, Mascow, 1981 (Russian). English transl.: Gordon and Breach, Amsterdam, 1986.

[89] REED, M, SIMON, B., *Methods of Modern Mathematical Physics, Vol. II: Fourier Analysis, Self-adjointness*, Academic Press, New York, 1975.

[90] ROBINSON, D.W., *Basic theory of one-parameter semigroups*, Proceedings of the Centre for Mathematical Analysis, Australian National University, Canberra, 1982.

[91] RUDIN, W., *Real and complex analysis*, third edition. McGraw-Hill Book Co., New York, 1987.

[92] SIMON, B., Schrödinger Semigroups, *Bull. Amer. Math. Soc.*, 7, 1982, 447–526

[93] SOBOL, Z., VOGT, H., On the L_p-theory of C_0 semigroups associated with second order elliptic operators. I, *J. Funct. Anal.*, 193, 2002, 24–54.

[94] STEIN, E.M., The characterization of functions arising as potentials, I, *Bull. Amer. Math. Soc.* 67, 1961, 102–104.

[95] STEIN, E.M., *Topics in harmonic analysis related to the Littlewood–Paley theory*, Ann. of Math. Stud. 63, Princeton Univ. Press, 1970.

[96] STEIN, E.M., *Singular Integrals and Differentiability Properties of Functions*, Princeton University Press, 1970.

[97] STRICHARTZ, R.S., Analysis of the Laplacian on a Complete Riemannian Manifold, *J. Funct. Anal.*, 52, 1983, 48–79.

[98] STRICHARTZ, R.S., L^p contractive projections and the heat semigroup for differential forms, *J. Funct. Anal.*, 65, 1986, 348–357.

[99] YOSIDA, K.. On the differentiability and the representation of one-parameter semigroup of linear operators, *PJ. Math. Soc. Japan* 1, 1948. 15–21.

Index

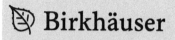 Birkhäuser | **www.birkhauser-science.com**

Operator Theory: Advances and Applications (OT)

This series is devoted to the publication of current research in operator theory, with particular emphasis on applications to classical analysis and the theory of integral equations, as well as to numerical analysis, mathematical physics and mathematical methods in electrical engineering.

Edited by
Joseph A. Ball (Blacksburg, VA, USA), Harry Dym (Rehovot, Israel),
Marinus A. Kaashoek (Amsterdam, The Netherlands), Heinz Langer (Vienna, Austria),
Christiane Tretter (Bern, Switzerland)

■ **OT 242: Bastos, A. / Lebre, A. / Samko, S. / Spitkovsky, I.M.** (Eds.), Operator Theory, Operator Algebras and Applications. IWOTA 12 (2014).
ISBN 978-3-319-06265-5

■ **OT 240: Ball, J.A. / Dritschel, M.A. / ter Elst, A.F.M. / Portal, P. / Potapov, D.** (Eds.), Operator Theory in Harmonic and Non-commutative Analysis (2014).
ISBN 978-3-319-06265-5

■ **OT 239: Denk, R. / Kaip, M.,** General Parabolic Mixed Order Systems in L_p and Applications (2013).
ISBN 978-3-319-01999-4

■ **OT 238: Edmunds, D.E. / Evans, W.D.,** Representations of Linear Operators Between Banach Spaces (2013).
ISBN 978-3-0348-0641-1

■ **OT 237: Kaashoek, M.A. / Rodman, L. / Woerdeman, H.J.** (Eds.), Advances in Structured Operator Theory and Related Areas. The Leonid Lerer Anniversary Volume (2013).
ISBN 978-3-0348-0638-1

■ **OT 236: Cepedello Boiso, M. / Hedenmalm, H. / Kaashoek, M.A. / Montes Rodríguez, A. / Treil, S.** (Eds.), Concrete Operators, Spectral Theory, Operators in Harmonic Analysis and Approximation. IWOTA11 (2013).
ISBN 978-3-0348-0647-3

■ **OT 234/OT235: Eidelman, Y. / Gohberg, I. / Haimovici, I.** (Eds.), Separable Type Representations of Matrices and Fast Algorithms.
Vol. 1. Basics. Completion problems. Multiplication and inversion algorithms (2013).
ISBN 978-3-0348-0605-3

Vol 2. Eigenvalue method (2013).
ISBN 978-3-0348-0611-4

■ **OT 233: Todorov, I.G. / Turowska, L.** (Eds.), Algebraic Methods in Functional Analysis. The Victor Shulman Anniversary Volume (2013).
ISBN 978-3-0348-0501-8

■ **OT 232: Demuth, M. / Kirsch, W.** (Eds.), Mathematical Physics, Spectral Theory and Stochastic Analysis (2013).
ISBN 978-3-0348-0590-2

■ **OT 231: Molahajloo, S. / Pilipović, S. / Toft, J. / Wong, M.W.** (Eds.), Pseudo-Differential Operators, Generalized Functions and Asymptotics (2013).
ISBN 978-3-0348-0584-1

■ **OT 230: Brown, B.M. / Eastham, M.S.P. / Schmidt, K.M.,** Periodic Differential Operators (2013).
ISBN 978-3-0348-0527-8

■ **OT 229: Almeida, A. / Castro, L. / Speck, F.-O.** (Eds.), Advances in Harmonic Analysis and Operator Theory. The Stefan Samko Anniversary Volume (2013).
ISBN 978-3-0348-0515-5

■ **OT 228: Karlovich, Y.I. / Rodino, L. / Silbermann, B. / Spitkovsky, I.M.** (Eds.), Operator Theory, Pseudo-Differential Equations, and Mathematical Physics (2013).
ISBN 978-3-0348-0536-0

■ **OT 227: Janas, J. / Kurasov, P. / Laptev, A. / Naboko, S.** (Eds.), Operator Methods in Mathematical Physics. Conference on Operator Theory, Analysis and Mathematical Physics (OTAMP) 2010, Bedlewo, Poland (2013).
ISBN 978-3-0348-0530-8

Printed in the United States
By Bookmasters